Psychology Revivals

Neurophysiology and Psychophysiology

Originally published in 1988, in several respects this book is a tribute to the outstanding career of the late Donald B. Lindsley (1907-2003) who, over a span of more than 55 years, had contributed greatly to the development of research in the fields of neurophysiology, psychophysiology, and experimental psychology. The impetus for the book was a conference held at UCLA to honor Professor Lindsley for his numerous and significant contributions to psychology. The chapters of this book have been written by Professor Lindsley's colleagues and co-workers, and by former students and postdoctoral fellows. The introductory chapter, written by Lindsley himself, tells of 2000 years of "pondering". The chapter is a discussion of the lengthy history of neurophysiology, psychophysiology, and behavior. Many of the topics mentioned in this chapter are subsequently presented in the book as reports of ongoing research in the field.

Neurophysiology and Psychophysiology

Experimental and Clinical Applications

Edited by

G. C. Galbraith, M. L. Kietzman and E. Donchin

Routledge
Taylor & Francis Group
LONDON AND NEW YORK

First published in 1988
by Lawrence Erlbaum Associates, Inc

This edition first published in 2021 by Routledge
2 Park Square, Milton Park, Abingdon, Oxon OX14 4RN

and by Routledge
605 Third Avenue, New York, NY 10017

Routledge is an imprint of the Taylor & Francis Group, an informa business

Publisher's Note

The publisher has gone to great lengths to ensure the quality of this reprint but points out that some imperfections in the original copies may be apparent.

Disclaimer

The publisher has made every effort to trace copyright holders and welcomes correspondence from those they have been unable to contact.

A Library of Congress record exists under ISBN: 0898599466

ISBN: 978-0-367-75923-0 (hbk)
ISBN: 978-1-003-16464-7 (ebk)

DOI: 10.4324/9781003164647

Neurophysiology and Psychophysiology: Experimental and Clinical Applications

Edited by

G. C. Galbraith
University of California at Los Angeles

M. L. Kietzman
Queens College of the City University of New York
and
New York State Psychiatric Institute

E. Donchin
University of Illinois at Urbana-Champaign

LEA LAWRENCE ERLBAUM ASSOCIATES, PUBLISHERS
1988 Hillsdale, New Jersey Hove and London

Lawrence Erlbaum Associates, Inc., Publishers
365 Broadway
Hillsdale, New Jersey 07642

Library of Congress Cataloging-in-Publication Data

Neurophysiology and psychophysiology.

A collection of papers in honor of Donald B. Lindsley.
Includes bibliographies and indexes.
1. Neurophysiology—Congresses. 2. Psychology, Physiological—Congresses. 3. Lindsley, Donald B.—Congresses. I. Galbraith, G. C., 1937-
II. Kietzman, Mitchell L. II. Doncin, Emanual.
IV. Lindsley, Donald B. [DNLM: 1. Neurophysiology.
2. Psychophysiology. WL 102N494]
QP360.N4933 1986 612'.8 87-24516
ISBN 0-89859-946-6

Printed in the United States of America
10 9 8 7 6 5 4 3 2 1

CONTENTS

CONTRIBUTORS

P. Apkarian
Netherlands Ophthalmic Research Institute
P.O. Box 6411
1005 EK Amsterdam, The Netherlands

Patricia R. Barchas
Departments of Sociology and
Psychiatry and Behavioral Sciences
Program in Sociophysiology
Building 120
Stanford University
Stanford, CA 94305

Jackson Beatty
Department of Psychology
University of California at Los Angeles
Los Angeles, CA 90024

Steven D. Berry
Department of Psychology
Miami University
Oxford, OH 45056

Warren Brown
Graduate School of Psychology
Fuller Theological Seminary
180 North Madison Avenue
Pasadena, CA 91182

Gerard E. Bruder
New York State Psychiatric Institute
722 West 168th Street
New York, NY 10032

Leo M. Chalupa
Department of Psychology
University of California at Davis
Davis, CA 95616

James R. Coleman
Department of Psychology
University of South Carolina
Columbia, SC 29208

Michael Coles
Cognitive Psychophysiology Laboratory
University of Illinois at Urbana-Champaign
Champaign, IL 61820

Emanuel Donchin
Cognitive Psychophysiology Laboratory
University of Illinois at Urbana-Champaign
Champaign, IL 61820

David Dupree
Cognitive Psychophysiology Laboratory
University of Illinois at Urbana-Champaign
Champaign, IL 61820

Dennis M. Feeney
Departments of Psychology and Physiology
University of New Mexico
Albuquerque, NM 87131

Itzhak Fried
Department of Psychology
University of California at Los Angeles
Los Angeles, CA 90024

Morton Friedman
Department of Psychology
University of California at Los Angeles
Los Angeles, CA 90024

Gary Galbraith
University of California at Los Angeles
Mental Retardation Research Group
Lanterman Developmental Center
P.O. Box 100-R
Pomona, CA 91769

K. M. Gerrity
Health Sciences Center
University of Oklahoma
Oklahoma City, OK 73190

J. B. Gliddon
Lanterman Developmental Center
P.O. Box 100
Pomona, CA 91769

Gabriele Gratton
Cognitive Psychophysiology Laboratory
University of Illinois at Urbana-Champaign
Champaign, IL 61820

Noriaki Hirasuna
Department of Psychology
University of California at Los Angeles
Los Angeles, CA 90024

Chuong C. Huang
Department of Psychiatry
and Mental Health Sciences
Medical College of Wisconsin
9455 Watertown Plank Road
Milwaukee, WI 53226

William S. Jose, II
Division of Computer-based Education
Control Data Corporation
Minneapolis, MN 55440

Mitchell L. Kietzman
Queens College
City University of New York
Flushing, NY 11367; and
New York State Psychiatric Institute
New York, NY 10032

B. S. Kopell
Department of Psychiatry and Behavioral Science
Stanford University School of Medicine
Stanford, CA 94305; and
Palo Alto Veterans Administration Medical Center
Palo Alto, CA 94304

Franklin B. Krasne
Department of Psychology
University of California at Los Angeles
Los Angeles, CA 90024

C. Kreinick
Laboratory of Neuropsychology
Division of Neurobiology
Barrow Neurological Institute
350 West Thomas Road
Phoenix, AZ 85013

Robert W. Lansing
Department of Psychology
University of Arizona
Tucson, AZ 85721

Donald B. Lindsley
Department of Psychology
University of California at Los Angeles
Los Angeles, CA 90024

M. Mackeben
Smith-Kettlewell Institute of Visual Sciences
Medical Research Institute of San Francisco at
Pacific Presbyterian Medical Center
2200 Webster Street
San Francisco, CA 94115

Irving Maltzman
Department of Psychology
University of California at Los Angeles
Los Angeles, CA 90024

P. Marchok
Laboratory of Neuropsychology
Division of Neurobiology
Barrow Neurological Institute
350 West Thomas Road
Phoenix, AZ 85013

George Marsh
Department of Psychology
California State University, Dominguez Hills
Dominguez Hills, CA 90747

James T. Marsh
Department of Psychiatry and Biobehavioral Sciences
University of California at Los Angeles
Los Angeles, CA 90024

Loren H. Meyerink
Denver Neuropsychological Associates
Aurora, CO 80010

Allan F. Mirsky
Laboratory of Psychology and Psychopathology
National Institute of Mental Health
Building 10, Room 4c-110
Bethesda, MD 20205

Risto Näätänen
Institute of Psychology
University of Helsinki
Ritarikatu 5,
SF - 00170 Helsinki 17, Finland

Ken Nakayama
Smith-Kettlewell Institute of Visual Sciences
Medical Research Institute of San Francisco at
Pacific Presbyterian Medical Center
2200 Webster Street
San Francisco, CA 94115

Barbara Payne
Division of Environmental Impact Studies
Argonne National Laboratory
Argonne, IL 60439

Carol K. Peck
School of Optometry
University of Missouri-St. Louis
8001 Natural Bridge Road
St. Louis, MI 63121

Mary Pendery
San Diego Veterans Administration Medical Center
San Diego, CA 92161; and
Department of Psychiatry
School of Medicine
University of California at San Diego
San Diego, CA 92161

Ronald E. Ponsford
Department of Psychology
Northwest Nazarene College
Nampa, ID 83651

Constance Ray
Laboratory of Neuropsychology
Boston University Medical Center
Boston, MA 02118

Robert W. Rhodes
Department of Anatomy
College of Medicine
and Dentistry of New Jersey
School of Osteopathic Medicine
University Heights, Box 55
Piscataway, NJ 08854

John W. Rohrbaugh
Laboratory of Clinical Studies
National Institute on Alcohol
Abuse and Alcoholism
9000 Rockville Pike
Bethesda, MD 20892

W. T. Roth
Palo Alto Veterans Administration
Medical Center
Palo Alto, CA 94304

Guenter H. Rose
Psychology Department
Bowdoin College
Brunswick, ME 04011

John Schlag
Department of Anatomy
and Brain Research Institute
School of Medicine
University of California at Los Angeles
Los Angeles, CA 90024

Madeleine Schlag-Rey
Department of Anatomy
and Brain Research Institute
School of Medicine
University of California at Los Angeles
Los Angeles, CA 90024

Arthur S. Schwartz
Laboratory of Neuropsychology
Division of Neurobiology
Barrow Neurological Institute
350 West Thomas Road
Phoenix, AZ 85013

Eleanor Shapiro
New York State Psychiatric Institute
722 West 168th Street
New York, NY 10032

P. V. Simonov
Institute of Higher Nervous Activity
and Neurophysiology
USSR Academy of Sciences
Butlerova str. 5a
117485 Moscow, USSR

James E. Skinner
Section of Neurophysiology
Department of Neurology
Baylor College of Medicine
Houston, TX 77030

Charles F. Sydnor
Department of Ophthalmology
Duke University Medical Center
Durham, NC 27710

John E. Thomas
Management Research Division
City of Tucson
Tucson, AZ 85701

Richard F. Thompson
Department of Psychology
University of Southern California
University Park
Los Angeles, CA 90089

Lee E. Travis (deceased)
Graduate School of Psychology
Fuller Theological Seminary
Pasadena, CA 91182

C. W. Tyler
Smith-Kettlewell Institute of Visual Sciences
Medical Research Institute of San Francisco at
Pacific Presbyterian Medical Center
2200 Webster Street
San Francisco, CA 94115

Francisco Velasco
Neurosurgery Service
General Hospital of Mexico, SS, and
Clinical Neurophysiology Section
National Medical Center IMSS
AP 73-032, Mexico DF 73

Marcos Velasco
Division of Neurophysiology
Clinical Neurophysiology Section
National Medical Center IMSS
AP 73-032, Mexico DF 73

Veronica Welch
Department of Psychology
University of California at Los Angeles
Los Angeles, CA 90024

Charles L. Wilson
Department of Neurology
Reed Neurological Center
University of California at Los Angeles
Los Angeles, CA 90024

Diana S. Woodruff-Pak
Department of Psychology
Temple University
Philadelphia, PA 19122

PREFACE

In several respects, this book is a tribute to the outstanding career of Donald B. Lindsley who, over a span of more than 55 years, has contributed greatly to the development of research in the fields of neurophysiology, psychophysiology, and experimental psychology. The impetus for the book was a conference held at UCLA to honor Professor Lindsley for his numerous and significant contributions to psychology. He has published over 230 articles including some 150 empirical studies, dozens of chapters in books (many of which are now considered classics), and 15 *Science* articles. In addition he has edited three books. And this productivity does not include his World War II activities, when he edited and wrote some 64 memoranda, research reports, and manuals.

The chapters of this book have been written by Professor Lindsley's colleagues and co-workers, and by former students and postdoctoral fellows.

The introductory chapter, written by Professor Lindsley, tells of 2000 years of "pondering." The chapter is a discussion of the lengthy history of neurophysiology, psychophysiology, and behavior. Many of the topics mentioned in this chapter are subsequently presented in the book as reports of ongoing research in the field.

The section on Neural Control has six chapters which reflect a wide range of topics and a diversity of experimental techniques, including the escape behavior of the lowly crayfish; the identification of the effects of lesions on EEG activity; the investigation of unitary sensory responses recorded extracellularly; the effects of electrical stimulation within the brainstem and hypothalamus of the cat; and the relationship between psychological stress and cardiac vulnerability.

The research reported here involving neural control reflects a potentially rich and productive impact on the fields of neurophysiology and psychophysiology. Throughout all of the chapters in this section, the influence of Donald Lindsley is clearly manifested. Many of the section's authors say so directly, while others do not need to say so, because their chapters are either co-authored with Lindsley or the references cited include a number of his publications.

The three chapters of Neural Substrates of Visual Discrimination have a number of interesting features. One chapter involves surgically induced, strabismic misalignment of the eyes in the cat. Two chapters use human visual evoked potential techniques, one to isolate cortical sub-populations of cells narrowly tuned for different spatial frequencies, the other to study the visual pathways of human albino subjects. The animal study and the study of human albino subjects have important direct implications for the study of abnormal visual function.

It is no surprise that the chapters of this section all deal with visual performance, visual evoked potentials, and both normal and abnormal vision. All are long-standing interests of Donald Lindsley. He sponsored the dissertations of the senior authors of the three chapters of this section. These chapters are essentially continuations of the interests held and pursued by Professor Lindsley over his long career as teacher and scientist. It should also be noted that many of the other chapters in other sections of this book reflect the same themes--normal and abnormal vision and evoked potentials.

The chapters entitled Neural Substrates of Attention nicely reflect the diversity of the concept of attention as well as providing examples of different approaches to studying attention. First is a chapter on vigilance or alertness as measured in both monkeys and humans by performance on the continuous performance test. There are also two chapters on selective attention, one measuring pupillary dilation, the other measuring evoked potentials and regional cerebral blood flow. Next is a clinical research study concerned with the relationship between unilateral neglect and sensory extinction. Finally, there is an investigation of visual fixation (the ability to direct gaze in space), which is tested in cats by the firing of neurons of the thalamic internal medullary lamina.

Over the years, the topic of attention is one that Professor Lindsley has investigated and discussed. His research has measured several types of attention. He also has introduced novel hypotheses about attention, which have influenced others to do the research to test these ideas. As early as the mid-1930s, Professor Lindsley began to investigate attention, even though at that time it was not fashionable. Beginning in the early 1950s, he introduced his activation theory of emotion and in the early 1960s, he began to theorize about such topics as consciousness, arousal, activation, and attention. These theories offer rich conceptual formulations not only about the behavioral aspects of attention, but also, and most importantly, about the neural substrates of attention. It is in the area of the neurophysiology of attention that Professor Lindsley has done some of his strongest empirical research and made some of his most profound observations.

The section entitled Latencies and Motor Performance is concerned with response latencies and motor behavior. The first paper is a combined electrophysiological-behavioral approach in which event-related potentials are used to study probability behavior. The next chapter is about the little known and infrequently investigated topic of human respiratory reaction times. Next is a paper investigating stimulus intensity-time reciprocity or temporal integration. Does temporal integration apply to the measurement of reaction time as it does for so many other electrophysiological and behavioral response measures? Finally, there is a report of a dual sensorimotor task (motor tapping with and without speech) with the concomitant measurement of the EEG. The purpose of this research is to measure time-shared performance and EEG asymmetry.

Measurement of the temporal properties of behavior, especially response latencies, has always been an important component of Professor Lindsley and his colleagues' research. Studies of simple and choice reaction times, more complex motor responses, forward and backward visual masking, latencies of event-related potential components, and alpha excitability relationships have all been topics of research in his laboratory. A unifying thesis of this research concerns the "when" of behavior, under the assumption that it is of value to learn when an event occurs, even though it is not always clear as to the cause or nature of the event. The chapters of this section fully illustrate some of the values of measuring motor behaviors, especially response latencies.

The three chapters of the section entitled Neural Substrates of Development represent a range of topics and subjects, both human and animal. The first chapter is about theory, research, and application in the field of developmental psychobiology, with a primary emphasis upon animal studies. The next chapter concerns the role of environmental influences, such as visual restrictions and visual deprivation, on the golden hamster's visual system and visual behavior. The third chapter examines the rapid changes occurring in the central nervous system of an infant during the first four months of life.

The chapters in this section have all been influenced by Professor Lindsley. His developmental research ranges from his early investigations in the 1930s when he studied his son's brain wave activity to the late 1930s and early 1940s when he studied the EEG brain waves of normal and abnormal children. During the 1960s, he studied the development of visual evoked potentials in kittens. In the field of developmental research, Professor Lindsley has contributed more than 25 publications over a period of more than 30 years. These have included articles concerned with research, theory, and applications.

The section entitled Higher Cognitive Processes covers a wide range of topics. The first chapter is a review of Professor Lindsley's contributions to physiological psychology in general and to the areas of learning and motivation in particular. The second chapter is a theoretical discussion of four main brain systems and their relationship to both the genesis of emotional states and the organization of purposeful behavior. The next report is a theoretical and empirical study of reading disabilities in the framework of contemporary neuropsychology and cognition. The next chapter utilizes the asymmetry of event-related potential differences between males and females as a measure of hemispheric asymmetry in language processing. Language dominance is studied in three groups: (a) normal right-handed males, (b) normal right-handed females, (c) male right-handed stutterers. Next is an event-related potential study involving confidence ratings for a line discrimination task of varying levels of difficulty. The next chapter is a classical conditioning experiment of the orienting response as indexed by the galvanic skin response. The final chapter investigates whether the contingent negative variation, a central nervous system measure, correlates with subject self-evaluation of their own competence.

It is appropriate that the papers of this section reflect such a wide number of topics, ranging from learning, emotion, and motivation to reading, language, laterality, and even social status. Professor Lindsley's contributions to the study of higher cognitive processes are well-known and are supported by a large body of empirical evidence collected by him, his colleagues and coworkers, and his former students. Professor Lindsley truly is a renaissance scientist with interests and research knowledge and skills ranging from cellular recordings and lesion studies to the "higher cognitive processes" described here.

In the preparation of this book we wish to acknowledge support of the University of California at Los Angeles, School of Medicine, Mental Retardation Research Center grant No. PHS-DHEW HD 04612. We appreciate the secretarial support of Karen Hochwater, Dorothy Trottier, and Celeste Hutchinson. We express special appreciation to Kathleen Schwarz, Carolyne Givens, and Terry Givens whose impressive talents and abilities were required to convert to a new word processor system and to deal with the problems of camera-ready copy.

DONALD BENJAMIN LINDLSEY
1907 -

SECTION I. HISTORICAL OVERVIEW

In this first section, Professor Lindsley introduces us to an extended overview of the topics that have motivated and guided his own professional career as a teacher and researcher, namely, brain and behavior, and especially the relationship between them. His "Two Thousand Years of Pondering" tells us that neurophysiology or psychophysiology, or whatever term is chosen to characterize this field of study, is not a new endeavor, but one that has a long and interesting history. About 500-300 B.C. was the earliest serious thought given to behavior by the famous Greek philosophers, Socrates, Plato, and Aristotle. They primarily used rational approaches based on reason and logic. It was probably Hippocrates (*ca.* 460-377 B.C.) who began the empirical study of human physiology as well as describing numerous medical conditions. Later, Herophilus (active *ca.* 300 B.C.) recognized the brain as the controlling center of the nervous system and the probable locus of intelligence. Subsequent contributions by such luminaries as Galen, Descartes, Galvani, Müller, and Helmholtz, among many others, bring the history of neurophysiology to the mid- and late 19th century, and to the threshold of rapid developments in neurophysiology and psychophysiology in the first half of the 20th century, followed by the neuroscience explosion commencing in the 1960s.

Important topics and themes are noted and presented by Lindsley in his chapter. Localization of function was, and still is, a topic of interest. Where exactly in the nervous system, and where especially in the brain, are certain functions centered? Such questions concerned many investigators: the phrenologists, Gall and Spurzheim; Flourens, famous for his ablation research; Broca, for his 30-year study of a brain-lesioned patient; and Fritsch and Hitzig, for their studies of electrical stimulation of the human cortex. Other theories that have been important in the study of brain and behavior relationships include the use of electrical recording techniques such as the electroencephalogram (EEG), the contingent negative variation (CNV), and the distinctive evoked/event-related potential (ERP) techniques. Other topics related to the book's theme include the neurohistological and architectonic studies of such notables as Golgi, Cajal, and Brodmann. Finally, the topics of behavior and conditional reflexes have an obvious relevance to the underlying theme of the book--the relationship between the brain and behavior.

 DOI: 10.4324/9781003164647- 1

1 · TWO THOUSAND YEARS OF PONDERING BRAIN AND BEHAVIOR[1]

D. B. Lindsley

University of California at Los Angeles

Man's awareness of his brain and that it has something to do with his behavior and mental life is a comparatively recent discovery relative to the estimated age of man on Earth. Practically no knowledge of the brain existed 2500 years ago, and only the mere rudiments of its form were known 1800 years ago at the time of the physician-anatomist, Galen. Only gradually did its gross form and structure come to be known reasonably accurately 400-500 years ago, and at that time almost nothing was known about its function. Two hundred years ago it was not yet known that nerves conduct electrical impulses, and it was thought that animal spirits flowed in tubular nerves. As recently as 125 years ago only the most elementary information was known about receptors and the way in which they transform physical energy into electrical nerve impulses, and how sensory nerves conduct these messages to the spinal cord and brain and motor nerves transmit impulses to muscles and glands where action takes place. However, since then, progress in the acquisition of knowledge and understanding about brain organization and behavior has been very rapid and extensive. But despite the remarkable advances that have been made, there is much to suggest that brain and behavior, like the universe, is an endless frontier awaiting new discoveries.

It is estimated that life on Earth began over two billion years ago. Simple primates first evolved about 60 million years ago. Hominid apes date from about one million years and *Homo sapiens* from 25,000 to 50,000 years. Figure 1.1 shows the evolutionary sequence believed to have led to man's present state of development, with gradual emergence of upright posture freeing the hands for tool making, shifts in the center of gravity with consequent changes in skeleton and musculature, changes in shape and size of skull, brain and lower jaw (see Figure 1.2), and the acquisition of speech and language.

---

[1]Reprinted from Lindsley, D. B. (1972). Two Thousand Years of Pondering Brain and Behavior. In H. Messel and S. T. Butler (Eds.), *Brain mechanisms and the control of behavior* (pp. 207-234). Sydney, Australia: Shakespeare Head Press.

DOI: 10.4324/9781003164647- 2

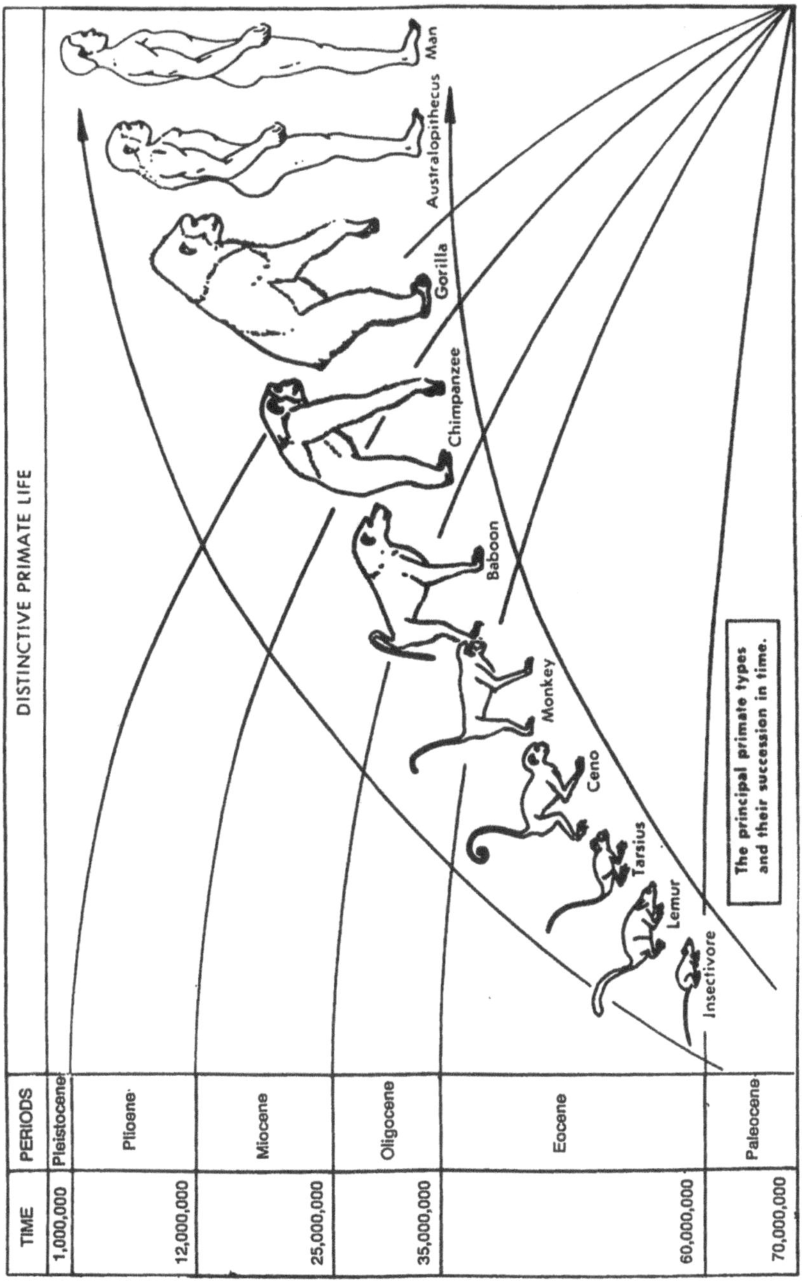

FIGURE 1.1. Evolutionary development with the Primate order of prosimians, monkeys, apes and man in relation to geologic periods (from Dart, 1959).

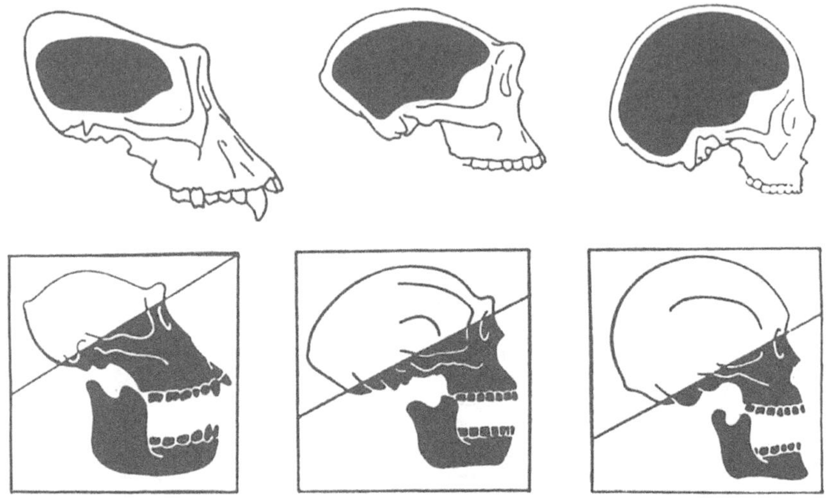

FIGURE 1.2. Changes in skull, brain and jaw configuration as man (right) evolved from Pithecanthropus (middle) and the gorilla (left), assuming erect posture with freedom of hands for manipulative and prehensile purposes. Note that large incisors of gorilla drop out, frontal and posterior association areas of brain increase, and jaws recede (from Moore, 1953).

Man's serious thought about himself and his behavior began a little over two thousand years ago. During the period from about 500 to 300 B.C. there was an overlapping succession of three famous Greek philosophers -- Socrates (469-399 B.C.), Plato (428-348 B.C.) and Aristotle (384-322 B.C.). The first two, in keeping with philosophic thought at that time, were concerned broadly with the nature of things in the world, including man. Their philosophy was one of rationalism based on reason and logic. They were more interested in the relationships which exist between men in a society than in the nature of man himself. Aristotle, on the other hand, was interested in the nature of man and other living organisms and his comparative approach reflected to some extent the beginning of a scientific attitude involving observation, description and classification. It is even said that his "ladder of nature," arranging animal structures from the lowest to the highest forms, might have become an evolutionary theory such as Darwin proposed 2200 years later had he been able to conceive of it in this way. But even Aristotle relied heavily upon philosophical rationalism and deduction rather than analytical observation and induction upon which the experimental sciences ultimately came to be based. All three of these philosophers were concerned with knowledge and ideas, and judgments about the ethics of man's behavior in society, subjects which contributed more to basic education and law than to a science of man and his behavior. There was very little thought about the role of the brain in man, although Plato's concept of a tripartite soul located one part in the head and the others in the body. Because of the vital nature of the heart, Aristotle located the soul there.

It is generally acknowledged that astronomy and physics were the earliest of the sciences to develop, one by observing, recording and analyzing the nature of the universe, the other the nature of matter. Biology and psychology, as sciences of animal life and behavior, have usually lagged far behind, due partly to the complexity of organic matter and the difficulty in studying and analyzing living things, and partly to man's personal reticence about studying himself or in accepting lower animals as surrogates. Nevertheless, empirical study of man and animals did occur during this early Greek period. Hippocrates (*ca.* 460-377 B.C.), regarded as the father of medicine, observed, described and classified many medical conditions. He established an ethics of medical practice, based on high standards, which is still honored in medicine today and known as the Hippocratic Oath. Herophilus (active *ca.* 300 B.C.), unable to do dissections in Greece, went to Egypt and studied at Alexandria where the dissection of human bodies was encouraged. He dissected human bodies and brains and became the greatest of Greek anatomists. He distinguished sensory and motor nerves and recognized the brain as the controlling center of the nervous system and the probable locus of intelligence. Erasistratus (*ca.* 304-250 B.C.), said to be the grandson of Aristotle, also dissected human bodies and compared their anatomy with that of animals, identifying and naming nerves and parts of the brain, although he did not understand their functions. He proposed that the complexity of the convolutions of the brain is related to intelligence, a view consistent with modern day thought.

At the time of Galen (*ca.* A.D. 129-200), some 400 years later, dissection of the human body was again forbidden in Greece and remained so in western countries for at least 12 centuries. Galen, a distinguished physician and anatomist, who systematized and unified Greek knowledge of anatomy and medicine, was forced to rely upon careful dissections of animals, including the Barbary ape, in order to draw conclusions about human anatomy. Unfortunately, because of this and his great prestige, a number of errors were perpetuated for some centuries to come. He did dissections of the brain, spinal cord and peripheral nerves and held that the mind is located in the brain.

Although there was some progress in science and medicine in Byzantine, Moslem and Chinese civilizations, there was very little progress in the study of man for the next 13 centuries in the West. This medieval period, often known as the "Dark Ages," was a time of military conquest, religious crusades and monastic scholasticism, the latter combining theology and philosophy but having little relevance to science. In the 15th century Leonardo da Vinci

(1452-1519), noted for his art and spirit of scientific inquiry, made remarkable anatomical illustrations of man. In the 16th century Vesalius (1514-1564) prepared the first comprehensive treatise on human anatomy and corrected the errors of Galen. In the 17th century Harvey (1578-1657) experimentally demonstrated the circulation of blood and diagrammed the circulatory system. Also in the 17th century René Descartes (1596-1650) was perhaps the first to leave out the concept of the soul and to think of the brain and nervous system in a mechanistic way. He conceived of a vital force which consisted of animal spirits flowing in the tubular nerves from sensory receptors to the ventricular cavities of the brain and from there to the muscles to produce action. The residual pineal gland, because of its central location in the brain, he thought served as a valve regulating the flux and flow of the spirits. This view, in a sense, was the first concept of a sensory-motor reflex arc. It is illustrated in Figure 1.3.

The next important step in the experimental study of the nervous system came when Galvani (1737-1798) discovered that an electric current generated by the contact of two dissimilar metals in the fluid surrounding nerves and

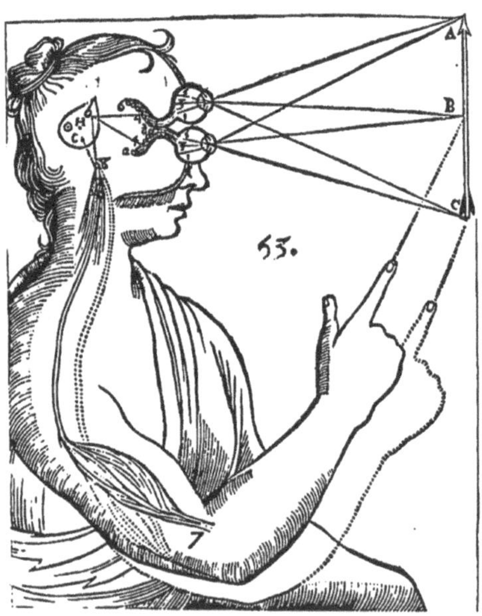

FIGURE 1.3. Visuo-motor reflex action from Descartes' *Treatise on man* (1677). Rays of light generate vital spirits which are directed to centrally located pineal body (H) from pores in the optic nerves and are received by pores of the motor nerve from whence transmitted to the arm muscles to effect action (from Magoun, 1963).

muscles excited them to action. He correctly concluded that nerves conduct electrical impulses. In the early part of the 19th century this opened up the possibility of using sensitive galvanometers to record electrical potentials generated in nerves and muscles and thus to learn more about their organization and action, either by recording from them or using a source of electrical current to stimulate them. This, and other developments of the early 19th century, soon enabled Matteucci to record the electromyogram; Du Bois-Reymond to study reflexes; Bell and Magendie to establish the law of independence of sensory and motor roots entering the spinal cord; Marshall Hall to distinguish between various kinds of action movements, such as voluntary and involuntary; Johannes Müller to define his law of specific energy of nerves; and Helmholtz to measure the velocity of the nerve impulse, showing that action resulting from excitation was delayed and not immediate as was supposed from observation. In gross outline, much of what we know today about the physiology of sensory receptors, sensory and motor nerves, muscles, the spinal cord and the functions of the brain was provided in a monumental series of volumes entitled, *Handbuch der Physiologie des Menschen* (1833-1840) by Johannes Müller, who was a great teacher and had many famous pupils such as Helmholtz, Du Bois-Reymond, Ludwig and others.

Localization of Function in the Brain

It is difficult to say precisely when the problem of localization of function in the brain first began or was first recognized. Aristotle located the soul or the seat of life in the heart; Plato located it in the head and considered the brain the "organ of the mind." The Alexandrian physicians of Egypt had discovered the ventricular cavities of the brain, and later Galen taught that animal spirits flow from the brain ventricles to the heart and are distributed to the body by the arteries. As already noted Descartes, centuries later, thought that tubular nerves served this function.

Albertus Magnus (1193-1280) prominently illustrated the three ventricular chambers of the brain, to which, like the Greek three-chambered courts of law, he and others attributed the functions of information gathering (sensory influx) in the foremost, consideration and judgment (integration) in the middle chamber, and recording and action (memory and motor outflow) in the posterior chamber. Numerous other similar diagrams persisted over the centuries in locating these functions in like manner (see Figure 1.4). Curiously, to the extent that such functions can be localized in the cerebral cortex of the brain today, they would be reversed, with memory and motor functions in the front half of the brain and sensory and integrative functions in the rear half.

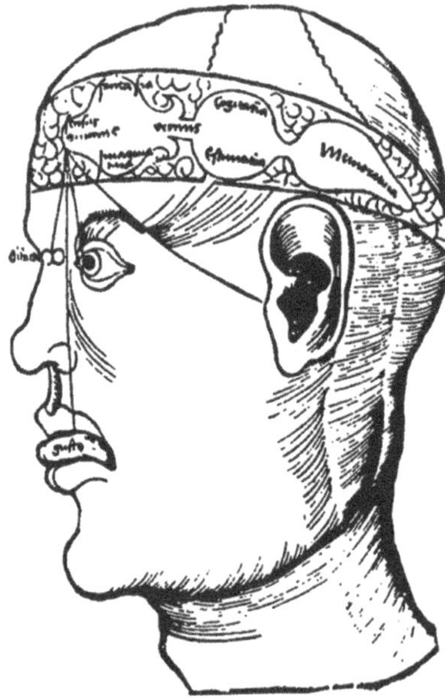

FIGURE 1.4. From an early 16th century printed woodcut (Reisch, 1504) this figure shows the three-chambered cerebral ventricle system known from the time of early Egyptian physicians, subsequently by Galen, and persisting through medieval times. Sensory information led into the anterior chamber - *sensus communis* - thence to the middle chamber for cogitation and estimation, and finally to the posterior chamber for memory and action (from Magoun, 1963).

Phrenology: Gall and Spurzheim. Around the beginning of the 19th century, Gall (1758-1828), a respectable anatomist, who concentrated on physiognomy, craniology and the anatomy of the brain, became convinced that the shape and size of the head, and particularly its protrusions or "bumps," could be correlated with personality and mental traits, such as he observed among criminals and mental patients. His studies of the anatomy of the brain and the topography of the head were considered reliable, but there was great controversy about the association of specific traits, both good and bad, with the variations of the surface of the skull. He was joined later by his pupil, Spurzheim (1776-1832), who turned out to be somewhat of an entrepreneur and propagandist. Although the church objected, scientists seemed to approve of the work and showed an interest in it initially; later Spurzheim elaborated the topography and the faculties or traits each bump was reputed to represent (see Figure 1.5). Spurzheim lectured extensively on the subject of phrenology in England and America and died in Boston in 1832 while advocating the doctrine. Periodically, for at least 100 years, there have been recurrences of efforts to utilize topography of the head to predict traits of various kinds.

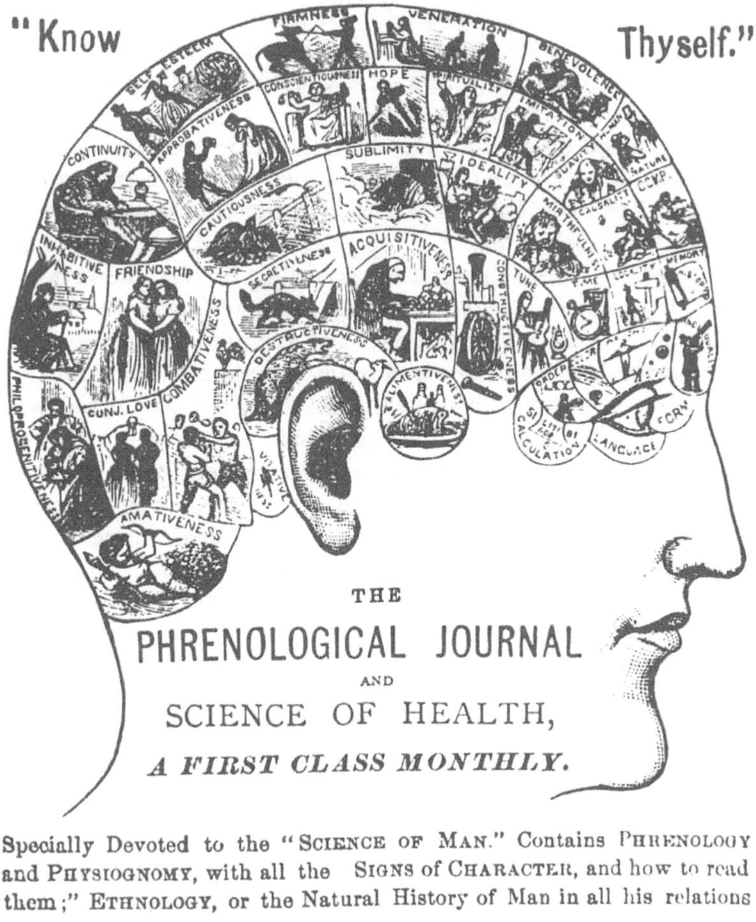

FIGURE 1.5. An illustrated phrenological brain map after Spurzheim, naming a great variety of traits supposedly localized on the brain as shown and represented by elevations on the skull at corresponding locations.

As recently as 1937, while in Cleveland, Ohio, a few of us from the medical school at Western Reserve university and from the Psychology Department were called upon by the Better Business Bureau to investigate a personnel organization located in a modern office building in the city, which was performing a kind of modern day "phrenology," with a machine called a Psychograph. A shiny aluminum bowl was positioned carefully over the head. Through it extended some 36 rods with contact points along the side. When all of the rods had been brought in contact with the skull, a switch was closed and their exact location in depth was registered on a chart containing 36 personality traits. A fee was charged for the making of the chart and another fee for an analysis of its meaning. Several respectable business firms were sending prospective employees to have evaluations made. One of us in the visiting delegation also served as a subject.

When we raised objections based on scientific grounds, such as, "It is well known that the surface of the brain does not conform exactly to the contour of the skull," or "There is no evidence that localization of function in the brain exists for such traits," the reply we received was, "That may be so according to your school of thought, but according to ours, predictions can be made from such measurements." This was fortified further by the assertion that thousands of these measurements and evaluations had been made in other cities of the United States and many people were satisfied with the service and the personnel selections made. Even the lawyer for the Better Business Bureau, who was skeptical and sought our consultation, indicated that a reading taken upon himself seemed to represent some of his traits as he knew them. Of course, fortune telling by gypsies or others has its adherents and sometimes if enough predictions are made some of them may turn out to be true, at least as interpreted by the subject. In this case we suspected that an experienced personnel man might, by his contact and interview with a subject, be able to make a reasonable selection of a man for a position without the aid of the machine, thus making the phrenological aspect somewhat superfluous. Nevertheless, we found it an interesting and illuminating experience in modern day phrenology. We later found out that there was no way the Better Business Bureau could prevent the selling of such a service, providing care was exercised in composing the advertising as to what the service claimed to do for a client. The reading of palms, fortune telling, the analysis of handwriting and other services of similar nature can often be performed for a fee in many localities.

Many modern day psychological tests have been developed for the purpose of assessing various abilities, capacities and traits. Usually these have been subjected to rigorous and arduous statistical tests of their reliability and validity after being administered to sizable populations of people and after appropriate criterion measures have been established. Such standardized tests, although never one hundred percent perfect in their appraisals and predictions for various reasons, can be characterized as having a certain probability of success. The main difficulty resides in defining the traits and capacities to be measured and in establishing suitable criteria against which to check the instruments designed to measure them.

Pierre Flourens and the Ablation Method. Flourens (1794-1867), a French physiologist, was stimulated by the work of Gall to test the validity of the claims of phrenology which depend upon the localization of functions over various parts of the brain. He utilized the method of *ablation*, that is, surgical removal of particular parts of the brain to determine whether functions attributed to that part were lost or showed a deficit after removal or inactivation. This method is still used today after many years of valuable service in experimentation on the brain. In general, it has been greatly improved over the years due to better surgery, to better understanding of the structures ablated and their relationship to other structures of the brain, both neuroanatomically and functionally, and to many other technical refinements. Nevertheless, the method is subject to criticism since it is often uncertain whether an area removed represents the function which shows a deficit or whether trauma to adjacent regions is responsible. In some instances a region ablated may sever fibers connecting two or more other regions of the brain, or interfere with the blood supply to other regions. Often scar tissue forms on the margins of lesions, causing irritation and abnormal discharge of neurons comprising the intact areas. Despite these and other limitations of the ablation method it has provided valuable information about various functions of the brain.

In birds, rabbits, cats and dogs, Flourens removed various parts of the brain. His approach was necessarily gross in nature, for the regions of the brain had not been specifically defined with respect to their functions in animals, although Gall and Spurzheim had presumed various small and localized regions of the brain of man to be endowed with specific and varied functions. He removed the cerebral hemispheres, the cerebellum, the corpora quadragemina, the medulla oblongata, the spinal cord and various peripheral nerves. He was able to draw certain general conclusions. Removal of the cerebral hemispheres causes an animal to be blind and deaf and to lack voluntary actions. From this he concluded that the cerebrum serves perception and volition, i.e., sensory and motor functions. Although the animal did not respond to light stimulation the pupil of its eye showed a reflex constriction. The cerebellum serves coordination of locomotor movements. The medulla has a conservative function of vital nature for when removed the animal dies. Thus he had shown that different parts of the brain in a gross sense have different functions; a dose of opium mimicked the results of removal of the hemispheres and alcohol produced incoordination, as did removal of the cerebellum. Removal of parts of the cerebrum did not cause loss of specific sensory function or motor control, or if it did interfere there was recovery of function.

Thus, in part, his results controverted the specific localization notions of Gall and Spurzheim, but did show differentiation of functions for removal of different gross parts of the brain. He maintained that there was unity of function within these gross parts. Thus his results speak partly in favor of localization and partly in favor of generalization of functions. It is of interest that subsequent work even unto the present day reflects these generalized conclusions, for as time went on one set of investigators after another would periodically demonstrate that something could be localized and another would find results suggesting generalization of function and lack of localization. This has led to repeated changes in emphasis of the theories of brain function, ranging from an extreme holistic notion that all functions are represented to some extent in all parts of the brain, to an extreme view with respect to highly localized and individualized functions. As we will see, the truth probably lies somewhere between these two extreme positions, but the problem has still not been resolved in its entirety.

Broca and the Neuropathological Method. Paul Broca (1824-1880), a French surgeon, examined a patient admitted to a hospital for the insane in 1831 because he could not talk. Broca examined the man carefully, including his larynx and vocal apparatus, and concluded that there was nothing wrong anatomically with the speech organs. Broca continued his study of the patient over a period of thirty years and when the man died in 1861 he performed an autopsy on his brain. He found that there was a lesion of the third frontal convolution on the left side of the brain, a region which has been named Broca's area. He published a description of the patient's symptoms and the region of neuropathology that he found. It is now well known that motor aphasia, such as this patient exhibited, occurs only when the lesion is in the left or dominant hemisphere of the brain in a right-handed person, or in the right hemisphere of a left-handed person.

Whereas Flourens' experimental ablations of various regions of the brain in animals were said to disprove the contentions of Gall and Spurzheim, that there are highly localized functions in the brain, Broca argued that his results pointed to a high degree of localization of the speech function and that this tended to controvert the conclusions of Flourens. Broca's argument was fortified by the electrical stimulation work of Fritsch and Hitzig, to be described later, in which they found specific points along the motor cortex which were excitable to electric current and elicited specific motor movements of the opposite side of the body.

The clinical neurological study and assessment of deficits in function exhibited by a patient, when correlated with the results of gross and microscopic study of the brain after the patient dies, is called the *neuropathological method*. This method has been responsible for a considerable part of our knowledge about the functions of the brain in cases where nature or accident has provided, through disease or trauma, destruction of brain tissue in certain regions. Although the neuropathological method has been of great value in identifying the functions of various regions of the human brain it is not without disadvantages. For example, in the case of Broca's patient, a 30-year period intervened between the recognition of the problem and the identification of the region of the brain which had been damaged. An additional difficulty is that having studied a patient for a number of years the neuropathologist may not be able to take advantage of his accumulated data, if the relatives of a patient do not permit an autopsy or give permission to section, stain and study the brain grossly and histologically. Other disadvantages are that nature does not plan the site, size and type of lesion, in contrast to lesions deliberately made in specific regions of the brain of animals which are of interest to study under experimental conditions. Furthermore, lesions may increase in size with time so that it is not always possible to relate precisely the changes in behavior which occurred earlier with the lesion which existed at that time. Also in the case of expanding lesions produced by growing tumors, the particular area involved may be less a determinant of the changes in behavior or loss of psychological functions than are the secondary effects created by pressure upon other, and sometimes remote, regions of the brain.

Fritsch and Hitzig and the Method of Electrical Stimulation. Following Galvani's discovery of the effect of an electrical current on nerve and muscle, subsequent investigators during the first half of the 19th century utilized the method to stimulate nerves and muscles and to study spinal reflexes, and other problems. Few attempts had been made to stimulate the surface of the exposed brain of animals, although Rolando, an Italian anatomist and pathologist, after whom the central sulcus of the brain is named, had stimulated the cerebellum and produced vigorous motor movements. Indeed, Hitzig first observed that electrical stimulation of the human cortex produced eye movements, which he then confirmed in the rabbit. Thereafter Fritsch and Hitzig undertook a systematic study of point by point stimulation of the cortex of the dog, concentrating upon the most anterior regions which were the principal areas from which motor movements could be elicited. They were able to show that various parts of the body could be made to move when an electrical stimulus of appropriate strength was applied to different localized points in what was to become known as the motor cortex. They published their results in 1870. Soon thereafter Ferrier in England published a monograph on *The Functions of the Brain* in 1876 in which he described in detail the results of stimulating much of the surface of the monkey brain and in which he plotted all of the excitable regions, especially those that gave rise to specific motor movements.

Subsequently, investigators extended the use of the stimulation method to other species of animals, including man, and more and more refined maps of the excitable motor cortex became available. The neurosurgeons Förster, in Germany, and Penfield and Boldrey in Canada, in 1936 and 1937, respectively, published extensive studies of the stimulation not only of motor regions of the human cortex, but of sensory areas as well. The stimulation in many instances was done with the patient under local anesthesia so that he was consciously aware of what was going on and could report any sensations or experiences that occurred as a result of the stimulation, especially in sensory zones. Still later Penfield and his collaborators explored by electrical stimulation regions of the temporal lobe where they were able to elicit memories of musical tunes and other experiences which had not been in the patient's thoughts for some time. This suggested that some parts of the temporal lobe, and perhaps the underlying hippocampus, may play a role in the storage of memories of past experiences.

In studying the nervous system, electrical stimulation has many other uses in addition to stimulating the brain directly. For example, electrical currents may be employed to excite a sensory receptor to action in lieu of its natural stimulus. A receptor may be severed from its sensory nerve and the nerve stimulated directly. Stimulation of a nerve pathway electrically, or the aggregation of nerve cells comprising a nucleus, may be carried out to determine where the pathway goes or what functions the nucleus serves. Modern day electrophysiology employs electrical stimulation in many other ways, but these applications require more sophisticated apparatus than was available 100 years ago, or even 50 years ago.

Initially, the voltaic pile or battery, developed by Volta at the time Galvani discovered the electrical excitability of nerves in the late 18th century, made a source of direct current available. This was called galvanic stimulation. The difference in potential between the two poles of the battery caused a current to flow from one to the other when they were connected. This made a continuous and direct flow of current, which had to be interrupted if the excitability of the nerve tissue was to be preserved. It was necessary to reverse the flow of current repeatedly if one wanted to observe the effect of frequently presented stimulation. An alternating current stimulator of this type produces faradic stimulation, after Faraday. Faraday's discovery that current could be induced in a circuit electromagnetically when the circuit was not connected to a battery led eventually to the development of an inductorium, some form of which was used for nearly 100 years, from about 1835 to 1935, when the electronic age ushered in electronic stimulators. Over the years electronic stimulators for physiological purposes have been greatly refined to a point where they can provide continuously varied frequency, duration of stimulus pulse, and magnitude of current or voltage. The manner of applying these stimulus parameters to the nerve or brain tissue has also been improved, and can be administered to even a single nerve cell by means of microelectrodes.

In the late 1930s electroshocks were applied to the brain on either side of the frontal lobe by large electrodes attached to the surface of the scalp at the temples. With voltage and current properly adjusted this produced an electroconvulsive seizure in the human patient which was thought at the time to have a beneficial effect in treating psychiatric patients by presumably breaking up aberrant habit formations thought to be associated with neurotic anxiety, depressed states and schizophrenic behavior. This method of treatment, like metrazol and insulin induced seizures, as well as frontal lobotomy, seems now to have fallen into disrepute since lasting cures were seldom effected and it was uncertain what other damages to the brain or body might occur. Furthermore, it was never clearly demonstrated why these methods might effect even temporary relief from the disorder, and since there were side effects and even possible brain damage, they were all abandoned during the 1950s in favor of tranquilizing drugs and other psychopharmacological agents.

Still another use to which electrical stimulation of the brain in animals has been put was discovered by Olds and Milner in 1954. Fine wire electrodes were implanted deep in the brain of a rat, in the hypothalamus, septum, or reticular formation. When these were connected to an appropriate strength of alternating current which could be applied when the rat pressed a lever with his paw, it was found that the rat either continued to press for more shocks, as if he enjoyed whatever effect it had, or else he refused to press and withdrew from the lever, as if the effect were an unpleasant one to be avoided. Interest has centered especially upon those points in the brain where the rat's self-stimulation was perpetuated even to the extent of being chosen in preference to the fulfillment of other basic drives such as hunger, thirst, and sex. Centers or regions of the brain which had a positive effect have been dubbed reward centers or even pleasure centers; other regions where the effect was negative have been called aversive or avoidance centers.

Electrical Recording in the Brain (Caton) and the Electroencephalogram (Berger)

It was discovered very early that when muscles contract they generate an electrical action potential. When recorded this is called an electromyogram (EMG). In the 1820s and 1830s, as soon as sensitive galvanometer-like equipment was available, investigators began to record muscle action currents or potentials. These potentials were of the order of millivolts, but even so required rather sensitive devices to detect them. It was some time before anyone tried, or was able, to record electrical potentials from the brain. In 1875 Caton, a British physician, in a very brief published communication, described the recording of electrical potentials from the exposed brain of rabbits, cats and monkeys. These potentials were of the order of microvolts to millivolts and hence needed even more sensitive instruments for their detection than for the muscle action potentials. Others verified Caton's findings during the next fifty years, but it was not until 1929 that Professor Hans Berger, a German neuropsychiatrist, described for the first time the recording of rhythmic oscillating potentials through the scalp and skull of human subjects.

These spontaneous, self-generated, electrical oscillations of the brain which did not require stimulation to evoke them, and were best observed when the subject was relaxed, quiet and unstimulated, Berger called the electroencephalogram (EEG). The principal and most prominent resting rhythm during wakefulness he called the alpha rhythm; it is comprised of trains of waves at about 10 per second on the average in older children and adults (see Figure 1.6). In most normal adults the alpha rhythm falls within the range of 8 to 12 waves per second. When a subject is relaxed and has his mind free enough from thoughts or concentration to generate rather steady alpha waves it was found that an unexpected stimulus such as a weak flash of light, a low intensity sound, or a touch upon the skin would block or desynchronize the alpha waves and the EEG would have a rather flat appearance (see Figure 1.7).

The ongoing electrical activity of the brain, comprised mainly of alpha rhythms, but including other types of rhythms, either slower or faster than alpha, came as a surprise and a shock to classical electrophysiologists for they thought that the only kind of electrical manifestation which could be recorded from nervous tissue was the well-known spike potential which accompanies the passage of an impulse in a nerve. This is a very brief and discrete electrical change having a definite pattern, whereas the alpha waves are relatively long-duration oscillations. Most electrophysiologists, when they heard about Berger's discovery, thought it was a fluke, or an artifact of some kind, but when distinguished neurophysiologists at the University of Cambridge in England -- Professors E. D. Adrian and B. H. Matthews -- confirmed Berger's findings in 1934, one could no longer ignore the electroencephalogram.

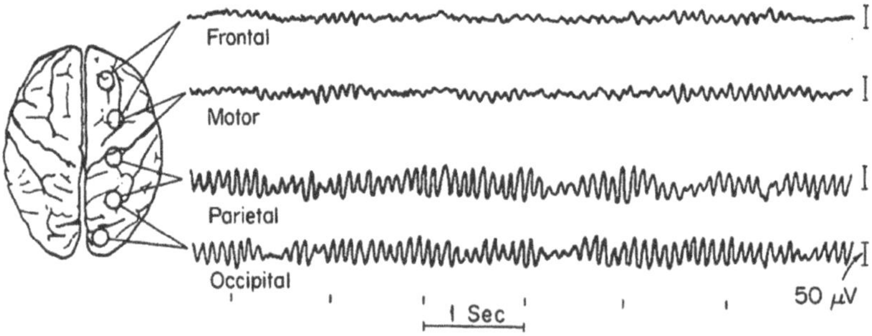

FIGURE 1.6. An electroencephalogram from a normal adult subject showing alpha wave oscillations at about 10 per second, typically more prominent over the sensory zones of the posterior half of the head as shown here (from Lindsley, 1948).

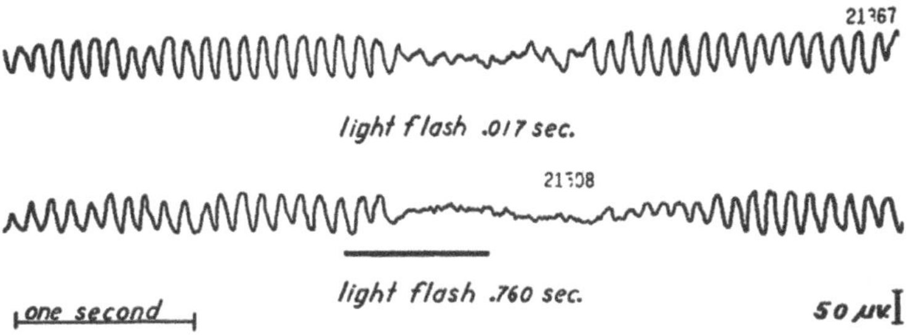

FIGURE 1.7. Occipital alpha waves blocked by light flashes of different durations (from Lindsley, 1944).

Soon quite a few investigators around the world began to record electrical activity from the brain, either through the scalp and skull of human subjects or by direct contact with the exposed cortex, or by needle-like electrodes inserted into the brain. In one respect, the importance of Berger's discovery of the electroencephalogram was the effect it had upon scientists' conceptions about the brain and the way it might work. It certainly had a marked influence in attracting investigators toward brain research, whether or not their methods were those of electrical recording or other quite different methods. In its own right the EEG has held considerable interest. Initially, many people thought it would correlate highly with higher level functions of the brain and mind, such as intelligence, perception, learning, thinking, problem solving and other psychological and cognitive functions. In this respect the EEG was initially a great disappointment since no important correlations with intelligence or personality were found; there were variations of the EEG related to physiological and psychological states and the alpha waves were responsive to stimulation in a negative way in that they broke up the rhythm.

It was soon discovered that certain neurological disorders such as epilepsy, brain tumors or abscesses, acute brain trauma and so forth often produced distinctive features in the EEG. It became an adjunct tool in neurological diagnosis, and even in the evaluation of treatment by drugs and other means. Today almost every major hospital in the world has an EEG laboratory where the electrical brain activity of its patients can be evaluated in relation to their other symptoms and in relation to other diagnostic methods.

The EEG is more than the spontaneous alpha and other ongoing rhythms; in the past twenty years it has been possible to record discrete short-duration evoked potentials over sensory areas when a stimulus of corresponding sense mode is presented. In order to do this through the scalp and skull it is necessary to use some kind of computer averaging technique and repeated stimulation so that the very small evoked potentials can be enhanced relative to the background spontaneous rhythms.

There is still a third type of electrical activity which can be recorded from the surface of the scalp in human subjects; it is a slow negative potential shift which W. Grey Walter in England has called the CNV (contingent negative variation) or expectancy wave. It was called contingent because after a warning stimulus it seemed to be contingent upon a set or expectancy that another stimulus would be presented, and indeed that is the experimental plan or paradigm under which it was first elicited and studied. Since 1964, when Walter and his collaborators first described the human CNV in the English science journal, *Nature*, there has been considerable study of this negative potential shift both in man and animals. Some feel that it reflects a kind of cortical "priming" such as may be required by motivation, by attention, by volition, or by anticipation. Others working with animals, and even some working with humans at the time of operation upon the brain, have thought of it as a temporary change in physiological state in the cortex which may have something to do with the excitatory or inhibitory state of the neural tissue. One thing that has seemed promising about the CNV is the possibility that it is associated in some fashion with the higher cognitive functions. Since these have generally been neglected it would be important to have some electrophysiological index of a changing psychological state or information processing condition.

Neurohistology and Architectonics of the Brain (Golgi, Cajal, Brodmann)

Among other developments, the latter half of the 19th century was a period of intensive activity on the part of many investigators whose interest focused on the internal structure of the cerebral cortex of the brain. The refinement of microscopes and the development of new methods for the sectioning and staining of brain tissues so as to bring out the details of nerve cells and their processes made it possible to differentiate the various regions of the cerebral mantle in terms of its structural features. It also made clear how numerous are the nerve cells comprising the cortex and how complex their matrix of interconnections. The monumental work of Golgi, an Italian, and Cajal, a Spaniard, in developing methods of impregnating neural tissue with silver to bring out the dendritic and axonic process of nerve cells as well as the outline of their cell bodies opened up a vast area of investigation. For the first time it became possible to describe and define objectively the types and distributions of nerve cells in the cortex. For their monumental contributions to the microscopic structure of the cortex, Golgi and Cajal were jointly awarded the Nobel Prize in 1906.

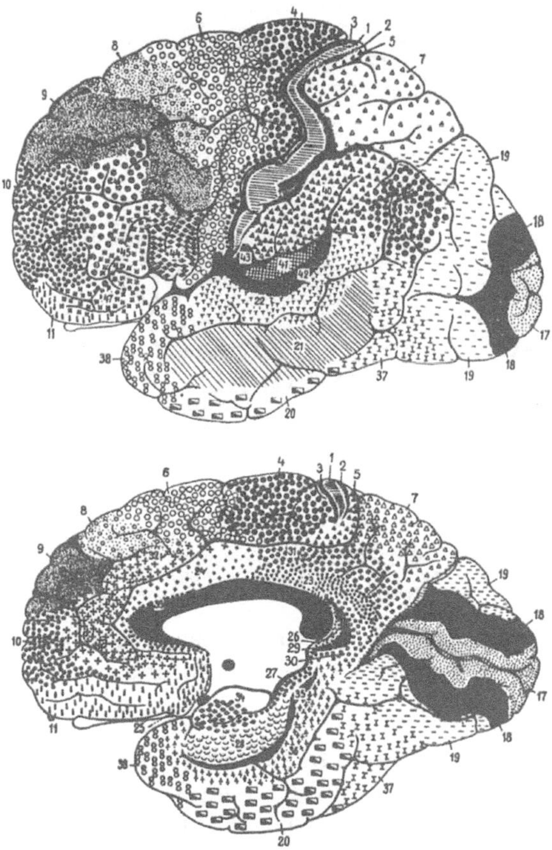

FIGURE 1.8. Human cortical architectonic brain map by Brodmann showing lateral view (top) and mesial view (bottom). Each numbered area was thought to be distinguishable in terms of detailed histological structure (from Brodmann, 1908).

Several investigators concentrated on the architecture of the cerebral cortex with respect to the mapping of different areas in terms of their respective structural features and the types of nerve cells found in each. Among these was Brodmann, whose famous architectonic map of the human cortex, first published in 1908, still remains as the standard basis of reference for the cortical areas. Figure 1.8 shows a lateral and a mesial view of Brodmann's maps with numbered areas designated up to forty-seven. Briefly, some of the areas are: primary visual receiving area (17), visual association areas (18, 19); auditory areas (41, 42, 22); somatosensory areas (3, 1, 2); motor area (4), premotor or extrapyramidal area (6), motor eye fields (8); prefrontal association cortex (9, 10); orbitofrontal cortex (11); cingulate gyrus (23, 24, 31). Figure 1.9 shows a brain map employing some of Brodmann's

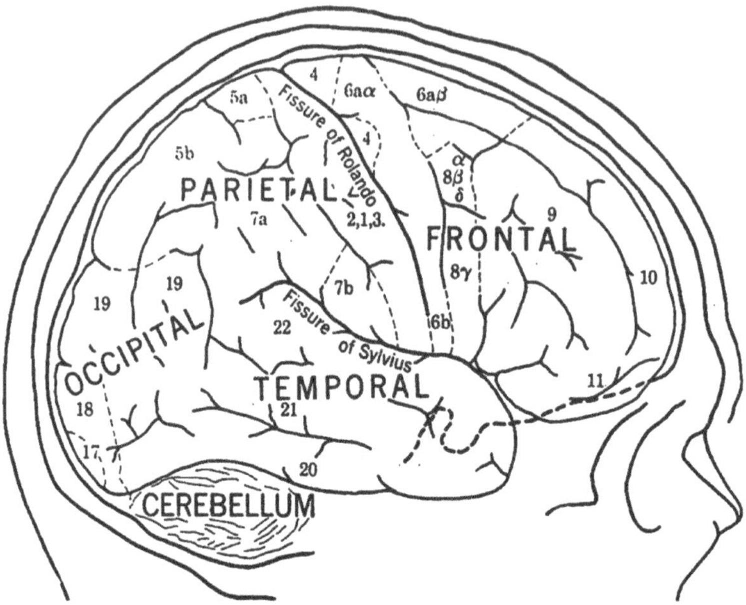

FIGURE 1.9. Brain map with labeled lobes, fitted to an x-ray of the skull, and with certain brain regions numbered according to Brodmann's map (from Lindsley, 1948).

numbers which has been fitted to an outline tracing from an X-ray of my skull. Among other things it provides some idea of the relative thickness of the scalp and skull through which the EEG must be recorded in the case of intact human subjects.

Experimental Study of Behavior and Conditioned Reflexes (Thorndike, Pavlov)

Very great progress was made during the second half of the 19th century in gaining a basic understanding of the gross and microscopic structure, and to some extent of the functioning of the brain of animals by stimulating, recording, and ablating portions and testing for loss of sensory and motor capability. Few, if any, attempts were made to study the behavior of animals in perceptual discrimination and learning situations. Thorndike must be credited with initiating, on an experimental and quantitative basis, the first thorough-going studies of behavioral learning. While still an undergraduate at Harvard University in 1897, Thorndike, under the aegis of the famous psychologist, William James, began his experiments with chicks in the cellar of James' house. In many respects this was the beginning of the rapid rise of American behavioristic psychology which flourished especially during the period from about 1930 to 1950 and still constitutes an important part of objective psychology. The behavioristic movement was given further impetus during the 1920s by E. B. Holt, John B. Watson and A. P. Weiss.

Thorndike received his Ph.D. degree from Columbia University in 1898 and his thesis was based on a series of comparative studies of behavior in animals. It was titled, *Animal Intelligence: An Experimental Study of the Associative Processes in Animals.* His experiments had important implications for human, as well as animal learning, and the basic data from his thesis were published in book form in 1911. For forty years or more his learning theory dominated all others in America, though vigorously attacked by others. He experimented with fish by making them learn to find and swim through an opening in a glass partition in an aquarium. He taught chicks to find their way through mazes containing cul-de-sacs or blind alleys. He experimented with cats and dogs in a puzzle box situation in which they had to learn, as he said, by "trial-and-error" to find the method of releasing the door to get out to get food. Monkeys were exposed to more complicated mechanical devices which they had to learn to manipulate in order to open a door. In all instances he felt that the animal did not "think its way out," but instead used what he called a trial-and-error procedure. This meant that the animal learned by making multiple responses from its instinctive or previously learned repertoire until one of them worked. He felt the animal had to have a set, attitude or disposition (internal drive), such as a state of hunger, which made it want to get out of the puzzle box to get food. As the learning proceeds the animal gradually eliminates useless or unsuccessful steps, but it does not, according to his views, suddenly develop "insight" as a human does, or perhaps the higher apes. Animals assimilate and utilize steps learned in previous problem situations, whether or not they are correct in the new situation. Finally, he thought that animals utilize what he called associative shifting, i.e., they will respond to a part of a general situation as if it were the whole, or they will shift a response to a stimulus which has been presented with an earlier rewarding stimulus if the rewarding stimulus is no longer present. This is in essence what Pavlov called a conditioned reflex.

Thus, Thorndike was of the opinion that animals do not think or reason or develop concepts as does man. Koffka and Köhler, German Gestalt psychologists, later contradicted Thorndike with respect to the matter of whether animals learn by "insight," i.e., the sudden solution of a problem or puzzle. Köhler's famous experiments with apes, while he was confined on the island of Tenerife off the coast of Africa during World War I, were thought to be examples of insightful behavior. Harlow also has expressed the opinion that the higher primates below man show evidence of reasoning. After the first ten years Thorndike devoted most of the rest of his life to human learning, but found that many of the basic principles upon which he built had been earlier derived, in part, from his animal studies.

Conditioned Reflexes. The distinguished Russian physiologist, I. P. Pavlov, who won a Nobel award in 1904 for his work on visceral function and digestive physiology, began a whole new investigative career at that time. In his earlier studies of digestive function in dogs he had pondered questions which related to the dog's salivary secretion when no food was actually presented, but when the attendant who usually fed the dog appeared with an empty dish in his hand. Pavlov concluded that some connection or association had been formed in the brain between two or more stimuli associated with the feeding situation. For the next 30 years he and his many students and collaborators did some remarkable experiments which confirmed this view and which extended the concept of the conditioned reflex to many aspects of behavior and its elaboration. The associations formed between two stimuli in the external environment he referred to as involving the *first signal system*. Associations formed between percepts, concepts and ideas belong to the *second signal system*, which is what we call today cognitive functions. This involves speech, language and communication functions, as well as thinking, reasoning and problem solving. One can see that the basic question raised by Thorndike in his early animal experiments as to whether animals can think and reason and form and store concepts is a very real one. Can dogs or other animals form concepts and can they forge associative or conditioned links between concepts as man does and as even young children do in simpler situations? This is a fundamental and persistent problem which always seems to defy a completely unequivocal positive answer.

Remarkable things have been accomplished experimentally by conditioning methods with animals, children and human adults. Conditioning procedures have been employed in various contexts known as classical conditioning, instrumental conditioning, and in the past 20 years or more as operant conditioning. In attempting to relate what goes on in the brain in classical conditioning, it has been postulated that an unconditioned stimulus (smell or sight of food), which is effective in causing the salivary glands to secrete (see Figure 1.10), becomes linked at a motor center for salivary secretion with an auditory or other stimulus which normally does not induce salivary secretion. When the unconditioned stimulus (UCS) occurs repeatedly in association with the ineffective or conditioned stimulus (CS) the latter will eventually elicit the response in the absence of the UCS. How does this happen? In connection with a motor response like a leg withdrawal to electric shock on the foot pad, it has been found possible to elicit this response by stimulating the motor cortex directly without the use of the foot shock. If now a sound or a light stimulus is given there will be no foot lifting. After many pairings of a stimulus, say the light, with the weak electric shock to the motor cortex, the light will eventually come to cause the foot to lift before the UCS occurs, or even if it does not occur.

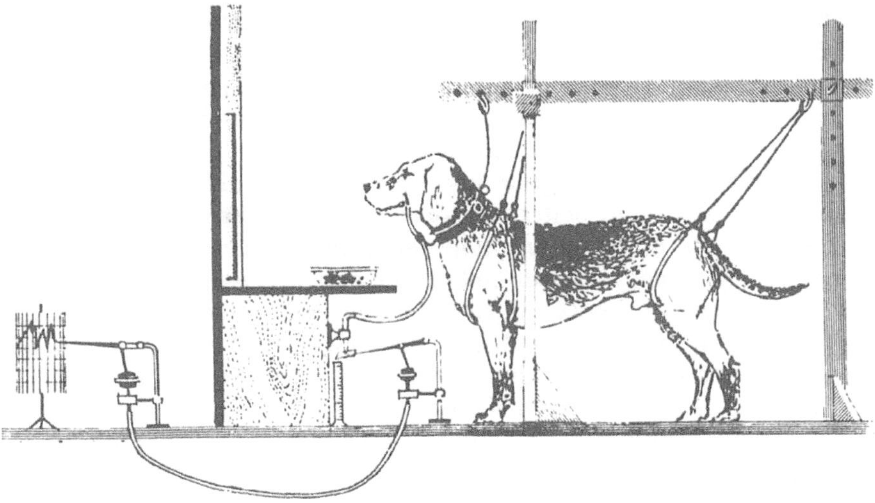

FIGURE 1.10. Typical Pavlovian dog stock restraining animal during classical Pavlovian salivary conditioning. A cannula in duct of salivary gland leads drops of saliva to delicately balance lever which activates pneumatic system causing stylus to write on drum with each increment in salivary output. Smell or sight of food and eating causes salivary flow. A neutral stimulus (tone) serving as a conditioned stimulus (CS), preceding or simultaneous with presentation of food, after a number of pairings with the UCS will cause salivation when presented without the UCS. This is a conditioned response (from Ruch, 1958).

Rusinov has demonstrated that applying a very weak DC polarizing current to the motor cortex, which does not elicit the response but apparently leaves a condition of sensitization there, creates a predisposing state at that motor site so that now a stimulus which did not cause the response can be applied and elicit the response. This type of experiment implies that the pairing of a CS with a UCS accomplishes somewhat the same type of thing. The initially effective UCS creates a negative polarization each time it utilizes these circuits or pathways leading from the cortex to the responding structure, and the radiation of impulses from the sensory cortex of the ineffective stimulus finds the threshold at the motor cortex lowered and the response is triggered by the originally ineffective stimulus. There are of course many aspects of the entire process which are not understandable in terms of this simple explanation, but it is a starting point.

We have seen how for 2000 years there has been concern about how man senses, perceives, reasons, forms judgments and responds. We have also seen that he has gradually learned a considerable amount about the structure of his sense organs, sensory pathways, sensory cortex, motor cortex and so forth, and that he is gradually learning in gross and microscopic detail about the functioning of these mechanisms. Several methods or approaches have been mentioned, but by no means all. It seems clear that much progress has been made by studying the neurophysiology and electrophysiology of various parts of a system, as well as whole, but experimentally isolated systems. It is also clear that much progress has been made in diversified studies of behavior, by developing better controlled stimulus presentation methods and better methods of observing and recording behavior objectively and quantitatively. Finer and finer units of the nervous system have come under study whether by stimulating single nerve cells or recording electrical potentials from single brain cells, and by finer control of drugs or chemical agents injected by micro-pipette into specific localized regions. These are only a part of the specialized methods and instrumentation which is available today. The problem seems to be not one of instrumentation and methodology, but the formulation of the problem in a logically appropriate way, that is, learning how to ask the right questions and then how to formulate a strategy which will produce the right answers.

In general, students today are much better and more broadly trained than many of us were; they have access to, and knowledge about, many new techniques for acquiring and treating data. With computers properly programmed it can all be done with much greater speed and less expenditure of energy. Hopefully, many of these ancient problems, thus far attacked by relatively crude methods and technologies, will give further ground as young bright minds, motivated by the vast unknown of the brain, sharpen the focus of their questions and apply new and powerful tools that are available.

REFERENCES

Brodmann, K. (1908). Beiträge zur histologischen Lokalisation der Grosshirnrinde. VI. Die Cortexgliederung des Menschen. *Journal für Psychologie und Neurologie (Leipzig), 10*, 231-246.

Dart, R. A. (1959). *Adventures with the missing link*. New York: Harper.

Lindsley, D. B. (1944). Electroencephalography. In J. McV. Hunt (Ed.), *Personality and the behavior disorders* (Vol. 2, pp. 1037-1103). New York: Ronald Press.

Lindsley, D. B. (1948). Studying neuropsychology and bodily functions. In T. G. Andrews (Ed.), *Methods of psychology* (pp. 417-458). New York: Wiley.

Magoun, H. W. (1963). *The waking brain*. Springfield,IL: Thomas.

Moore, R. (1953). *Man, time and fossils*. New York: Knopf.

Ruch, F. L. (1958). *Psychology and life*. Chicago: Scott, Foresman.

SECTION II. NEURAL CONTROL

The six chapters of this section are all concerned with neural control, and thus can lead to a better understanding of brain-behavior relationships. By neural control is meant the automatic regulation of bodily functions by the nervous system, such as occurs in homeostasis. The different investigations of neural control involve a variety of experimental techniques such as intracellular unit recordings (Huang, Krasne); selected lesions (Feeney, Velasco & Velasco); electrical stimulation (Krasne, Skinner, Wilson); cryogenic blockade (Skinner); and a number of unique behavioral techniques used by almost all of these investigators.

In his chapter, Professor Krasne nicely illustrates how the topic of brain-behavior relationships can be better understood through an examination of the escape behavior of the lowly crayfish, an organism with a relatively simple nervous system. In summarizing his research on crayfish, Krasne concludes that the purpose of his research has "graduated from analyzing the circuits which directly *mediate* a behavioral reaction," to studying the network of neurons that *control* the excitability of the mediational circuitry. In short, his research is no longer primarily concerned with studying the afferent and efferent pathways of the crayfish as much as it is concerned with studying the crayfish's neural control mechanisms--its "reticular activating system," as it were.

In their chapter, the Velascos continue along similar lines, describing how stereotaxic lesions in the rostral reticular formation of the monkey suppress experimental tremors. On two occasions, separated by an interval of three months, rhesus monkeys were lesioned in the mesencephalon. The first lesion was to produce tremors; the second was to suppress them. Regions of both tremor production and tremor suppression were identified in the mesencephalon. Such lesion research has potential implications for a better understanding of Parkinson's disease in humans.

Feeney's chapter describes the investigation of the effects of rostral thalamic and/or basal forebrain lesions on the EEG activity of cats before and after the inducement of seizure activity by administering either penicillin or Metrazol. The results showed that lesions of either the rostral thalamus or the basal forebrain suppressed several forms of synchronous EEG rhythms: spontaneous sleep spindles, barbiturate spindles, recruiting responses, and seizure activity. These results support the concept of a forebrain mechanism that regulates synchronous neural discharges.

Huang's chapter is a follow-up report on a systematic study using electrophysiological methods. The study investigated unitary sensory

 DOI: 10.4324/9781003164647- 3

responses recorded extracellularly in the pulvinar and other posterior association nuclei of the cat. Responses were recorded during visual, auditory, and somatosensory stimulation. Ninety-two per cent of the units responded to flash stimulation, but far fewer responded to somatosensory and auditory stimulation. The latencies of the unit responses were divided into two groups, early and late unit responses, with the visual units responding earlier than the auditory and somatosensory units.

The chapter by Wilson, Hirasuna, and Lindsley summarizes prior research describing the effects of electrical stimulation of functional systems within the brainstem and hypothalamus of the cat. These systems were studied in order to determine their influence upon behavior and upon electrical activity recorded in the hippocampus and brainstem. Three basic questions were posed and partially answered. It is concluded that the brainstem regions investigated may represent some of the origins of the hypothalamic-hippocampal systems described previously by Lindsley and his colleagues. Also considered are the relationships between questions addressed by this study and more general issues that have been discussed previously--namely, are there brainstem-limbic systems by which certain behavioral manifestations are expressed?

Skinner's chapter also is concerned with regulation and control, in this case the relationship between factors of psychological stress and cardiac vulnerability. Specifically, Skinner argues that cardiac disease, and death due to cardiac illness, may be related to both the autonomic nervous system and the frontal cortex. In a careful and scholarly fashion, he marshals together an abundance of evidence supporting this thesis. In essence, Skinner's theory, as he notes, is an interesting update of Cannon's cerebral defense system, which postulated that the frontal cortex subserves the function of the simultaneous regulation of both sensory and autonomic channels of information processing.

2·MODES OF CONTROL OVER SENSORY AND MOTOR EVENTS IN A SIMPLE SYSTEM

F. B. Krasne

University of California at Los Angeles

Viewed in a broad sense, the greater portion of Dr. Donald Lindsley's research career has focused on those control processes which turn us from reflexive, stimulus-response automatons into variable, capricious organisms. For what after all are attention, arousal, emotion and motivation but constructs that we use to explain why a stimulus that successfully reaches a receptive surface *may* or may *not* evoke awareness, perception, learning or response.

I was asked to present my research in this festschrift, not because my research interests were viewed as close to those of Professor Lindsley's, but because I happen to be the current chairman of the UCLA Psychology Department's Physiological Psychology Group, a group to which Donald Lindsley has belonged for the greater part of his research career and a group of which he is surely the most illustrious member. Actually, my work, which has all been done on the central nervous system of the crayfish, seems on the face of it to be exceptionally *inappropriate* to this *festschrift*. However, I do not believe that this is so. For leaving the animal aside for the moment, my research interests, like Professor Lindsley's, are in those processes which make behavior variable and uncertain and which are responsible for the impression we all have that our friends and even some of our laboratory animals are possessed of consciousness and a free will.

Unlike Donald Lindsley, however, I have chosen not to attack these problems frontally in the most sophisticated of organisms, but rather have chosen to study them in one of the simplest organisms where they appear as though they might be operating. This animal, which is shown in Figures 2.1 and 2.2, is the lowly crayfish. The behavior pattern that I shall discuss is one of the animal's escape responses. Crayfish escape from threats by rapid flexions of the abdomen which drive the animal rapidly through the water in a more or less backward direction. There are two different fixed patterns of escape that are mediated by nerve cells with giant axons (Krasne & Wine, 1977; Wiersma, 1947). *Lateral giant fiber* mediated reactions are produced by sudden tactile stimulation of the abdomen and pitch the animal upward as shown in Figure 2.2, while *medial giant fiber* mediated reactions are produced by sudden rostral stimulation and drive the animal directly backwards. The

 DOI: 10.4324/9781003164647- 4

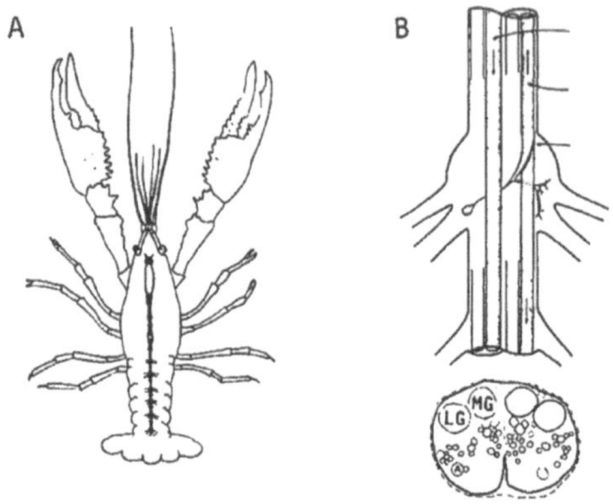

FIGURE 2.1. The crayfish giant fiber system. A. The crayfish nerve cord is diagrammed as it is situated in the animal; it is a chain of ganglia which contain cell bodies and synaptic integrating regions, joined by large tracts, called connectives, of interconnecting axons. B. The lateral giant (LG) and medial giant (MG) command neurons are diagrammed and their location in the connectives between ganglia shown. The location of the largest of the sensory interneurons of the circuit in Figure 2.3 below, interneuron A, is also labeled.

research discussed here all involves lateral giant fiber mediated escape reactions. Anyone who has set out to study these reactions will assure you that the crayfish, if not "conscious", at least does have a mind of its own. And it is this fickleness of crayfish escape behavior which caused me to select it for study.

When my work was started about 20 years ago the nerve cell circuitry for lateral giant mediated escape was only slightly charted; however now, as the result of work in a number of laboratories, the broad outlines of the circuitry are well established (Figure 2.3) (see Krasne, 1969, and especially Zucker, 1972a). The circuit is in essence a four-neuron arc. The sensory neurons are sensitive tactile hairs located all over the dorsal surface of the abdomen. These project to a population of sensory interneurons which in turn project to the dendrites of the neurons that form the lateral giant axons. Finally, the lateral giant neurons make a host of divergent connections to the many motoneurons that are involved in producing lateral giant type abdominal flexions. In Figure 2.3 two categories of motoneurons, the non-giant fast flexors and the giant fast flexors, which will be differentiated below, are indicated separately. There are some additional known connections which are not indicated in this diagram, but the circuit given here will be adequate for matters to be discussed. In general the exact locations and characteristics of the neurons that comprise the wiring diagram of Figure 2.3 are sufficiently constant that one can go to them for stimulation or recording at will in any

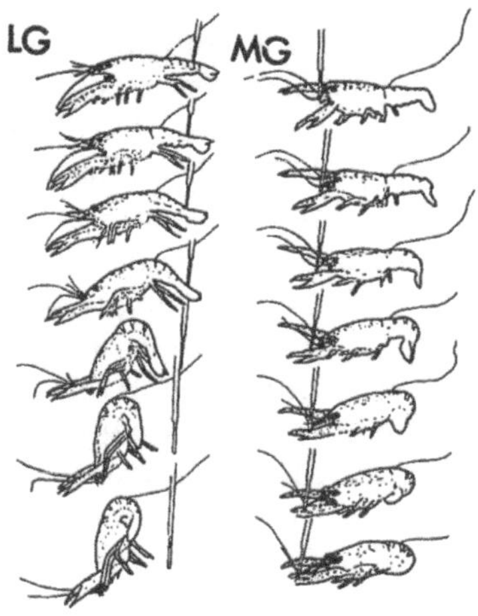

FIGURE 2.2. Forms of escape behavior. MG: A medial giant tailflip, initiated by a tap to the rostrum. Note that all abdominal segments flex and the animal darts backward along a low trajectory. LG: A lateral giant tailflip, initiated by a tap to the abdomen. Note that only the anterior segments flex, so that the force of the caudal abdomen is directed downward, causing the animal to pitch forward.

animal (Figure 2.1). That these neurons can be identified specifically in this way and that one can draw such an explicit wiring diagram as Figure 2.3 for a real behavior is a remarkable thing, and it is one of the primary reasons for studying simple behavior in simple animals. In the present instance, armed with the circuit diagram of Figure 2.3 and a behavior pattern that is fickle, one is in a position to investigate the basis for the capriciousness of the behavior in a way that we cannot yet even approach for the interesting, variable behaviors of higher animals.

Just as with the psychologically interesting behaviors of higher animals, the apparent capriciousness of lateral giant escape results from the influence of experience, behavioral context, psychological set, the occurrence of other on-going responses, and the animal's state of arousal. However, what is relatively unique in the present instance is that by recording from nerve cells in the circuit of Figure 2.3, we and others have begun to determine the routes by which these various factors affect the behavior.

Perhaps the first thing to be said, and one of the most important findings to have emerged, is that most of the control over the excitability of lateral giant escape is exerted by imposing decreases of excitability on an intrinsically very excitable reflex, rather than by augmenting the excitability of an inherently weak reflex (however, see Krasne & Glanzman, 1986 and Miller & Krasne, 1985, for recent evidence of augmentations). This is true for a number of kinds of control (Figure 2.4).

For example, if one restrains an animal so that it cannot get away, the LG escape reaction is profoundly depressed, becoming virtually impossible to evoke even with the strongest stimuli. This loss of reactivity is due to inhibitory influences which play directly on the LG command neuron itself and perhaps also on motoneurons but not on the initial synapses of the escape reflex pathway (Krasne & Wine, 1975). By contrast if we test the escape reflex while an animal is locomoting in ways other than by tailflips, such as by forward walking, we find that the reflex is again suppressed (though less profoundly than before) -- but in this case by an inhibitory action that is at least in part directed to the *first* synapse (Fricke, Block, & Kennedy, 1982; Fricke & Kennedy, 1983; Kandel, Krasne, Strumwasser, & Truman, 1979). If we present stimuli for escape repeatedly, we find that reflex excitability is again lower, but in this case the cause is not inhibition but, rather, an intrinsic depression (or "tiring") of the initial, and only the initial, synapses in the escape reflex pathway (Krasne, 1969; Wine, Krasne, & Chen, 1975; Zucker, 1972b).

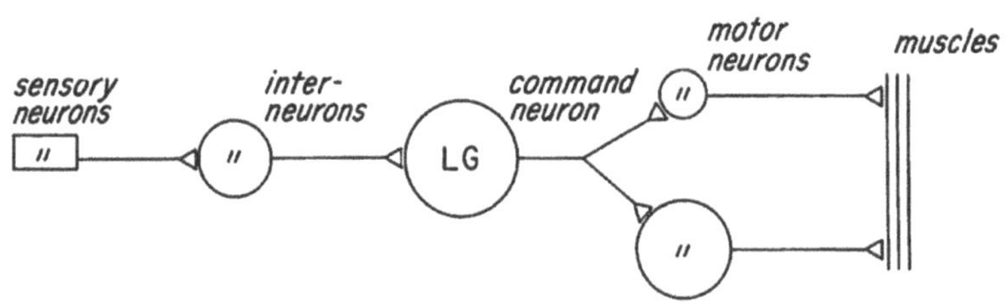

FIGURE 2.3. Circuit mediating the lateral giant escape reaction. Ditto marks indicate that there are a number of neurons of this type in parallel. Only prominent connections relevant to the present discussion are shown.

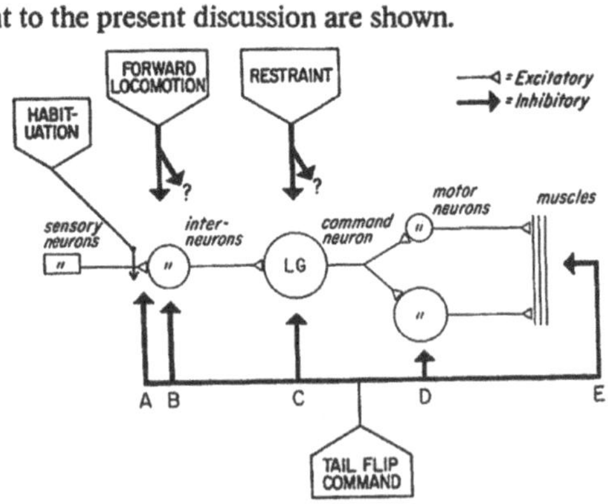

FIGURE 2.4. Controlling influences on the lateral giant escape reaction. Explanation in text.

Such information on modes and loci of control at once raises questions about *why* control is exerted as it is in each case. Attempts to answer this question can in turn begin to give us new insights into the strategies and techniques used by nervous systems for achieving integrated action. Let me illustrate the heuristic value of the kind of information that a picture such as Figure 2.4 provides with an example that will bring us to one of Professor Lindsley's favorite topics, the control of information flow through sensory pathways.

One of the circumstances under which the lateral giant escape reaction is most powerfully suppressed is when the tailflip motor apparatus is operating due to the issuance of a tailflip command by any of the several systems that can issue such commands (Figure 2.4). Analysis shows that shortly after a tailflip command is issued by any system there is direct inhibitory action on virtually every possible target in the LG escape circuit (Krasne & Wine, 1977): this includes presynaptic inhibition of the terminals of the primary afferent fibers (Figure 2.4A), postsynaptic inhibition of the sensory interneurons (Figure 2.4B), postsynaptic inhibition of the lateral giant command neurons (Figure 2.4C), inhibition of tailflip motoneurons (Figure 2.4D), and even direct postsynaptic inhibition of the muscle fibers themselves (Figure 2.4E). When this widespread inhibition was first discovered, it seemed to make sense as a way to prevent a new reaction from starting while an old one was already in progress, and we at first supposed that it occurred at so many levels in order to insure absolutely that a competing reaction would not occur. However, detailed analysis showed that this inhibition of the escape reflex pathway was far from a unitary phenomenon.

As Figure 2.5 shows, the inhibition at the lateral giant command neuron itself starts first; it is followed only later by the two initial synapse inhibitory actions and much later by inhibition to the muscles. The giant flexor motoneurons are inhibited at about the same time as the command neurons, but the non-giant motoneurons are not inhibited at all. The various inhibitions are also of very different durations from one another. In other words, the inhibition is not a unitary phenomenon but, rather, a carefully orchestrated spatial and temporal pattern of potentially independent modulatory effects. One, therefore, presumes that each action has its own special function (Bryan & Krasne, 1977b; Wine, 1977; Wine & Mistick, 1977).

In the remainder of this essay I will focus on the first of the inhibitory actions (A), the presynaptic inhibition at the terminals of the sensory neurons. It is of particular interest here, first of all because it is the sort of centrifugal control of sensory inflow that has played such a large part in the thinking of Professor Lindsley and others in the UCLA Brain Research Institute, secondly because it appears to be such an expensive way to do a simple thing (effective presynaptic inhibition requires that a very large number of terminals be inhibited whereas inhibition of the pathway

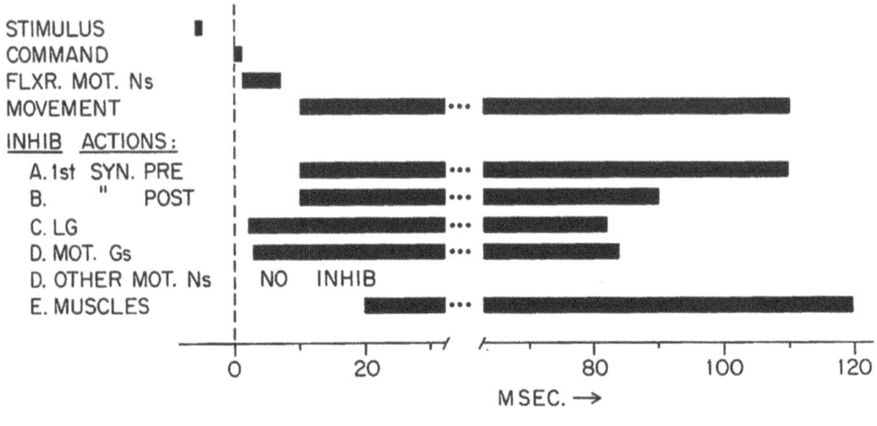

FIGURE 2.5. Time courses of inhibitory actions that follow a tailflip command. The moment of firing of the command neuron is taken as time zero.

postsynaptically, after convergence, requires inhibition of a relatively much smaller population of targets), and thirdly because the investigation of this presynaptic inhibition turns out to tell us something quite new and interesting.

The first synapse inhibition that is under discussion is illustrated in Figure 2.6. At the left is shown an EPSP produced in interneuron A, the largest of the first order sensory interneurons in the escape reflex pathway, by a brief shock to a sensory nerve (given at the square dot). At the right the same stimulus was given about 20 msec after activation of tailflip motor circuitry by a shock to a tailflip command neuron, indicated by the arrow. Obviously the inhibition is remarkably powerful. The evidence that this inhibition is in substantial measure presynaptic stems partly from experiments I am about to describe and partly from intracellular recordings from the terminal regions of afferents, which show that the terminals become depolarized and spikes in them become shunted during this period of inhibition (Kennedy, Calabrese, & Wine, 1974).

At issue is the question of just what the purpose of this inhibition might be. What is it that can be achieved here by presynaptic inhibition that is not adequately achieved by the postsynaptic inhibition that is also present? A possible answer comes when we think about the way that this nervous system achieves habituation to innocuous mechanical stimuli. This habituation is due to intrinsic depression of the synapses between the primary afferents and the first order interneurons, and a variety of lines of evidence indicate that such depression is in large measure due to decreased transmitter release from the presynaptic neuron. When we think about this situation a bit, it becomes apparent that the animal has a real problem on its hands. The tactile

FIGURE 2.6. Inhibition of transmission between tactile afferents and sensory interneuron A. In both traces the square dot marks afferent nerve stimulation. In the second trace an escape command neuron was stimulated at the point indicated by the triangle. Calibration 2 mV and 20 msec.

receptors which are the normal input pathway to the lateral giant escape reflex are extremely sensitive and respond not only to touch but also to vibratory stimuli and water currents. Now, crayfish perform tailflip locomotion not only in the context of lateral giant fiber mediated escape but also for many other reasons -- for example, to reach food that is suspended in the water high above the substrate. Whenever a crayfish flips its tail for any reason it will necessarily excite the very sensitive tactile receptors that feed into the lateral giant escape reflex pathway, and given the mechanism for habituation that we have indicated, one would expect that the firing of the tactile receptors produced by the animal's own tailflip movements would lead to habituation of the lateral giant escape reflex. This would obviously be a most unfortunate state of affairs for the animal; for it would mean that whenever an animal flipped its tail for any reason its lateral giant escape reflex would subsequently become habituated, and the animal would become temporarily unable to respond to threats to its abdomen.

However, notice in Figure 2.5 the particular time in the sequence of inhibitory actions at which the afferent terminals are inhibited; their inhibition starts precisely at the time when tailflip movements begin. This suggests the possibility that their inhibition could function to protect the afferent terminals from habituation during periods when the animal is actually flipping its tail. This can be tested by a simple experiment (Bryan & Krasne, 1977a, 1977b). Using a de-efferented central nervous system to avoid complications due to feedback from movement, we have two conditions (Figure 2.7, top). In *control* series we give a sequence of stimuli to afferent nerves that will cause depression of synaptic transmission to interneuron A, which in turn contributes to habituation of the lateral giant escape reaction. In *experimental* series we give precisely the same sequence of stimuli except that each of the "training" stimuli is preceded by activation of tailflip circuits

TRIAL NUMBER:	1	2	. . .	10	11
CONTROL SERIES:	S	S	. . .	S	S
EXPTL SERIES:	M→S	M→S	. . .	M→S	S

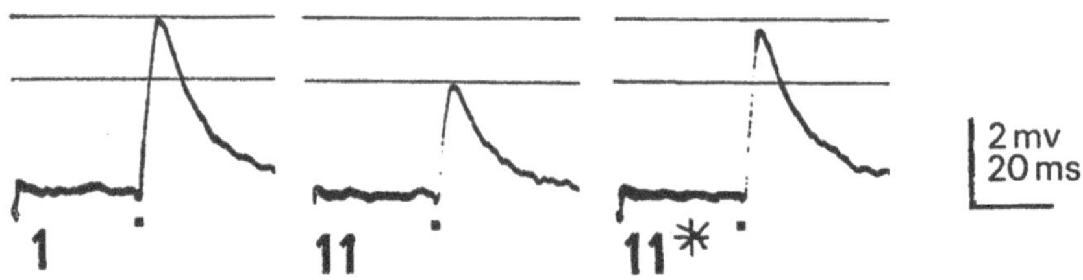

FIGURE 2.7. Protection from synaptic depression at sensory fiber - interneuron A synapses. Explanation in text.

at a time such that the stimulus deliveries come when the animal, were it intact, would be making a tailflip movement. In both sorts of series a final test trial is given to evaluate the ultimate excitability of the escape reflex. The bottom part of Figure 2.7 shows the results of such an experiment. The first two traces show the reduction in the EPSP in interneuron A resulting from a control series of repetitive stimulations; the first and the last trial are shown. The third trace shows the final test trial in an experimental series in the same animal; though there has been a slight reduction from the initial naive size of the EPSP, the reduction has been very slight. In other words when stimuli are presented during the time when the animal would be flipping its tail they do not lead to depression of the first synapses.

Figure 2.8 shows the relationship between the time courses of protection from habituation, inhibition at the first synapse (this is the combined effect of pre- and postsynaptic inhibition), and postsynaptic inhibition alone as indicated by the duration of the IPSP produced in interneuron A following a tailflip command. What is clear is that there is a very good correspondence between inhibition (though not postsynaptic inhibition) and protection. Given these results it seems reasonable to argue that the purpose of pathway A in Figure 2.4 is not to control input per se, but rather to control the impact of stimuli on the animal's plastic synapses -- i.e., on its *memory* system.

I believe that this is more than a clever adaptation of one lowly species. Plastic synapses that alter their efficacy with use are used for all kinds of jobs in the central nervous system and spurious presynaptic activity must be a common problem at many of these synapses. It, therefore, seems to me quite likely that special control circuitry that makes use of presynaptic inhibition to protect plastic synapses from undesirable change will be found to be a common feature in many nervous systems, and such protection may in fact be

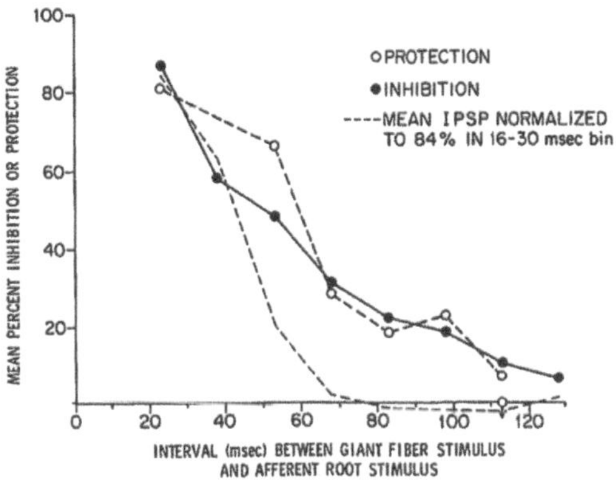

FIGURE 2.8. Time courses of protection from depression, total inhibition, and postsynaptic inhibition. Percentage protection is computed by the formula (1 - % decline in EPSP due to 10 experimental trials/% decline due to 10 control trials) X 100.

one of the paramount services that presynaptic inhibition provides to nervous systems (Krasne, 1978). Of course, this is only a hypothesis. The important point right now is that it shows how analysis of integration in simple behavioral systems can lead to ideas of great potential generality.

Let me conclude by returning briefly to the overall pattern of inhibition produced during tailflips. We and Jeffrey Wine of Stanford have been successful in beginning to locate and determine the connections of neurons that orchestrate the inhibitory pattern of Figure 2.5. Our findings are shown in Figure 2.9 (see Wine & Krasne, 1982). Above is again shown the basic LG escape reaction pathway; below is shown such inhibitory circuitry as has been discovered. What is beginning to become clear is that the inhibitory control circuitry is probably going to be very much more complicated than the circuitry of the excitatory pathway itself. While this work is just beginning, I think that it is very exciting. For it means that at least in this case we have graduated from analyzing the circuits which directly *mediate* a behavioral reaction, and we are beginning to work our way into the network of neurons that *control* the excitability of the mediational circuitry. And it is, of course, these control neurons that are responsible for much of the caprice of behavior that prompted our original investigation. In a certain sense it might be said that we have graduated from the study of the classical afferent and efferent pathways of our little animal to the study of the animal's "reticular activating system" -- but with a possibility for incisive analysis hardly achievable in complex higher animals.

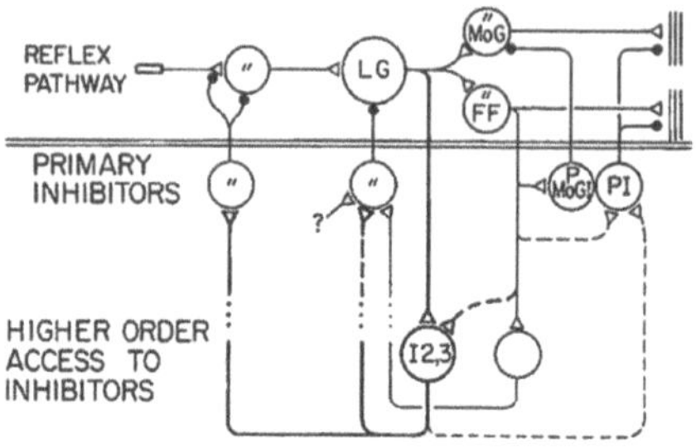

FIGURE 2.9. Some of the circuitry mediating the inhibitory effects that are produced by command neuron firing. Above the double line the basic LG circuit is recapitulated. Here MoG refers to the giant motor neurons that were indicated by large circles in previous figures. Below the lines are neurons that have been identified as being involved in mediating inhibition. I2 and I3 also receive excitatory input from other escape command neurons (in addition to the lateral giants themselves). Solid pathways are strong; dashed ones weak. Based on material in Krasne and Wine, 1977; Wine, 1977; and unpublished material.

REFERENCES

Bryan, J. S., & Krasne, F. B. (1977a). Protection from habituation of the crayfish lateral giant fibre escape response. *Journal of Physiology, 271*, 351-368.

Bryan, J. S., & Krasne, F. B. (1977b). Presynaptic inhibition: The mechanism of protection from habituation of the crayfish lateral giant fibre escape response. *Journal of Physiology, 271*, 369-390.

Fricke, R. A., Block, G. D., & Kennedy, D. (1982). Inhibition of mechanosensory neurons in the crayfish. II. Inhibition associated with proprioceptive feedback from locomotion. *Journal of Comparative Physiology, 149*, 251-262.

Fricke, R. A., & Kennedy, D. (1983). Inhibition of mechanosensory neurons in the crayfish. III. Presynaptic inhibition of primary afferents by a central proprioceptive tract. *Journal of Comparative Physiology, 153*, 443-453.

Kandel, E. R., Krasne, F. B.,Strumwasser, F., & Truman, J. W. (1979). Cellular mechanisms in the selection and modulation of behavior. *Neurosciences Research Program Bulletin, 17*, 523-710.

Kennedy, D., Calabrese, R. L., & Wine, J. J. (1974). Presynaptic inhibition: Primary afferent depolarization in crayfish neurons. *Science, 186*, 451-454.

Krasne, F. B. (1969). Excitation and habituation of the crayfish escape reflex: The depolarizing response in lateral giant fibres of the isolated abdomen. *Journal of Experimental Biology, 50,* 29-46.

Krasne, F. B. (1978). Extrinsic control of intrinsic neuronal plasticity: An hypothesis from work on simple systems. *Brain Research, 140,* 197-216.

Krasne, F. B., & Glanzman, D. L. (1986). Sensitization of the crayfish lateral giant escape reaction. *Journal of Neurosciences, 6,* 1013-1020.

Krasne, F. B., & Wine, J. J. (1975). Extrinsic modulation of crayfish escape behavior. *Journal of Experimental Biology, 63,* 433-450.

Krasne, F. B., & Wine, J. J. (1977). Control of crayfish escape behavior. In G. Hoyle (Ed.), *Identified neurons and behavior of arthropods* (pp. 275-292). New York: Plenum.

Miller, M. W., & Krasne, F. B. (1985). Long-lasting stimulation-induced enhancement of synaptic transmission in the crayfish lateral giant escape circuit. *Society for Neuroscience Abstracts, 11,* 795.

Wiersma, C. A. G. (1947). Giant nerve fiber system of the crayfish: A contribution to comparative physiology of synapse. *Journal of Neurophysiology, 10,* 23-38.

Wine, J. J. (1977). Neuronal organization of crayfish escape behavior: Inhibition of the giant motoneuron via a disynaptic pathway from other motoneurons. *Journal of Neurophysiology, 40,* 1078-1097.

Wine, J. J., & Krasne, F. B. (1982). The cellular organization of crayfish escape behavior. In D. C. Sandeman & H. L. Atwood (Eds.), *The biology of crustacea: Vol. 4. Neural integration and behavior* (pp. 29-35). New York: Academic Press.

Wine, J. J., Krasne, F. B., & Chen, L. (1975). Habituation and inhibition of the crayfish lateral giant fibre escape response. *Journal of Experimental Biology, 62,* 771-782.

Wine, J. J., & Mistick, D. C. (1977). Temporal organization of crayfish escape behavior: Delayed recruitment of peripheral inhibition. *Journal of Neurophysiology, 40,* 904-925.

Zucker, R. S. (1972a). Crayfish escape behavior and central synapses. I. Neural circuit exciting lateral giant fiber. *Journal of Neurophysiology, 35,* 599-620.

Zucker, R. S. (1972b). Crayfish escape behavior and central synapses. II. Physiological mechanisms underlying behavioral habituation. *Journal of Neurophysiology, 35,* 621-637.

3 · RETICULAR LESIONS SUPPRESS EXPERIMENTAL TREMOR IN MONKEYS

F. Velasco and M. Velasco

National Medical Center IMSS, Mexico City

It has been shown that a small subthalamic lesion suppresses tremor in the contralateral extremities of patients with Parkinson's disease (Bertrand, Hardy, Molina-Negro, & Martinez, 1969). Although the precise anatomical structures involved are unknown, it has been postulated that such lesions interfere with the prelemniscal radiations, a system of fibers extending from the mesencephalic reticular formation to the nucleus ventralis oralis anterior thalami, based upon a series of indirect radiological, electrophysiological and computer vectorial tests (Velasco & Molina-Negro, 1973; Velasco, Molina-Negro, Bertrand, & Hardy, 1972; Velasco, Velasco, & Machado, 1975; Velasco, Velasco, & Maldonado, 1976; Velasco, Velasco, Maldonado, & Machado, 1975; Velasco & Velasco, 1975, 1979). Presumably, tremorogenic impulses originating in the mesencephalic reticular formation reach the pyramidal system through a subthalamic-thalamic-cortical pathway. In order to test this hypothesis, an attempt to suppress experimental tremor in monkeys by means of lesions in the rostral reticular formation was made.

METHODS

Experiments were performed on a group of rhesus monkeys. Two consecutive stereotaxic lesions were placed in the mesencephalon of each monkey with an interval of 3 months between lesions. A first lesion for tremor production aimed at the pars compacta of the substantia nigra (A = 6.0 to 9.0; L = 2.5, H = -2.0) and a second lesion for tremor suppression aimed at the rostral portion of nucleus reticularis mesencephali (A = 5.0 to 7.0; L = 3.0, H = 2.0 to 3.0). Surgical procedures were done aseptically under light barbiturate anesthesia (Nembutal 30 mg/kg) combined with local anesthesia of fixation points and surgical incisions (xylocaine 1%). The surgery consisted of a small incision and burrhole at the parasagittal parietal region which permitted the sequential penetration of two electrodes into the brain: a first exploratory electrode (tip size 10 microns, resistance 40 Kohms) was normally oriented to the stereotaxic target, recording the cellular activity while simultaneously observing the behavioral response to electrical stimulation (60/sec, 1 msec duration, 50-400 μA) of each thalamic and subthalamic region found along its vertical trajectory. Once the targets were

DOI: 10.4324/9781003164647- 5

electrophysiologically identified, a second lesioning electrode (Radionics Model RFG-4) replaced the first one for the production of a radiofrequency lesion (80° C, duration 1 to 2 min). After each lesion, animals were allowed to recover and presence or absence of tremor was detected by means of a single plane accelerometer (Grass Mod. SPA) and electromyography of flexor and extensor muscles in the forearm at least once every week during 3 months. At the completion of the experiments, the site and extent of cerebral lesions were histologically determined by means of serial coronal sections (10 microns thick, taken every 100 microns) from the diencephalon and mesencephalon stained with the Klüver-Barrera technique.

RESULTS

The area for tremor production was identified before lesioning by a spontaneous rhythmical 4 to 6 c/sec cellular activity with a mean interburst interval of 197 msec. Electrical stimulation of this area induced ipsilateral oculomotor signs consisting in eyelid retraction, internal rotation of the eye and pupillary constriction. The effectiveness of the lesion for tremor production depended on its critical site and volume: lesions involving the pars compacta of the substantia nigra, the magnocellular portion of the red nucleus and intermediate zone between these two structures produced tremor (Figure 3.1, stripped ventral areas). All 17 monkeys in which the above-mentioned electrophysiological criteria for lesioning the ventromedial mesencephalon were fulfilled developed persistent tremor that appeared within 15 days: 11 monkeys developed postural type, and 6 monkeys "intentional" type tremors. Besides the trembling extremities, 8 animals presented a fixed posture in flexion and 4 a mild incoordination. No other motor disturbances were noticed.

The area for tremor suppression was identified before lesioning by non-rhythmical cellular activity enhanced by nociceptive stimulation applied anywhere on the face and extremities. Electrical stimulation of this area induced bilateral oculomotor signs consisting of eyelid retraction, pupillary dilatation, facial gesticulation, and body movements resembling an arousal response (Magoun, 1958). The effectiveness of the lesion for tremor suppression also depended on its critical site and volume: lesions which involved the subthalamic area of nucleus reticularis and extended backwards to its mesencephalic portion (A = 3.0) and which compromised small marginal parts of the brachium conjunctivum and red nucleus suppressed postural and intentional tremor (Figure 3.1, crossed dorsal areas). On the contrary, lesions involving only small portions of nucleus reticularis mesencephali, even when they were accompanied by involvement of large portions of contiguous structures (red nucleus, brachium conjunctivum, medial lemniscus, midline thalamic nuclei, etc.), produced no modification or

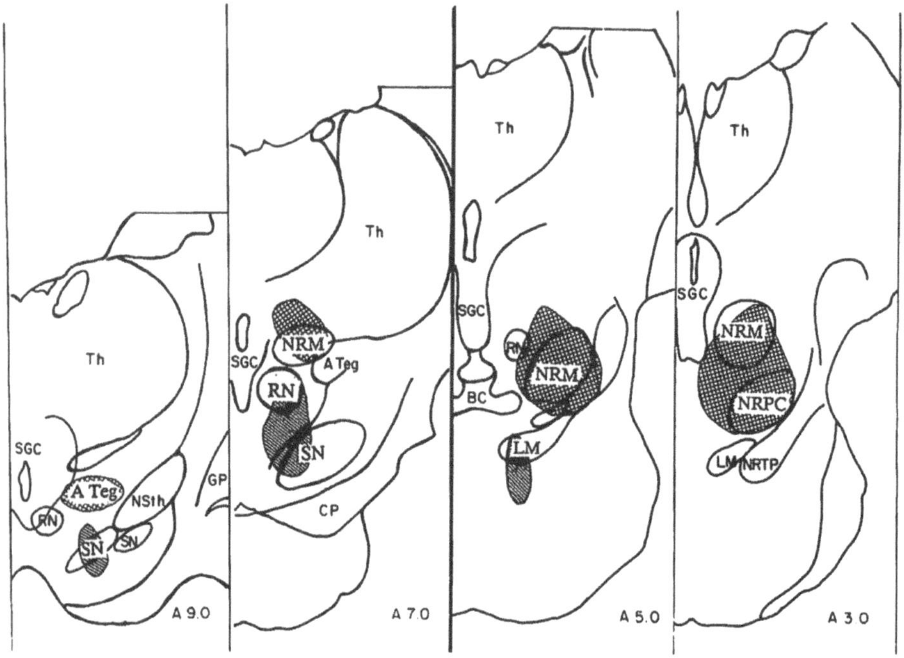

FIGURE 3.1. Diagrams of the monkey diencephalon and mesencephalon at frontal levels 9.0, 7.0, 5.0 and 3.0 showing the site and extent of lesions which produced (stripped ventral areas) and suppressed tremor (crossed dorsal areas). Abbreviations: A Teg = area tegmentalis, BC = brachium conjunctivum, CP = cerebral peduncle, LM = medial lemniscus, NR = red nucleus, NRM = nucleus reticularis mesencephali, NRPC = nucleus reticularis pontis oralis, NSth = nucleus subthalamicus, SGC = periaqueductal gray, SN = substantia nigra, Th = thalamus.

transient decrease in tremor amplitude.

Reticular lesions which suppressed tremor induced also a transient state of "inattention" of the contralateral extremities, similar to that found in humans after subthalamic lesions (Velasco, Velasco, Maldonado, & Machado, 1975). In the monkey it consisted of a preferred use of the extremities ipsilateral to the lesion side over the contralateral ones for spontaneous movements, such as feeding, scratching or tactile exploration. However, when ipsilateral extremities were held from being used the animals used the contralateral extremities with good muscular strength and coordination. The animals also showed postural defects and a tendency to fall to the side contralateral to the lesion. As in the humans, these motor abnormalities vanished in 2 to 10 days and the tremor did not reappear.

3. RETICULAR LESIONS

DISCUSSION

Present results confirm those of previous investigations establishing that lesions in the nigro-rubral area produce tremor. Whether or not the lesions interfere with an ascending nigro striatal (Poirier, 1960; Poirier et al., 1969) or a cerebello-rubro-thalamic system (Carrea & Mettler, 1947) cannot be assessed by the present experiments. On the other hand, the presence of rhythmic 4-6/sec spike burst activity in the nigro-rubral area of an intact monkey and the fact that lesions in this area produced (instead of suppressing) tremor are contrary to the idea that rhythmical activity is the electrophysiological manifestation of a brain stem tremorogenic generator (Lamarre, 1975; Lamarre, De Montigny, Dumont, & Wess, 1971). In contrast, present experiments showed that lesions in the nucleus reticularis mesencephali were critical for the suppression of tremor at the contralateral extremities and although lesions to neighboring structures were unavoidable in most cases, we found that lesions destroying these structures but sparing nucleus reticularis mesencephali did not have the same effect.

The pathways through which the nucleus reticularis mesencephali influences the presence of tremor in the contralateral extremities remain to be studied. However, since the effects found with lesions in this area are similar to those found in humans after lesions in the subthalamic and ventrolateral thalamus (Velasco, Velasco, & Machado, 1975; Velasco et al., 1976; Velasco, Velasco, Maldonado, & Machado, 1975; Velasco & Velasco, 1975, 1979), the pathways most likely ascend to the motor cortex. Anatomical, physiological and clinical data of other investigators support this idea. Anatomically, reticulo-thalamo-cortical fibers originating or passing through this reticular area have been described (Nauta & Kuypers, 1958; Scheibel & Scheibel, 1967). Physiologically, tremor has been produced by electrical stimulation of the reticular formation (Ward, McCulloch, & Magoun, 1948) and clinically it has been traditionally found that tremor is present only during wakefulness, a state which depends on activation of the reticular activating system (Lindsley, Bowden, & Magoun, 1949; Magoun, 1958; Moruzzi & Magoun, 1949). Furthermore, the electrophysiological characteristics of this area before, and the neurological signs appearing after lesioning, suggest that this region is associated with an ascending activating system regulating the excitability of the motor cortex and related to the process of selective attention (Velasco & Velasco, 1979; Watson, Valenstein, & Heilman, 1981).

Recent experiments suggest that the mesencephalic reticular formation may also exert a cholinergic control on the caudate nucleus activity related to motor control and the development of tremor (Velasco, Velasco, & Romo, 1982).

REFERENCES

Bertrand, C., Hardy, J., Molina-Negro, P., & Martinez, N. (1969). Optimum physiological target for the arrest of tremor. In F. J. Gullingham & I. M. Donaldson (Eds.), *Third symposium on Parkinson's disease* (pp. 251-262). Edinburgh: Livingstone.

Carrea, R. M. E., & Mettler, F. A. (1947). Physiological consequences following extensive removals of cerebellar cortex and deep cerebellar nuclei and effect of secondary cerebral ablations in primate. *Journal of Comparative Neurology, 87*, 169-288.

Lamarre, Y. (1975). Tremorogenic mechanisms in primates. In B. S. Meldrum & C. D. Mardsen (Eds.), *Primate models of neurological disorders: Advances in neurology* (Vol. 10, pp. 23-24). New York: Raven Press.

Lamarre, Y., De Montigny, C., Dumont, M., & Wess, M. (1971). Harmaline induced rhythmic activity of cerebellar and lower brain stem neurons. *Brain Research, 32*, 246-250.

Lindsley, D. B., Bowden, J. W., & Magoun, H. W. (1949). Effect upon the EEG of acute injury to the brain stem activating system. *Electroencephalography and Clinical Neurophysiology, 1*, 475-486.

Magoun, H. W. (1958). *The waking brain* (pp. 91-114). Springfield, IL: Thomas.

Moruzzi, G., & Magoun, H. W. (1949). Brain stem reticular formation and activation of the EEG. *Electroencephalography and Clinical Neurophysiology, 1*, 455-473.

Nauta, W. J. H., & Kuypers, J. G. J. M. (1958). Some ascending pathways in the brain stem reticular formation. In H. H. Jasper et al. (Eds.), *Henry Ford symposium on reticular formation of the brain* (pp. 3-30). Boston: Little, Brown.

Poirier, L. J. (1960). Experimental and histological study of midbrain diskinesias. *Journal of Neurophysiology, 23*, 534-551.

Poirier, L. J., Bouvier, G., Bedard, P., Boucher, R., Larochelle, L., Olivier, A., & Singh, P. (1969). Essai sur les circuits neuronaux impliques dans le treblement postural et l'hypokinesie. *Revue Neurologique (Paris), 120*, 15-40.

Scheibel, M. E., & Scheibel, A. B. (1967). Structural organization of nonspecific thalamic nuclei and their projection toward cortex. *Brain Research, 6*, 60-94.

Velasco, F., & Molina-Negro, P. (1973). Electrophysiological topography of the human diencephalon. *Journal of Neurosurgery, 38*, 204-214.

Velasco, F., Molina-Negro, P., Bertrand, C., & Hardy, J. (1972). Further definition of the subthalamic target for the arrest of tremor. *Journal of Neurosurgery, 36*, 184-191.

Velasco, F., & Velasco, M. (1979). A reticulo-thalamic system mediating proprioceptive attention and tremor in man. *Neurosurgery, 4*, 30-36.
Velasco, F., Velasco, M., & Romo, R. (1982). Effect of the perfusion of carbachol and atropine in the mesencephalic tegmentum and caudate nucleus on experimental tremor in monkeys. *Experimental Neurology, 78*, 450-460.
Velasco, M., & Velasco, F. (1975). Differential effect of task relevance on early and late components of cortical and subcortical somatic evoked potentials in man. *Electroencephalography and Clinical Neurophysiology, 39*, 353-364.
Velasco, F., Velasco, M., & Machado, J. P. (1975). A statistical outline of the subthalamic target for the arrest of tremor. *Applied Neurophysiology, 38*, 38-46.
Velasco, F., Velasco, M., & Maldonado, H. (1976). Identification y lesion de las radiaciones prelemniscales en el tratamiento quirurgico del temblor. *Archivos de Investigacion Medica (Mexico), 7*, 29-42.
Velasco, M., Velasco, F., Maldonado, H., & Machado, J. P. (1975). Differential effect of thalamic and subthalamic lesions on early and late components of the somatic evoked potentials in man. *Electroencephalography and Clinical Neurophysiology, 39*, 163-171.
Ward, A. A., & McCulloch, W. S., & Magoun, H. W. (1948). Production of alternating tremor at rest in monkeys. *Journal of Neurophysiology, 11*, 317-330.
Watson, R. T., Valenstein, E., & Heilman, K. M. (1981). Thalamic neglect, possible role of the medial thalamus and nucleus reticularis in behavior. *Archives of Neurology, 38*, 501-506.

4 · FOREBRAIN LESIONS PRODUCE SENSORY NEGLECT AND SUPPRESS SPINDLES AND SEIZURES

D. M. Feeney

University of New Mexico

Since the early work of Lindsley and his colleagues (Skinner & Lindsley, 1967; Velasco & Lindsley, 1965; Velasco, Skinner, Asaro, & Lindsley, 1968) there has been interest in rostral thalamic and forebrain mechanisms involved in attention and the regulation of synchronous EEG activity. However, the control of synchronous EEG discharges, recruiting responses, and sleep spindles by neural mechanisms in the forebrain has been disputed. The reports of suppression of spindles after orbital cortex lesions have not been replicated (Dahl, Gjerstad, & Skrede, 1972; Robertson & Lynch, 1971; Staunton & Sassasaki, 1971). Experiments in my laboratory with lesions of the basal forebrain-preoptic area, which is adjacent to orbital cortex, have demonstrated a clear suppression of spindles and recruiting waves. Additionally, we have observed that these same lesions also suppress or block another type of synchronous discharge--epileptic seizures--and also produce a syndrome of contralateral sensory neglect.

METHOD

In these experiments two types of preparations were studied in 41 cats weighing 2.0 - 4.0 kg, and details of these procedures are presented elsewhere (Feeney, Gullotta, & Pittman, 1977). Thirty-four acute animals were surgically prepared under short-lasting general anesthesia (ether and Brevital, 0.55 mg/kg). A spinal transection (encéphale isolé) and wide craniotomy were performed and local anesthetic was administered at wound and pressure points. Throughout the experiment freedom from pain was verified by the presence of slow rhythms in the EEG and pupillary constriction. The animal was paralyzed with gallamine triethiodide (Flaxadil) and respirated. The EEG was amplified and displayed using conventional procedures.

In 13 of these acute preparations, seizures were induced by intravenous administration of pentylenetetrazol (Metrazol) at a dosage of 80-100 mg/kg; and an additional 10 mg/kg given every 45 seconds until a generalized seizure was observed in the EEG. In three of these experiments evoked potentials were also recorded from the marginal gyrus to flash or optic tract stimulation and from the coronal gyrus to stimulation of the nucleus ventralis posterior

DOI: 10.4324/9781003164647- 6

lateralis (VPL). Also, recruiting responses evoked by 10/sec stimulation of nucleus parafasicularis (Pf) were recorded. In the other 21 acute preparations seizures were evoked by injection of 0.1-0.2 ml (20,000 - 40,000 units) of buffered K penicillin G into cortex. Areas of injection included anterior sigmoid, mid-suprasylvian and midmarginal gyri. Multiple injections and hyperventilation were sometimes necessary to produce seizures. Once evoked, seizures recurred spontaneously 10 to 20 times per hour and were followed for as long as three hours without significant change of amplitude. It is quite possible that in some cases this state of status epilepticus was a result of a general diffusion of the penicillin.

After recording spindle and seizure baseline control amplitudes, unilateral electrolytic lesions were made in rostral thalamus or the basal forebrain area. In a few experiments the lesions were extended bilaterally. To control for nonspecific effects of the lesions, control lesions were made in either the lateral hypothalamus, fields of Forel, lenticular nuclei or the head of the caudate nucleus.

The behavioral and chronic effects of the lesions were studied in seven animals. These subjects were anesthetized with promazine plus pentobarbital (Sparine, 5.5 mg/kg followed by Nembutal at 23.3 mg/kg intraperontoneal). The animals were paralyzed with Flaxadil and respirated via an endotracheal tube. The EEG was recorded from the skull surface using saline-soaked wick electrodes. After recording a baseline of barbiturate spindles, a unilateral electrolytic lesion was made in the rostral thalamus or basal forebrain area and Metrazol seizures evoked as described above. The electrodes were withdrawn, the wounds closed and antibiotics and Flaxadil antagonist (endrophonium chloride; Tensilon, 0.4 mg/kg intravenous) administered.

Systematic behavioral observations were conducted twice a week on these animals and included visual and tactile placing, blink to threat, and the withdrawal of the paws from water. After 6-8 months of behavioral testing, acute electrophysiological experiments were repeated and the amplitude of barbiturate spindles and Metrazol seizures re-examined.

On completion of the experiments, the cats were administered an overdose of Nembutal and then perfused with saline and formalin. The lesions were examined from frozen coronal sections stained with thionine.

In all of these experiments the effects of lesions on the EEG were compared both before and after the lesion as well as between the hemisphere ipsilateral to the lesion and the contralateral control hemisphere. This latter comparison is important. Changes observed in the ipsilateral hemisphere without changes in the contralateral hemisphere after the lesion cannot be attributed to nonspecific effects such as general diffusion of the local anesthetic, oxygen deficit or blood pH. Rather, selective ipsilateral effects suggest a direct neural interaction between the basal forebrain and thalamus or cortex. Such unilateral effects were readily observed since pre-lesion spindles and seizures were of symmetrical amplitude and duration in homotopical cortical areas.

RESULTS

In a total of 15 of 18 experiments unilateral or bilateral lesions of the basal forebrain or rostral thalamus suppressed or blocked seizures evoked by application of penicillin to the cortex. In 10 of 13 of these experiments with unilateral lesions of rostral thalamus, basal forebrain, or both, there was a clear ipsilateral reduction of seizure amplitude compared to pre-lesion and to the contralateral control hemisphere. A reduction of seizure amplitude on the ipsilateral cortex following each lesion was observed in five other experiments in which bilateral lesions were carried out in basal forebrain and/or rostral thalamus. This was most pronounced in anterior cortical areas and for the tonic phase of the seizure although the effect could be seen in posterior areas and in the clonic phase as well. This ipsilateral seizure amplitude reduction usually ranged from 20% to 60% of the baseline control level. In one experiment a complete cessation of seizures was produced by a unilateral lesion of the rostral thalamus and underlying area ipsilateral to the focus. This effect is illustrated in Figure 4.1. Since the seizures had been recurring at a rate of 10-30 per hour before the lesion, this result was most dramatic. In three control preparations these seizures had been followed for up to 12 hours and were never observed to spontaneously abate. In only one other experiment was a complete suppression of seizures produced, and this was after a bilateral lesion of the ventral rostral thalamus. A reduction of seizure amplitude on the ipsilateral cortex following each lesion was observed in five other experiments in which bilateral lesions were carried out in basal forebrain and/or rostral thalamus.

In these penicillin seizure experiments it was difficult to observe sleep spindles since they were usually obscured by the recurring seizures. But when observable they were only on the cortex contralateral to the lesion and suppressed ipsilaterally, as were the seizures.

In the acute experiments with Metrazol seizures, the lesion effect on spindles was readily observed and was dramatic, usually producing a complete abolition of the spindles ipsilateral to lesions of the rostral thalamus or basal forebrain. This is illustrated in Figure 4.2, which also shows a blockage of both cortical and thalamic recruiting responses. However, to obtain a reduction of the Metrazol seizures larger lesions often had to be made than with the penicillin model. In only one preparation did a small lesion of the rostral thalamus reduce the amplitude of ipsilateral Metrazol seizures. In six other experiments the lesions had to be extended below and anterior to the rostral thalamus to obtain an effect on Metrazol seizures. In two cases lesions of only the basal forebrain were successful in suppressing Metrazol

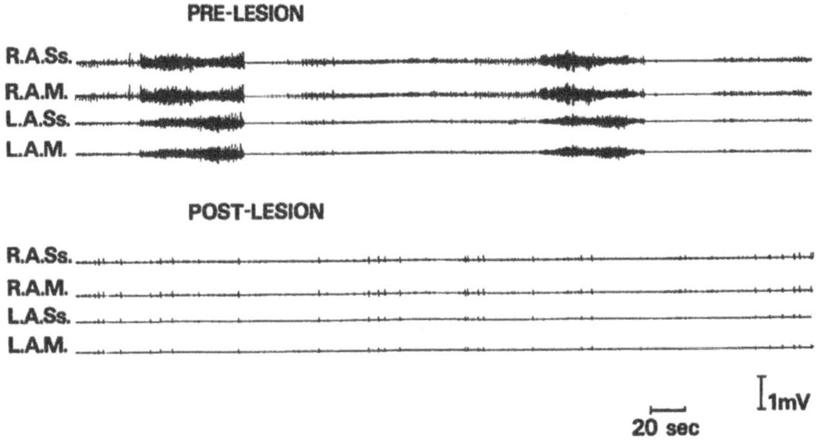

FIGURE 4.1. Blockade of recurring seizures by a lesion of the rostral thalamus. The lesion was ipsilateral to a penicillin focus in the right anterior sigmoid gyrus. Detailed descriptions of the lesions are available (Feeney, Gullotta, & Pittman, 1977). Note that in the post-lesion record seizures are absent but that interictal spiking persists. This indicates continued focal activity but a blockade of the spread. Abbreviations: R.A.Ss., right anterior suprasylvian gyrus; R.A.M., right anterior marginal; L.A.Ss., left anterior suprasylvian; L.A.M., left anterior marginal.

seizure amplitude on the ipsilateral cortex. However, in another preparation a very large lesion including rostral thalamus and basal forebrain blocked ipsilateral spindles but was not successful in reducing Metrazol seizure amplitude. In three experiments in which rostral thalamus-basal forebrain lesions successfully reduced seizure and spindle amplitude, evoked potentials were recorded before and after the lesion. In two of these three preparations the evoked responses from the marginal gyrus were enhanced, and that from the coronal gyrus unchanged on the hemisphere ipsilateral to the lesion. This rules out depression of the hemisphere as a mechanism for seizure and spindle suppression. In six experiments the effects of lesions of neighboring areas (caudate nucleus, lenticular nuclei, lateral hypothalamus) were studied. No effects were observed on sleep spindles or penicillin or Metrazol seizure amplitude.

In the seven chronic animals, barbiturate spindles and Metrazol seizures were recorded immediately after and again 6-8 months after a unilateral lesion of the rostral thalamus, basal forebrain, or a control lesion of the lateral hypothalamus. Of these animals, two died during the second surgery and one preparation showed no effect either immediately after, or seven months after the lesion. In this latter case histological analysis revealed that this lesion primarily involved the lateral hypothalamus. In the remaining four

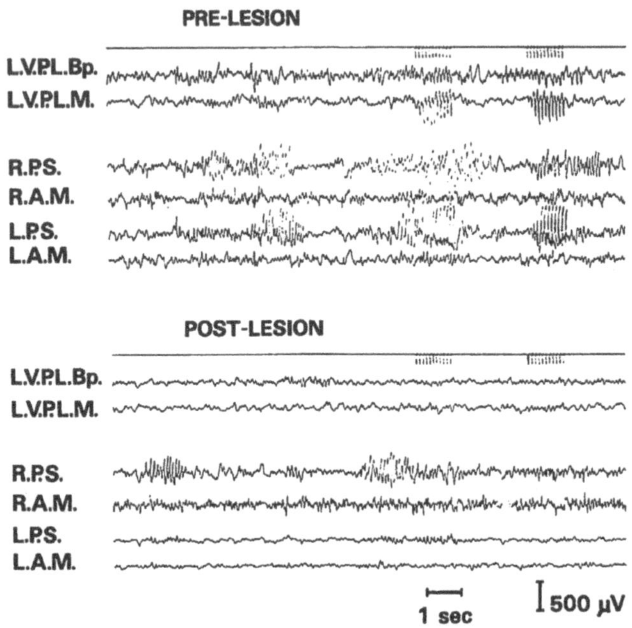

FIGURE 4.2. Suppression of barbiturate spindles and recruiting responses by a lesion of the left rostral thalamus. Recruiting was evoked by 10 Hz stimulation of the left nucleus parafasicularis (at marker on upper trace of each record) and is clear in the left ventralis posterior lateralis (L.V.P.L.M.) and left posterior sigmoid gyrus. Note the spontaneous spindle waves in the right and left posterior sigmoid gyri (R.P.S. & L.P.S.) in the pre-lesion record before the recruiting stimuli. The post-lesion record shows a suppression of spindles on the left hemisphere ipsilateral to the lesion. Recruiting is also abolished both at the cortex and within the thalamus. Abbreviations: L.V.P.L.Bp., left ventralis posterior-bipolar; L.V.P.L.M., left ventralis posterior lateralis-monopolar and others as in Figure 4.1.

cats the lesions reduced the amplitude of both seizures and spindles immediately after the lesion. Upon retesting 6-8 months later spindle amplitude suppression was as complete as immediately after the lesion. There was some recovery of Metrazol seizure amplitude; however, it remained 30% to 60% below the amplitude of the contralateral control hemisphere. Histological analysis of these successful chronic experiments indicated that the most effective lesions involved both the rostral thalamus and preoptic region. However, in two cases large lesions of the basal forebrain alone reduced spindles and seizure amplitude.

Behaviorally, these lesions induced a transient aphagia and adipsia, hypo- and hyperthermia and ipsiversive circling all of which recovered after a few weeks. Additionally, the behavioral testing indicated that these subjects displayed a "neglect" of visual and tactile stimuli presented to the sensory fields contralateral to the lesion. The subject with a control lesion of the lateral hypothalamus recovered completely in one month. However, for the other subjects there was only slight recovery over 6-8 months.

DISCUSSION

It is clear from these studies that lesions of either the rostral thalamus or basal forebrain can suppress several forms of synchronous EEG rhythms: spontaneous sleep spindles, barbiturate spindles, recruiting responses and seizure activity. This supports the concept of a forebrain mechanism regulating synchronous neuronal discharge and several specific hypotheses have been proposed to account of this action (Feeney, Gullotta, & Pittman, 1977). The conflicting reports concerning the effects of orbital cortex lesions on synchronous EEG noted above may be clarified by our observations. Perhaps those orbital cortex ablations which suppressed spindles also infringed on the neighboring basal forebrain or interrupted its blood supply. Those orbital lesions which did not affect spindles may have spared this region. Anatomical studies have indicated that there are direct ipsilateral projections from the basal forebrain to thalamus (Crosby & Woodburne, 1951) and to cortex (Divac, 1975; Whitehouse, Struble, Hedreen, Clark, & Price, 1985) either or both of which could support the regulation of synchronous EEG activity. The role of the basal forebrain and rostral thalamus in the propagation of epileptiform activity should be further studied since the cortical manifestations of penicillin or Metrazol seizures can be produced after removal of the *entire* rhombencephalon (Walker, Feeney, & Hovda, 1984).

Different approaches and models support the hypothesis of a provocative relationship between spindles and epileptiform discharges. A synchronization between spindles and the interictal spiking produced by topical application of penicillin to the cortex has been documented (McLachlan, Kaibara, & Girvin, 1983). After high intramuscular doses of penicillin a smooth transition to spike and wave discharges occurs (Kostopoulos, Gloor, Pelligrini, & Siatitsas, 1981). As in the experiments described above, sleep spindles, recruiting responses and the spike and wave epileptiform discharges are reduced or abolished by microinjection of KCl into the thalamus, which produces a functional thalamic lesion (Avoli & Gloor, 1981). Similarly, other experimental models of spike and wave discharges have been blocked by lesions of the rostral thalamus and basal forebrain (Villablanca, Schlag, & Marcus, 1970). Only the work of Shouse and Sterman (1979) challenges the hypothesis of a provocative relationship between spindles and seizures. These conflicting data appear to be only in the definition of spindle waves. They conclude that "sleep spindles do not facilitate seizure activation." However their thalamic lesions significantly reduced EEG activity in the 8-11 Hz sleep spindle range and also elevated monomethylhydrazine seizure threshold. They counted spindles as activity in the 12-15 Hz range, which was increased by their lesions, but sleep spindles in the cat are generally accepted as approximately 10 Hz (Anderson & Anderson, 1968; Anderson & Sears, 1964). Therefore, their work supports the proposition that spindles are epileptogenic and treatments which shift the EEG away from this frequency are anticonvulsant (Feeney, 1979). While spindles may provoke or enhance epileptiform discharges, and are of thalamic origination (Anderson & Anderson, 1968), grand mal type seizures can occur after total thalamic ablation (Bach-y-Rita, Poncet, & Naquet, 1969). Also, Metrazol seizures cannot result from brainstem activity as has been proposed (Velasco, Velasco, & Romo, 1980), since these seizures are normal in form even after complete removal of the brainstem (Walker, Feeney, & Hovda, 1984).

The enduring contralateral sensory neglect after basal forebrain lesions was surprising. Other studies of neglect induced in cats by diencephalic lesions describe a more rapid recovery (Feeney & Wier, 1979). Reeves and Hagman (1971) report that neglect is accompanied by sleep spindles in the hemisphere ipsilateral to the lesion and contralateral to the multimodal sensory loss but this was not confirmed by Feeney and Wier (1979). Furthermore, the lesions in the present study blocked spindles and produced a profound neglect so these phenomena are not causally related. The lack of complete EEG and sensory recovery after 6-8 months indicates a powerful control of the ipsilateral hemisphere by the basal forebrain.

REFERENCES

Anderson, P., & Anderson, S. A. (1968). *Physiological basis of the alpha rhythm*. New York: Appleton-Century Crofts.

Anderson, P., & Sears, T. A. (1964). The role of inhibition in phasing of spontaneous thalamocortical discharge. *Journal of Physiology, 173*, 459-480.

Avoli, M., & Gloor, P. (1981). The effects of transient functional depression of the thalamus on spindles and on bilateral synchronous epileptic discharges of feline generalized penicillin epilepsy. *Epilepsia, 22*, 443-452.

Bach-y-Rita, P., Poncet, M., & Naquet, R. (1969). Morphology and spatiotemporal evolution of ictal discharges induced by cardiazol in the presence of various cortico-diencephalic lesions in the cat. In H. Gastaut (Ed.), *Pathophysiology of the epilepsies* (pp. 255-267). Springfield, IL: C. C. Thomas.

Crosby, E. C., & Woodburne, T. T. (1951). The mammalian midbrain and isthmus regions. II. The fiber connections. *Journal of Comparative Neurology, 94*, 1-32.

Dahl, D. L., Gjerstad, I., & Skrede, K. K. (1972). Persistent thalamic and cortical barbiturate spindle activity after ablation of the orbital cortex in cats. *Electroencephalography and Clinical Neurophysiology, 33*, 483-496.

Divac, I. (1975). Magnocellular nuclei of the basal forebrain project to neocortex, brainstem and olfactory bulb. Review of some functional correlates. *Brain Research, 93*, 385-398.

Feeney, D. M. (1979). Marijuana and epilepsy: Paradoxical anticonvulsant and convulsant effects. In *Marijuana: Biological effects*. G. G. Nahas & W. D. M. Paton (Eds.), *Advances in the biosciences, 22 & 23*, 643-657.

Feeney, D. M., Gullotta, F. P., & Pittman, J. C. (1977). Slow-wave sleep and epilepsy: Rostral thalamus and basal forebrain lesions suppress spindles and seizures. *Experimental Neurology, 56*, 212-226.

Feeney, D. M., & Wier, C. S. (1979). Sensory neglect after lesions of substantia nigra or lateral hypothalamus: Differential severity and recovery of function. *Brain Research, 178*, 329-346.

Kostopoulos, G., Gloor, P., Pelligrini, A., & Siatitsas, I. (1981). A study of the transition from spindles to spike and wave discharges in feline generalized epilepsy: EEG features. *Experimental Neurology, 73*, 43-54.

McLachlan, R. S., Kaibara, M., & Girvin, J. P. (1983). Relationship between focal penicillin spikes and cortical spindles in the cerveau isolé. *Experimental Neurology, 79*, 38-53.

Reeves, A. G., & Hagman, W. D. (1971). Behavioral and EEG asymmetry following unilateral lesions of the forebrain and midbrain in cats. *Electroencephalography and Clinical Neurophysiology, 30*, 83-86.

Robertson, R. T., & Lynch, G. S. (1971). Orbitofrontal modulation of EEG spindles. *Brain Research, 28*, 562-566.

Shouse, M. N., & Sterman, M. B. (1979). Changes in seizure susceptibility, sleep time and sleep spindles following thalamic and cerebellar lesions. *Electroencephalography and Clinical Neurophysiology, 46*, 1-12.

Skinner, J. E., & Lindsley, D. B. (1967). Electrophysiological and behavioral effects of blockade of the nonspecific thalamocortical system. *Brain Research, 6*, 95-118.

Staunton, H. P., & Sassasaki, K. (1971). Recruiting responses not dependent on orbitofrontal cortex. *Brain Research, 30*, 415-418.

Velasco, M., & Lindsley, D. B. (1965). Role of orbital cortex in regulation of thalamocortical electrical activity. *Science, 149*, 1375-1377.

Velasco, M. E., Skinner, J. E., Asaro, K. D., & Lindsley, D. B. (1968). Thalamocortical systems regulating spindle bursts and recruiting responses. I. Effect of cortical ablations. *Electroencephalography and Clinical Neurophysiology, 25*, 463-470.

Velasco, F., Velasco, M., & Romo, R. (1980). Specific and nonspecific multiple unit activities during pentylenetetrazol. I. Animals with encéphale isolé. *Electroencephalography and Clinical Neurophysiology, 49*, 600-605.

Villablanca, J., Schlag, J., & Marcus, R. (1970). Blocking of experimental spike and wave by a localized forebrain lesion. *Epilepsia, 11*, 163-167.

Walker, A. E., Feeney, D. M., & Hovda, D. A. (1984). The electroencephalographic characteristics of the rhombencephalectomized cat. *Electroencephalography and Clinical Neurophysiology, 57*, 156-165.

Whitehouse, P. J., Struble, R. G., Hedreen, J. C., Clark, A. W., & Price, D. L. (1985). Alzheimer's disease and related dementias: Selective involvement of specific neuronal systems. *CRC critical reviews in clinical neurobiology, 1*, 319-339.

5 · FURTHER INVESTIGATION OF POLYSENSORY RESPONSES IN THE POSTERIOR THALAMIC ASSOCIATION NUCLEI OF CATS

C. C. Huang

Medical College of Wisconsin

The polysensory responses in pulvinar and other posterior association nuclei were studied systematically using electrophysiological methods by Huang and Lindsley (1973). An extensive mapping of evoked potentials to visual (V), auditory (A) and somatosensory (S) stimuli under chloralose anesthesia and in the unanesthetized brain was performed. Intermodal sensory interactions of the evoked potentials in this area were analyzed. The present study investigates the unitary sensory responses recorded extracellularly in this area.

METHOD

Nine chloralose anesthetized (70 mg/kg) cats were used. The surgical procedure, anesthesia method, stimulating and recording apparatus and the stereotaxic equipment used are similar to those described in Huang and Lindsley (1973). The microelectrodes for unit recording were electrolytically sharpened insect pins insulated with Insul-X and had a bare tip diameter of about 2-3μ. Upon completion of the preparation of the animal, the dura was opened and the microelectrode was inserted to the posterior thalamic regions to record evoked unit responses to flash, click and mild electric shock to the contralateral forepaw. After the tip of the electrode had reached H = 8.5 stereotaxically, the microelectrode was then lowered in small steps (0.1 mm) to record all possible sensory evoked units. Unit activity of the neurons was fed into a negative-capacitance high-impedance amplifier and then displayed on a Tektronix 502 oscilloscope. Photographs of unit responses were obtained with a Grass C-4 kymograph camera. The reference electrode was a silver plate wrapped in Ringer soaked cotton placed at the wound margin. At the end of each electrode recording, 5-10 μamps of positive current was applied through the electrode tip for 15 sec so that all locations of recording sites could be carefully determined histologically and retrospectively.

RESULTS

One hundred fifteen unit responses were recorded. These unit responses were recorded extracellularly from pulvinar, n. posterior, n.

 DOI: 10.4324/9781003164647- 7

lateralis posterior and n. suprageniculatus, especially from the pulvinar nucleus at A = 4.0 and 6.0 of the Horsley-Clarke stereotaxic coordinates. Very few evoked units were recorded from A = 8.0. The location of these units is shown in Figure 5.1. The percentages of the distribution of unit responses to the different stimuli are as follows: V = 28.7%; (V+A+S) = 40.9%; (V+S) = 18.2%; (V+A) = 4.3%; (S+A) = 0.9%; S = 6.1%; A = 0.9%. Ninety-two percent of the units responded to flash stimulation. The above findings are in agreement with the results of previous studies in mapping, recovery cycle, and intermodality interaction tests, which demonstrated that pulvinar is more specific in response to visual stimuli (Chalupa, Anchel, & Lindsley, 1972; Huang & Lindsley, 1973; Suzuki & Kato, 1969).

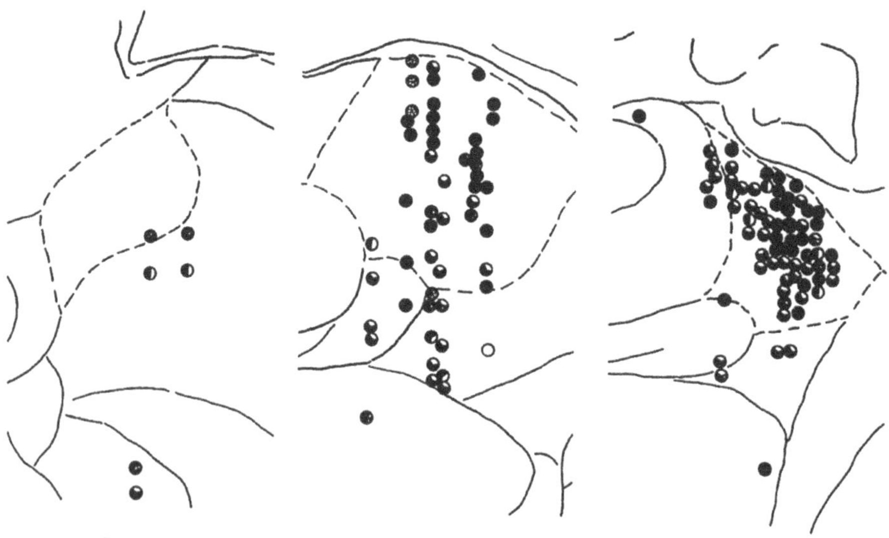

FIGURE 5.1. The locations of the evoked units studied in pulvinar and associated posterior thalamic nuclei; diagrams from Jasper and Ajmone-Marsan (1973). ● : visual (V) evoked unit; O : auditory (A) evoked unit; ⊛ : somatosensory (S) evoked unit; ◒ : V,A and S evoked unit; ◐ : V and A evoked unit.

The histograms of the latencies of the unit responses (Figure 5.2) show that there are two groups of unit responses to visual, auditory and somatosensory stimuli. The latencies of the early response group range from 15 to 150 msec and those of the late response groups range from 340 to 450 msec. The histogram of visual unit responses is different from those of the auditory and somatosensory responses with most of the unit responses having latencies ranging from 25 to 65 msec and peaking at 45 msec. The histograms of the auditory and somatosensory unit responses are similar, with most of

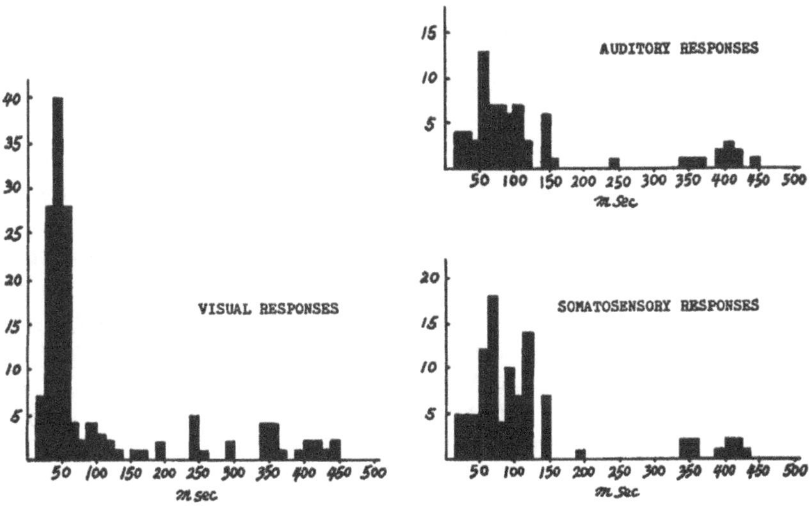

FIGURE 5.2. The histograms indicate the latency distributions of the unit responses evoked by visual, auditory and somatosensory stimuli. Ordinate: number of responses. Abscissa: time interval between the start of the stimuli and the start of the responses.

the unit responses having latencies ranging from 25 to 150 msec. The peaks of the auditory and somatosensory responses are at 60 msec and 75 msec, respectively.

DISCUSSION

It might be suggested that the long latency unit responses are polysynaptic in nature involving various structures of the brain and the short latency responses are from n. lateralis geniculatus, n. medialis geniculatus and n. ventralis basalis. Using a cryogenic cooling method to block the visual evoked potentials in pulvinar, Chalupa, Anchel, and Lindsley (1972) reported that the visual inputs to the pulvinar are via lateral geniculate, superior colliculus and/or visual cortex. The Chalupa et al. (1972) report is in agreement with the present microelectrode findings.

The unit responses to visual, auditory and somatosensory stimulation usually have the same patterns at each recording site. This indicates that the polysensory convergence and integration occur on the single cells in these association nuclei. However, occasionally the unit responses show different patterns in the responses to the different stimulation modalities at the same recording site. This probably indicates that the unit discharges were recorded from different nerve fibers rather than from the same cell. Further unitary studies in this area with more sophisticated stimulating apparatus or experimental design will be an interesting topic.

REFERENCES

Chalupa, L. M., Anchel, H., & Lindsley, D. B. (1972). Visual input to the pulvinar via lateral geniculate, superior colliculus and visual cortex in the cat. *Experimental Neurology, 36*, 449-462.

Huang, C. C., & Lindsley, D. B. (1973). Polysensory responses and sensory interaction in pulvinar and related posterolateral thalamic nuclei in cat. *Electroencephalography and Clinical Neurophysiology, 34*, 265-280.

Jasper, H. H., & Ajmone-Marsan, C. (1954). *A stereotaxic atlas of the diencephalon of the cat.* Ottawa: National Research Council of Canada.

Suzuki, H., & Kato, H. (1969). Neurons with visual properties in the posterior group of the thalamic nuclei. *Experimental Neurology, 23*, 353-365.

6 BRAINSTEM-LIMBIC SYSTEMS AND BEHAVIOR

C. L. Wilson, N. Hirasuna, and D. B. Lindsley

University of California at Los Angeles

In a series of studies carried out in this laboratory over the past several years, the effects of electrical stimulation of functional systems within the brainstem and hypothalamus of the cat have been investigated in order to determine the influences of these systems upon behavior and upon electrical activity recorded in the hippocampus and septum. Results of earlier investigations in this series have been reviewed in some detail by Lindsley and Wilson (1975). Here we will briefly recapitulate our principal findings and then relate them to the results of our more recent work and the work of other laboratories which have addressed this problem.

In their initial study, Anchel and Lindsley (1972) described two systems (functional tracts or pathways) extending from midbrain through medial and lateral hypothalamus which, when stimulated electrically at 100 Hz, could be differentiated in terms of their contrasting effects upon hippocampal electrical activity (synchronized theta rhythm vs. desynchronized fast activity) and behavior (alert, attentive, scanning, exploring vs. fixation of posture and gaze). The two systems are considered to consist of fiber pathways ascending from the lower brainstem via the hypothalamus and septum to the hippocampus. At the hypothalamic level, it was found that 100 Hz electrical stimulation in the medial hypothalamus (MH) was effective in eliciting the synchronous, rhythmic, slow-wave activity in hippocampus known as "theta rhythm," while stimulation in the region of the medial forebrain bundle in the lateral hypothalamus (LH) caused lower amplitude desynchronized electrical activity.

Intrigued by the earlier results of Petsche, Gogolak, and Van Zwieten (1965), Petsche, Stumpf, and Gogolak (1962) and Stumpf, Petsche, and Gogolak (1962) with respect to the role of the septum in relation to hippocampal theta rhythm, we began to investigate these two systems to determine the mechanisms by which they influence hippocampal activity. The ascending effects of MH and LH stimulation upon hippocampus were found to be mediated by a group of cells in the diagonal band of Broca within the septal nucleus (Wilson, Motter, & Lindsley, 1976). Stimulation of MH caused these cells to fire in rhythmic bursts in phase with hippocampal theta rhythm, while stimulation of LH either disrupted the rhythmicity of diagonal band cells, or inhibited their activity. Furthermore, surgical section

 DOI: 10.4324/9781003164647- 8

or cryogenic blockade of dorsal fornix and fimbria, thus separating septum from hippocampus, blocked the appearance of theta rhythm in hippocampus and influenced the firing pattern of many diagonal band cells. Significantly, rhythmic patterns of discharge remained intact in other cells, indicating that hippocampal rhythmicity is dependent upon septum, but not vice versa (Wilson, Motter, & Lindsley, 1975).

Macadar, Chalupa, and Lindsley (1974) sought the origin of these systems in the lower brainstem, and found that hippocampal theta rhythm could be elicited during stimulation of nucleus locus coeruleus, nucleus reticularis pontis oralis and the ventrolateral periaqueductal gray substance, while stimulation of nucleus centralis superior of the raphé and nucleus reticularis pontis caudalis desynchronized hippocampal activity. Stimulation of these structures had no effect upon hippocampal activity during cryogenic blockade of their ascending pathways in the hypothalamus (Wilson, Hirasuna, & Lindsley, 1976).

Having investigated some of the mechanisms by which hippocampal and septal activity is modified by brainstem-hypothalamic systems, and having delineated the extension of these systems from pontine and midbrain regions through hypothalamic and septal areas to the hippocampus, we began to examine the effects of stimulation of these systems upon behavior. Anchel and Lindsley (1972) had previously found that in addition to the contrasting electrophysiological changes resulting from stimulation of medial and lateral hypothalamus, there were also contrasting behavioral effects. Stimulation of the MH theta system, or cryogenic blockade of the LH desynchronizing system, produced orienting or searching behavior, while stimulation of the LH system, or cryogenic blockade of the MH system produced an arrest of ongoing behavior and caused fixation of gaze. The behavioral consequences of stimulation of MH and LH upon spontaneous, ongoing and learned behaviors, as well as hippocampal electrical activity, were studied by Coleman and Lindsley (1975, 1977), who found that MH stimulation producing theta rhythm sometimes led to cessation of bar-pressing for water reinforcement for extended periods ranging from 2 to 40 min. This response did not appear to be associated with any aversive effects of stimulation. LH stimulation, which desynchronized hippocampal activity, was also found to disrupt behavior, but only for a few seconds rather than many minutes. These findings were consistent with effects of MH stimulation which led animals to exhibit orienting responses or visual scanning of the environment. Such shifts of attention to the external environment were clearly incompatible with ongoing bar-pressing behavior. On the other hand, the disruption of bar-pressing by LH stimulation was largely limited to a brief arrest of behavior which resumed soon after stimulation.

After determining the behavioral consequences of hypothalamic stimulation, the next step was to study the behavioral effects of stimulating the midbrain and pontine structures mapped by Macadar, Chalupa, and Lindsley (1974), and it is to this topic that the remainder of this presentation is devoted. The questions investigated include (1) Are similar behaviors elicited by stimulation of the brainstem sites from which hippocampal theta rhythm can be evoked? (2) Do any brainstem theta sites evoke behavioral responses similar to those obtained from stimulation of MH or LH? (3) Does hippocampal desynchronization and its correlated behavioral changes resulting from stimulation of LH have corresponding electrical and behavioral effects during stimulation of specific brainstem sites?

PROCEDURE AND RESULTS

In order to answer these questions, electrodes were chronically implanted in cats in nucleus locus coeruleus (NLC), nucleus reticularis pontis oralis (RPO), the ventrolateral periaqueductal gray substance (PAG) and the nucleus centralis superior of the raphé (CSR). Six animals with stimulating electrodes in these areas and with recording electrodes in dorsal hippocampus and neocortex were tested during stimulation. The test chamber, containing a bar-press lever and above it a food reinforcement well, was spacious enough to allow a fairly large repertoire of behaviors to occur either spontaneously or during 8-sec stimulus trains of 0.15-msec duration biphasic pulses at 100 Hz with currents ranging from 15 to 400 μA. Hippocampal, neocortical, EOG and EMG activity was recorded conventionally on paper and magnetic tape, and behavior was continuously monitored and taped with a video system that allowed simultaneous viewing of hippocampal electrical activity and the animal on the monitor screen in order to better assess concomitant evoked brain activity and behavioral responses. Video-taping also allowed slow-motion and stopped-frame analysis of behavior-EEG relations.

Three brainstem sites which Macadar et al. (1974) described in immobilized cats under nitrous oxide, as areas where stimulation was capable of eliciting theta rhythm in hippocampus (NLC, RPO, ventrolateral PAG), also produced rhythmic hippocampal theta activity in the unanesthetized cat. Behaviorally, orienting and visual scanning responses were elicited at the same current thresholds required to evoke theta rhythm. At higher current levels, while theta rhythm became more prominent and showed increased frequency, behavior consisted only of automatic motor sequences. At theta suprathreshold stimulation levels, 10 out of 10 sites in NLC and 7 out of 9 sites in RPO caused ipsiversive circling. In PAG, 4 of 7 sites resulted in contraversive head turning and in all PAG sites agonistic responses were elicited, including both defensive and attack-like posturing, such as baring the teeth, and hissing.

In behaving animals, electrical stimulation of CSR produced

electrographic results differing from those reported by Macadar et al. (1974) in cats under nitrous oxide. While they found stimulation of CSR sites caused desynchronization of hippocampal electrical activity, we found in these awake behaving animals that CSR stimulation evoked hippocampal theta rhythm. In the majority of instances, stimulation in CSR produced (along with hippocampal theta rhythm) orienting responses, punctuated occasionally by brief fixation of posture and gaze.

Since most sites were capable of eliciting hippocampal theta rhythm at current levels below that producing motor behavior, current levels were chosen for those sites which elicited theta rhythm without eliciting involuntary motor responses. All animals were trained to bar-press on a continuous reinforcement (CRF) schedule for a liquid-food reward. Five animals were further trained to perform under a variable interval (VI-8) schedule, in which 2 to 20 sec (mean 8 sec) intervened between a bar-press and a reinforcement. In this situation animals characteristically press at high rates, and four of the cats trained on this schedule bar-pressed at rates near or above one per second. The effect of stimulation upon bar-pressing was then determined by recording the number of bar-presses during an 8-sec period immediately before stimulation (PRE), an 8-sec period during stimulation (STIM), and an 8-sec period immediately after stimulation (POST). Latencies from the termination of the stimulation train to the first bar-press were also recorded.

Examples of the effects of stimulation of two different brainstem sites are shown in Figure 6.1, as displayed on a video monitor with split-screen to show bar-pressing and other behavior together with the changes in patterns of the electrical activity of the hippocampus associated with stimulation of a particular brainstem site. Panel A shows a cat bar-pressing about once per second until electrical stimulation of CSR at 100 Hz induced rhythmic hippocampal theta rhythm which was accompanied by momentary (about 2 sec) cessation of bar-pressing that continued again during the latter half of the stimulation period. The theta rhythm continued throughout the stimulation period. Panel B shows the hippocampal activity of another cat which was bar-pressing at the onset of stimulation of NLC, but 3 sec later stopped bar-pressing as a burst of rhythmic theta activity was elicited. A few seconds after the end of the stimulus period the cat resumed bar-pressing at its regular rate. This result contrasts with the effect of stimulation of the MH system observed by Coleman and Lindsley (1977), which produced hippocampal theta rhythm but interrupted bar-pressing many minutes while the cat seemed to be distracted from the task and engaged in orienting, scanning and exploratory behavior.

Figure 6.2 shows the effect of stimulation of four brainstem sites upon bar-pressing before, during and after stimulation. Data points for all brainstem stimulation sites represent the mean number of bar-presses

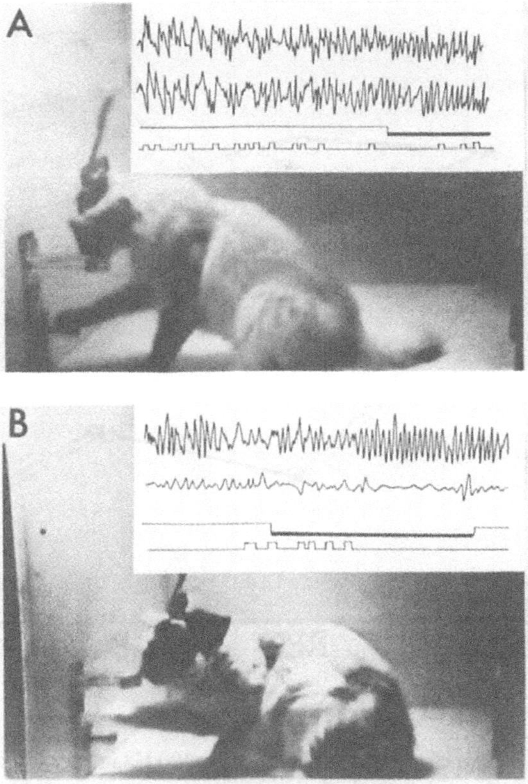

FIGURE 6.1. Electrophysiological and behavioral correlations during stimulation of brainstem sites which influence hippocampal electrical activity and trained bar-pressing behavior for liquid-food reward. Photos of split-screen video presentations of polygraph records and a cat's behavior are shown. Polygraph tracings from top to bottom: Panel A: Left and right dorsal hippocampus, signal for electrical stimulation of nucleus centralis superior of raphé (CSR) at 100 Hz, signal for bar-presses (up), for reinforcement (down); panel B: Left dorsal hippocampus, eye and head movement tracing, signal for electrical stimulation of nucleus locus coeruleus (NLC), signal for bar-presses (up). Panel A shows rhythmic hippocampal theta waves induced by stimulation of CSR, a 2-sec interruption in bar-pressing at onset of stimulation, and an immediate renewal of pressing during stimulation, but after a reinforcement; the cat's pressing and eating behavior illustrated occurs near end of the stimulation period. Panel B shows rhythmic theta beginning about 3 sec after onset of stimulation of NLC at which time bar-pressing was interrupted, cat looked up, oriented and scanned visually, but was reaching to press the bar again when photo was taken near end of polygraph record after offset of stimulation. No reinforcement occurred on this trial. Head and eye movement activity (second tracing) was suppressed during height of theta activity in hippocampus. Time is shown by the 8-sec reduced stimulation deflection in panel B.

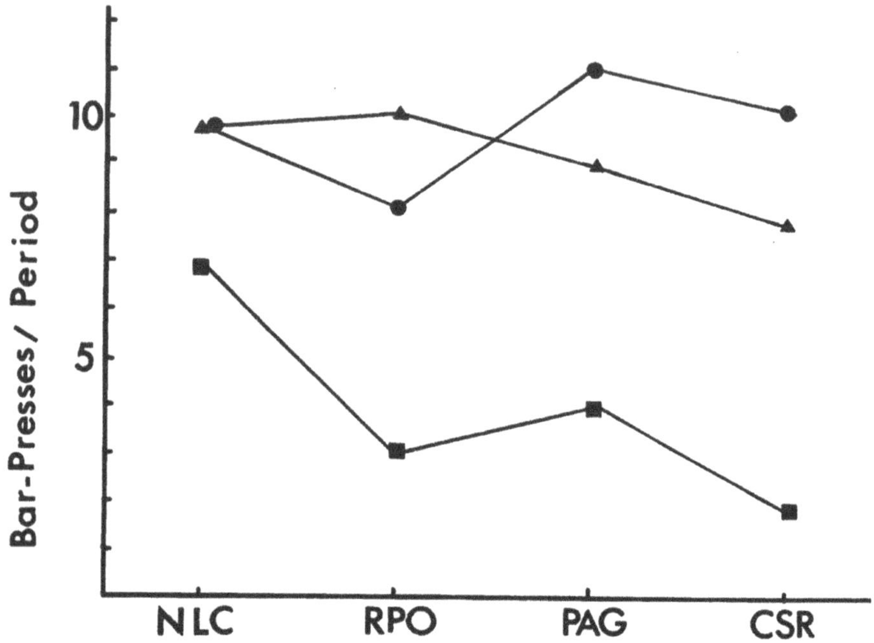

FIGURE 6.2. Effect of electrical stimulation at 100 Hz in four brainstem sites upon trained bar-pressing behavior for liquid-food reward under variable interval (VI-8) schedule. Mean number of bar-presses occurring in 8-sec periods: PRE-stimulation (triangles), POST-stimulation (circles), STIMulation (squares) are plotted according to the brainstem sites stimulated. NLC - nucleus locus coeruleus; RPO - nucleus reticularis pontis oralis; PAG - periaqueductal gray substance; CSR - nucleus centralis superior of the raphé.

recorded during six stimulus periods in a 30-min session and are subdivided according to each of the four brainstem areas studied. Means were based on stimulation of three to six sites in each area. There is no difference between PRE and POST bar-press means, but the number of bar-presses was clearly reduced during stimulation periods. PRE- and POST-stimulation periods of 8 sec each for the four different brainstem sites in all animals averaged about 10 presses per period; during "STIM" bar-pressing was significantly reduced for all sites, but the reduction was least for NLC and greatest for CSR. Stimulation at the other two sites, RPO and the ventrolateral PAG, was significantly less effective than the CSR but more effective than NLC (Mann-Whitney U, two-tailed: NLC vs. PAG: n1=36, n2=36, z=3.78, $p<.001$: PAG vs. CSR: n1=36, n2=18, z=2.52, $p<.006$).

In summary, stimulation at all four of these sites elicited hippocampal theta rhythm and inhibited bar-pressing. Except for CSR, the reduction in bar-pressing as observed during video monitoring was due to performing of incompatible behaviors such as orienting, visual scanning and exploratory behavior similar to that occurring during spontaneous theta rhythm. During stimulation of CSR, orienting also occurred, but it was followed by a brief arrest in behavior or by fixed gaze. It was noted that spontaneously occurring periods of immobility and fixed gaze were often accompanied by high amplitude low frequency (3-4 Hz) theta rhythm.

Stimulation of one additional site not studied by Macadar et al. (1974) showed a marked contrast to those sites just described in terms of behavioral response elicited. This midbrain site was studied in two cats and was identified histologically as nucleus linearis intermedius of the raphé (LIR). Even though stimulation of the site evoked prominent theta rhythm, the response during LIR stimulation was complete behavioral arrest for periods of stimulation up to 30 sec. If an animal was in the process of bar-pressing for water reward, for example, its paw would remain frozen in mid-air during stimulation. At the end of the usual 8-sec stimulation trains, the bar-press would be completed, and the animal would continue its reinforced behavior as if no interruption had occurred. Orientation to peripheral sensory stimuli during trains of LIR stimulation was absent to all sensory modalities (with the exception of noxious stimulation, e.g., strong tail pinch), but major postural responses such as the righting reflex were intact. Since the arrest during LIR stimulation resulted in the maintenance of position, a certain degree of tonicity was present, so that external manipulation of a limb often resulted in its return to the original position after its release. Therefore the term "tonic immobility" is a better description than "waxy flexibility." In contrast to stimulation of PAG, which evoked an emotive behavioral response at higher current levels, stimulation of LIR evoked no behavioral sign of fear or aggressiveness or any evidence that the stimulation was aversive.

DISCUSSION

These results provide some answers to the three questions posed earlier regarding the behavioral and electrographic effects of stimulation of the sites studied by Macadar et al. (1974). The question of whether hippocampal desynchronization is produced in unanesthetized cats by stimulation of the four brainstem sites studied by Macadar et al. was answered in the negative: all sites tested, including the raphé nuclei, CSR and LIR, evoked hippocampal theta rhythm. Two possible reasons for this difference are (1) the use of recording sites in the hippocampus of the animals in the present study which were less prone to desynchronization than the previous study, and/or (2) the nitrous oxide anesthesia used in the previous study. In regard to the first reason, the amplitude of the spontaneous theta rhythm recorded in the present study was greater than that in the Macadar et al. study. Also, some of the CSR sites of the present study showed a decrease in theta amplitude and increase in theta frequency during stronger stimulation. If the prestimulation theta amplitude had been less, the resulting pattern could have appeared desynchronized, but the electrode tips in the present study were positioned above and below the CA1 stratum pyramidale so the phase reversed signals summed to produce theta amplitudes near or greater than 1.0 mV, which may have prevented observation of the desynchronization response reported by Macadar et al., during stimulation of CSR. (For further discussion of this factor see Robinson, 1980.)

A second aspect of importance is the question of recordings obtained under anesthesia versus those obtained from freely behaving animals. Controversy has existed on this question since some investigators have reported absence of hippocampal desynchronization and elicitation of theta rhythm during stimulation of brainstem sites which include the raphé nuclei (Robinson & Vanderwolf, 1978). Others report hippocampal desynchronization occurs during stimulation of raphé sites and their rostral projection pathways (Vertes, 1981, 1982), in agreement with Macadar et al. (1974). The animals in the present study clearly show theta rhythm during stimulation of CSR while engaged in voluntary, automatic, or learned behaviors. It is important to note, however, that many CSR sites tested during implantation of electrodes under nitrous oxide in these and other animals studied in our laboratory have produced either desynchronization of hippocampal activity (even using threshold current levels), or else a marked suppression of theta amplitude and increase in frequency with increases in strength of stimulation. Since Vertes' studies were done under urethane anesthesia, it appears that the brainstem sites effective in producing hippocampal desynchronization may be unmasked by certain anesthetics, and are therefore not observable in freely behaving animals.

The answer to the second question, whether behavioral responses during stimulation of brainstem sites were similar to those obtained during stimulation of MH and LH by Coleman and Lindsley (1974), was both yes and no. Orienting, visual scanning and fixation of posture and gaze were elicited, but no long-lasting, post-stimulation disruption of behavior occurred as during MH stimulation, even when suprathreshold stimulation caused motor responses or defensive posturing. Short duration arrest responses accompanied CSR stimulation, resulting in a greater degree of bar-press interruption during raphé stimulation than any other site.

The most remarkable effect upon behavior was during stimulation of LIR at a point just dorsal to the interpeduncular nucleus in the midbrain. Although this site was not studied by Macadar et al. (1974), similar effects have been previously described by Kuroki (1961), who also reported arrest during raphé stimulation. It is also possible that the ventral tegmental bundle could have mediated the arrest effect (Vertes, 1982).

The final question, whether similar behaviors accompanied stimulation of brainstem sites evoking hippocampal theta rhythm, has a three-part answer. First, at theta rhythm threshold stimulation levels, all four sites produced a significant quantitative decrease in bar-pressing, but the behaviors which accompanied the reduction in bar-pressing differed qualitatively from site to site as detailed above. Second, NLC was significantly less effective and CSR significantly more effective in disrupting bar-pressing than either RPO and PAG. Third, when stronger stimulation eliciting incompatible motor activity was employed, different qualitative behavioral differences occurred, probably because adjacent motor nuclei or pathways were being activated, at which point the theta rhythm in hippocampus may have been only an epiphenomenon. As noted above, all quantitative behavioral measures were carried out at levels of stimulation below the threshold for motor activation. Although all sites produced qualitative behavioral differences at theta rhythm threshold, the sites which produced the clearest differences at theta rhythm threshold levels were in LIR.

The specific questions addressed by this study bear on more general issues that have been discussed previously (see Lindsley & Wilson, 1975; Vertes, 1982), namely, are there brainstem-limbic systems via which certain behavioral manifestations are expressed? What are the brainstem substrates of such systems? Can they be subdivided on the basis of synchronizing and desynchronizing influences which are correlated with behavior? At the hypothalamic level such systems can be nicely subdivided as medial and lateral, based on recordings at both septum and hippocampus (Wilson, Motter, & Lindsley, 1976), as well as upon behavioral differences (Coleman & Lindsley, 1974). In the brainstem, perhaps because nuclei rather than fiber pathways are being stimulated, the results are not as clear cut. Electrographic differentiation is lost since only theta rhythm can be evoked, but stimulation evoked behavioral differences do exist among NLC, PAG, RPO, and CSR. Attempts by others to separate the effect of stimulation of these sites upon hippocampal activity using pharmacological techniques have not been successful (Robinson & Vanderwolf, 1978), since only hippocampal theta and no desynchronization could be elicited. Behavioral differences were reported but are difficult to relate to the behavioral measures of the present study.

In conclusion, the evidence obtained in this study supports the hypothesis that the brainstem regions investigated may represent some of the origins of the hypothalamic-hippocampal systems described by Anchel and Lindsley (1972) and Coleman and Lindsley (1974). The theta system appears to be most strongly activated during stimulation of RPO, ventrolateral to PAG, and of NLC, while behavior elicited from these sites appears to resemble that of the medial hypothalamic system. Although the behavior elicited during CSR stimulation is more similar to that produced during lateral hypothalamic stimulation, no hippocampal desynchronization was obtained from these sites. It therefore may require either activation of a fiber pathway (such as the MFB) or facilitation of desynchronizing influences by pharmacological means to detect the raphé contribution to this system. Finally, further study of the nucleus linearis intermedius of the raphé or of the ascending tracts in that region is warranted on the basis of the striking behavioral arrest response observed during stimulation of this area.

ACKNOWLEDGMENT

We wish to acknowledge the assistance of Mr. Patrick M. Fleming in the conduct of this study.

REFERENCES

Anchel, H., & Lindsley, D. B. (1972). Differentiation of two reticulo-hypothalamic systems regulating hippocampal activity. *Electroencephalography and Clinical Neurophysiology, 32,* 209-226.

Coleman, J. R., & Lindsley, D. B. (1974). Hippocampal and neocortical EEG changes during free and operant behavior. *Society for Neuroscience Abstracts,* 175.

Coleman, J. R., & Lindsley, D. B. (1975). Hippocampal electrical correlates of free behavior and behavior induced by stimulation of two hypothalamic-hippocampal systems in the cat. *Experimental Neurology, 49,* 506-528.

Coleman, J. R., & Lindsley, D. B. (1977). Behavioral and hippocampal electrical changes during operant learning in cats and effects of stimulating two hypothalamic-hippocampal systems. *Electroencephalography and Clinical Neurophysiology, 42,* 309-331.

Kuroki, T. (1961). Arrest reaction elicited from the brain stem. *Folia Psychiatrica Neurologica Japan, 12,* 317-340.

Lindsley, D. B., & Wilson, C. L. (1975). Brain stem-hypothalamic systems influencing hippocampal activity and behavior. In R. L. Isaacson & K. L. Pribram (Eds.), *The hippocampus* (Vol. 2, pp. 247-278). New York: Plenum.

Macadar, A. W., Chalupa, L. M., & Lindsley, D. B. (1974). Differentiation of brainstem loci which affect hippocampal and neocortical electrical activity. *Experimental Neurology, 43,* 499-514.

Petsche, H., Gogolak, G., & Van Zwieten, P. A. (1965). Rhythmicity of septal cell discharges at various levels of reticular excitation. *Electroencephalography and Clinical Neurophysiology, 19,* 25-33.

Petsche, H., Stumpf, C., & Gogolak, G. (1962). The significance of the rabbit's septum as a relay station between the midbrain and the hippocampus. I. The control of hippocampal arousal activity by the septum cells. *Electroencephalography and Clinical Neurophysiology, 14*, 202-211.

Robinson, T. E. (1980). Hippocampal rhythmic slow activity (RSA; Theta): A critical analysis of selected studies and discussion of possible species differences. *Brain Research Reviews, 2*, 69-101.

Robinson, T. E., & Vanderwolf, C. H. (1978). Electrical stimulation of the brain stem in freely moving rats: II. Effects on hippocampal and neocortical electrical activity, and relations to behavior. *Experimental Neurology, 61*, 485-515.

Stumpf, C., Petsche, H., & Gogolak, G. (1962). The significance of the rabbit's septum as a relay station between the midbrain and the hippocampus. II. The differential influence of drugs upon the septal cell firing pattern and the hippocampus theta activity. *Electroencephalography and Clinical Neurophysiology, 14*, 212-219.

Vertes, R. P. (1981). An analysis of ascending brain stem systems involved in hippocampal synchronization and desynchronization. *Journal of Neurophysiology, 42*, 214-228.

Vertes, R. P. (1982). Brain stem generation of the hippocampal EEG. *Progress in Neurobiology, 19*, 159-186.

Wilson, C. L., Hirasuna, N., & Lindsley, D. B. (1976). Brain stem sources of hippocampal theta rhythm investigated by cryogenic blockade of hypothalamic pathways in the cat. *Neuroscience Abstracts (Part I), 2*, 399.

Wilson, C. L., Motter, B. C., & Lindsley, D. B. (1975). Effects of cryogenic blockade of dorsal fornix and fimbria upon hippocampal theta rhythm and septal cellular activity. *Neuroscience Abstracts, 1*, 544.

Wilson, C. L., Motter, B. C., & Lindsley, D. B. (1976). Influences of hypothalamic stimulation upon septal and hippocampal electrical activity in the cat. *Brain Research, 107*, 55-68.

7· REGULATION OF CARDIAC VULNERABILITY BY THE FRONTAL CORTEX: A NEW CONCEPT OF CANNON'S CEREBRAL DEFENSE MECHANISM

J. E. Skinner

Baylor College of Medicine

Several independent studies of the pathology of persons who died of "heart attack" while living in a modern urban setting have established consistently that 13-16% of all these deaths are sudden (unexpected) and are *unexplained* following extensive post-mortem examination (Kuller & Lilienfeld, 1966; Moritz & Zamcheck, 1946; Schwartz & Gerrity, 1975; Spiekerman, Brandenburg, Achor, & Edwards, 1962). That is, the hearts of members of this sudden unexplained death group showed no sign of coronary artery block, myocardial infarction, or the presence of any arteriosclerotic heart disease that could have caused cardiac ischemia and the onset of ventricular fibrillation. The only cause known for rapid death without prior symptoms is ventricular fibrillation and it has been confirmed in eight out of eight cases witnessed by medical experts as the condition present in victims who unexpectedly collapse and are in the process of sudden death (Friedman et al., 1973). The inescapable conclusion of these combined studies is that coronary artery blockade is not a necessary condition for evoking ventricular fibrillation and sudden death. Therefore other causal factors involved in the initiation of this lethal arrhythmia must be sought.

Rahe, Bennett, Romo, Seltanen, and Arthur (1973) have demonstrated a psychologic stress factor for sudden death that is as statistically significant as the well-known physical risk factors, hypertension and high serum cholesterol. They showed that individuals who scored high on a scale of life change, a scale characterized by "death of a spouse, divorce, concern over the health of a family member, recently out of work, etc.," are at high risk for sudden death. It has been known for some time that bereavement is associated with an increase in mortality in persons of western cultures (Parkes, 1967; Rees & Lutkins, 1967), and knowledge of a "voodoo" curse seems to precipitate sudden death in members of certain primitive cultures (Burrell, 1963; Cannon, 1942). Extensive post-mortem examinations of victims of Pygmy curse-death (James, 1971, personal communication) or of Bantu willed-death (Burrell, 1963) have revealed no detectable underlying pathology to explain their sudden demise. In conclusion it appears that recently precipitated states of psychologic stress can be a *sufficient* condition for sudden cardiac death.

Myocardial ischemia and sudden death are variables that cannot be

DOI: 10.4324/9781003164647- 9

causally related because they are doubly dissociated; not all persons who die suddenly have coronary obstructions and not all persons with massive coronary obstructions die. Stress-evoked potentiation of arrhythmogenesis in hearts already made ischemic by heart disease or myocardial infarction could account for the high incidence of sudden death in cardiac patients. This potentiation hypothesis is supported by such findings as the early observation by Järvinen (1955) that the presence of unfamiliar staff making rounds in a coronary care unit (i.e., psychologic stressors) was associated with a five-fold greater incidence of sudden death than was expected when the usual staff alone made the rounds. Mild stressors produce no detectable pathology, but severe ones, such as nontraumatic child abuse, robbery, assault, etc., can result in myofibrillar degeneration (Cebelin & Hirsch, 1980). Pavlov was the first to observe this phenomenon in animals (see Pavlov, 1951), an observation which more recently has been confirmed by Johansson et al. (1974).

A. Psychologic Stress: A Causal Factor for the Initiation of Ventricular Fibrillation

Animal studies offer the hope of unraveling the neural mechanism that intervenes between the psychologic stressors in the environment and the sudden death precipitated by ventricular fibrillation. Skinner, Lie, and Entman (1975) have shown that psychologic stress is a *necessary* factor for the precipitation of ventricular fibrillation following coronary occlusion in the conscious pig. If the pig was adapted to the unfamiliar staff and surroundings of the laboratory, then coronary occlusion did not result in ventricular fibrillation; in unadapted or stressed animals the lethal arrhythmia was always produced. That stress can by itself be a *sufficient* factor for the initiation of ventricular fibrillation in the conscious pig is suggested by the study of Johansson et al. (1974), which showed that the psychologic stress of immobilization plus the physical stress of cutaneous shock together were sufficient to produce cardiac arrhythmias, permanent myocardial damage, and ventricular fibrillation. Using another method to assess cardiac vulnerability, Lown, Verrier, and Corbalan (1973) showed that less current to the myocardium was required to produce a series of extrasystolies when a conscious dog was in a part of the laboratory where it was accustomed to receiving cutaneous shocks, than that which was required when the animal was located in its home cage.

That a brain factor alone is sufficient to produce ventricular fibrillation is confirmed by a study in which direct electric stimulation of the posterior hypothalamus produced this lethal arrhythmia (Garvey & Melville, 1969). Lown and associates were unable to replicate this result, perhaps because of different electrode placement, but they did show potentiation effects of the brain when combined with myocardial ischemia (Satinsky, Kosowsky, Lown, & Kerzner, 1971). Hall and associates have found that stimulation of either the frontal cortex (Hall, Livingston, & Bloor, 1977) or the posterior region of the hypothalamus (Hall, Sybers, Greenhoot, & Bloor, 1974) will produce spotty myocardial necrosis similar to that produced by coronary artery injections of catecholamines. This same pattern of pathological insult is also produced by stress alone (Cebelin & Hirsch, 1980; Johansson et al., 1974; Pavlov, 1951).

Skinner and Reed (1981) found that cryogenic blockade in a frontocortical-brainstem pathway, first described anatomically by Nauta (1964), will prevent the occurrence of ventricular fibrillation in the ischemic heart of an unadapted and highly stressed conscious pig. Figure 7.1 shows schematically the points of effective blockade (filled bars) in this pathway (FC-BS) as well as the control sites that were ineffective (open bars). It thus appears that direct electric activation of an anatomically defined frontocortical-brainstem pathway is sufficient to cause cardiac fibrillation or permanent myocardial damage, and that blockade in this pathway will prevent the deleterious effects of psychologic stress on the vulnerability of the myocardium, even in the presence of massive ischemia.

B. Psychologic Stress: Cerebral Responses

That the frontal cortex and its related structures respond specifically to psychologically and physically stressful stimuli (i.e., novel, conditioned, and noxious) has been demonstrated in an electrophysiological study by Skinner and Yingling (1976). After acquisition of learning in a classical tone-shock conditioning paradigm, the psychologically stressful stimulus (tone) produced an extracellular slow potential shift in the frontal cortex, parietal association cortex, mesencephalic reticular formation, and thalamic reticular nucleus. Cryogenic blockade in the pathway between the frontal cortex and thalamic reticular nucleus (i.e., the inferior thalamic peduncle, ITP, Figure 7.1) prevented cortical and thalamic reticular responses to the conditioned tone, but it did not prevent thalamic reticular responses to the physical stress (cutaneous shock) or to direct stimulation of the mesencephalic reticular formation. These results suggest a separation of cerebral systems reactive to psychologic and physical stressors, that is, the frontal cortex and the mesencephalic reticular formation, respectively.

The electrogenic basis for the cortical event-related slow potential (ERSP) response to psychologic stressors is beginning to be understood. To begin, the ERSP is not the only type of event-related cerebral response that occurs in response to a stressful stimulus. Three types of bioelectric events can be recorded simultaneously from the human scalp or animal cortex in response to meaningful stimuli: (1) the EEG desynchronizes, that is, changes

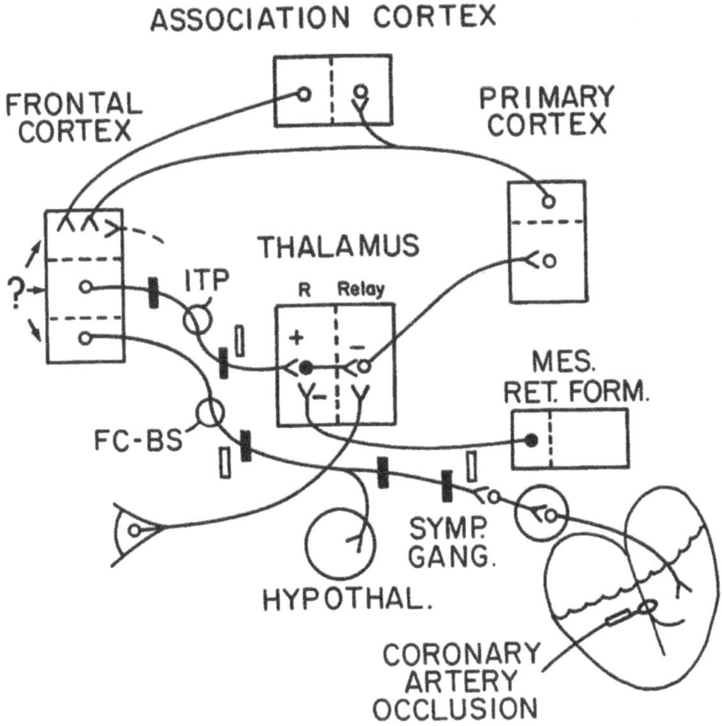

FIGURE 7.1. A neural model for the regulation of cardiac vulnerability to ventricular fibrillation. Two pathways descend from the frontal cortex: (1) the inferior thalamic peduncle, ITP, which has an excitatory effect on the sensory gating neurons located in the thalamic reticular nucleus, R; and (2) the frontocortical-brainstem pathway FC-BS, which has collateral branches into the hypothalamus and eventually effects activity in the sympathetic ganglia that project to the heart. Cryogenic blockade in the critical region of the ITP (filled bars) will prevent selective gating of sensory input to primary cortex, but blockade in nearby control structures has no effect (open bars). Such blockade also prevents conditioned cardiac responses in a classical tone-shock paradigm. Cryogenic blockade in the FC-BS (filled bars), but not in nearby control sites (open bars), will prevent the initiation of ventricular fibrillation following coronary artery occlusion in psychologically stressed animals. The integrative activities in the frontal cortex that couple the sensory and autonomic control are unknown (?), but are thought to be related to event-related slow potential processes that are perhaps mediated by norepinephrine-release, cyclic AMP activation and outward potassium conductance decrease.

from a predominant 10 Hz high-voltage pattern of activity characteristic of inattention and drowsiness to one of high frequency and low voltage; (2) evoked potentials elicited by attended stimuli are larger in amplitude than those resulting when the same stimulus is not attended; and (3) slow onset sustained potentials are elicited by warning signals, novel stimuli or any other similar type of meaningful stimulus. Each of these electrocortical responses can occur in the absence of a physical stimulus as long as the event of the absence has meaning (e.g., nonoccurrence of a repetitive expected stimulus). It is the *context* of the evoking stimulus, not its physical attributes, that governs the amplitude and waveform of these endogenous event-related responses.

A specific set of cerebral structures underlies the joint regulation of the three types of event-related cerebral responses. Rossi and Zanchetti (1957), in their early classical review, showed that all three were related to the functioning of the mesencephalic reticular formation. Later, Skinner and Lindsley (1973) demonstrated that the frontal granular cortex was also involved in their regulation. A neurophysiological model has been constructed by Skinner and Yingling (1977) to explain how the mesencephalic reticular formation and the frontal cortex together regulate the three types of cerebral activity. Skinner and King (1980a, 1980b) showed that some neurons in the frontal cortex manifest slow changes in membrane potential when the conscious animal responds to a tone that forewarns cutaneous shock. This result suggests that the extracellular ERSP is produced when these neurons manifest their slow event-related *membrane* potentials.

Libet (1978) suggested that the neurochemical mechanisms underlying the regulation of the cortical ERSP may be the same as those described in the sympathetic ganglion for the slow excitatory post synaptic potential (slow EPSP). This theory is supported by pharmacologic evidence in both humans (Ashton, Millman, Telford, & Thompson, 1974; Teece & Cole, 1974; Thompson et al., 1978) and animals (Marczynski, 1978). Cellular evidence from an *in vitro* cortical system (i.e., hippocampal slice) has also been supportive of the theory. The slow EPSP is present in the hippocampal slice and both cholinergic and catecholaminergic systems regulate the *inactivation* of a resting potassium current (Brown & Adams, 1980; Cole & Nicoll, 1983; Haas & Konnerth, 1983; Halliwell & Adams, 1982; Hashiguchi, Kobayashi, Tosaka, & Libet, 1982; Madison & Nicoll, 1982), just as Libet's slow ERSP theory predicts (Hashiguchi et al., 1982).

Early evidence by Krnjević's laboratory (Krnjević, Pumain, & Renaud, 1971) suggests that some cholinoceptive neurons in the cat frontal cortex manifest potassium-inactivation. Heinemann and Lux (1975) showed that synaptic stimulation in this same tissue can produce a reduction in the extracellular potassium activity, a response that could be explained by the inactivation of a potassium current. Skinner and Molnar (1983) have shown in this same frontocortical tissue that the temporal pattern of an event-related reduction in potassium activity is correlated with the *waveform* of the local EPSP.

The release of norepinephrine (NE) and its consequent stimulation of cyclic AMP is produced in the cortex by the same stimulus-events that evoke the local ERSP (Skinner, 1984a; Skinner, Reed, Welch, & Nell, 1978; Skinner, Welch, Reed, & Nell, 1978). What role the cortical noradrenergic release has in cerebral functioning is not yet clear. Daw, Rader, Robertson, and Ariel (1983), Kasamatsu and Pettigrew (1976), and Pettigrew and Kasamatsu (1978) have shown that cortical noradrenergic and beta-receptor functions are necessary for visual cortex neurons, commonly called "feature detectors," to be physiologically reprogrammed by a change in visual input. Long-term potentiation (LTP) of excitatory synaptic inputs is currently considered to be one of the more important cellular models of learning and memory (Bliss, Goddard, & Rives, 1983; Cotman & McGaugh, 1980; Thompson, 1980). Goddard (1981) has shown in rats that depletion of NE will block LTP. Hopkins and Johnston (1984) have shown that the addition of NE to the bath of a hippocampal slice will increase the amplitude and duration of LTP and that the addition of propranolol (a beta-blocker) will prevent its occurrence. Thus, NE has been implicated in the neuromodulation of at least three independent types of cortical phenomena: feature detection, change in synaptic efficacy and ERSP-generation.

The observation of a common noradrenergic mechanisms underlying these higher cerebral processes has important implications for interpretations directed at the processing of information in the higher centers of the brain. Further understanding of this common mechanism may eventually help us understand why the increase in cardiac vulnerability by a constant environmental stressor varies from subject to subject, as well as within a subject during *learned behavioral adaptation.*

It is to be emphasized that it is the endogenous *learning-related* aspect of an environmental stimulus, not its exogenous physical attributes, that determines whether or not it is a psychologic stressor. For example, it is the learning-related familiarization to the abundance of novel stimuli that prevents the occurrence of ventricular fibrillation after coronary artery occlusion (Skinner & Reed, 1981). To demonstrate that both the ERSP and a cerebral noradrenergic mechanism are indeed involved in determining cardiac vulnerability, Skinner (1984b) recently reported data which show that (1) in cardiac patients with high arrhythmia rates, ERSP amplitude and cardiac ectopy rates are *linearly* correlated, and (2) in unadapted psychologically stressed pigs, intracerebral injection of 0.01 mg/kg of levo-propranolol will *prevent* the lethal consequence of acute coronary artery occlusion.

C. Psychologic Stress: Mediation by Cannon's Cerebral Defense System

Cannon and de la Paz (1911) first showed that when a cat is frightened or angered by a dog, not only does it orient its exteroceptors toward this meaningful stimulus, but the autonomic nervous system is also prepared to support anticipated behavioral responses to the situation. Cannon (1931) thought that there must be simultaneous cerebral orchestration of sensory input channels and autonomic output channels when the animal orients to attended objects (i.e., he postulated that a cerebral defense system exists in higher structures of the brain). Skinner and Yingling (1977) have shown that frontal cortex, in its projection through the inferior thalamic peduncle (ITP) to the thalamic reticular nucleus, exerts selective control over sensory channels, as they relay in the thalamus (Figure 7.1). Stimulation of different parts of the thalamic reticular nucleus produces inhibition of transmission in the adjacent sensory relay nucleus (Yingling & Skinner, 1976). Graded cryogenic blockade in ITP produces graded facilitation of transmission of irrelevant stimulus-information in various sensory channels (Skinner & Lindsley, 1971), presumably by selectively reducing excitatory input to specific regions of thalamic reticular cells. Stimulation of the mesencephalic reticular formation, however, caused intense inhibition of all thalamic reticular cells and a generalized opening of all sensory gates (Yingling & Skinner, 1975, 1976). Skinner and Yingling (1977) concluded that frontocortical regulation of thalamic reticular cells provides the neurophysiological basis for *selective attention*, whereas mesencephalic reticular control mediates a more generalized process, such as *arousal or orienting reactions*.

Cryogenic blockade in the vicinity of the ITP not only blocks regulation in sensory input channels, but also prevents the cardiac response to a tone that forewarns the onset of electric shock (Skinner & Yingling, 1977). Ablation of the frontal cortex produces a similar effect on the cardiac conditioned response (Smith, Nathan, & Clarke, 1968). Thus the frontal cortex appears to subserve the function of simultaneous regulation of sensory and autonomic channels, as was hypothesized by Cannon in his early description of a cerebral defense mechanism.

How the frontal cortex functions to produce a particular pattern of output to the thalamic reticular nucleus to provide selective sensory gating and how it simultaneously activates output in the frontocortical-brainstem system to prepare the viscera to respond to attended objects remain goals for future research. Anatomic and physiologic evidence shows that the frontal cortex is the recipient of cortico-cortical input from virtually all parts of the cerebral mantel (Imbert, Bignall, & Buser, 1966). It also receives input from all parts of the medial thalamic nuclear group (Kuypers, 1977). This convergent input is somehow transformed by frontal cortex into the simultaneous regulation of both the sensory and autonomic channels.

The deleterious descending autonomic projection to the heart during states of increased vulnerability to arrhythmogenesis seems to be simultaneous activation of *both* sympathetic and parasympathetic nerves (Manning & Cotten, 1962; Skinner, Beder, & Entman, 1983; Skinner, Mohr, & Kellaway, 1975). These data could explain why the *peripheral* administration of propranolol (Beta HAT Research Group, 1982) only reduces mortality by 26% in cardiac patients with ischemic myocardial injury (i.e., because it only blocks sympathomimetic projections to the heart), whereas the *intracerebral* injection in pigs (Skinner, 1984b, 1985) reduces mortality by 75% (i.e., because it blocks the central mechanisms that give rise to the dual autonomic outflows). The peripheral propranolol may actually achieve its mortality-effect by its central action, as it is known to cross the blood-brain barrier.

The continued investigations of the neurochemical and biological mechanisms that give rise to activity in the frontocortical-brainstem pathway may lead to a significant pharmacological intervention in the mechanism of sudden cardiac death, for the physical blockade of this pathway leads to 100% reduction in mortality (Skinner & Reed, 1981). Clearly, the role of the brain in the mechanism of ventricular fibrillation, the number one cause of death in industrialized countries, should become a focus for research in the future. Given that coronary artery disease alone is not sufficient to cause ventricular fibrillation, that coronary bypass surgery in such diseased individuals has had only a slight impact on mortality, and that psychologic stress itself may have a role in the pathogenesis of arteriosclerotic disease, the National Heart, Lung, and Blood Institute has announced in its 5-year plan (Tenth Report NHLBI, 1982) that "Basic biobehavioral research has been refocused on defining the influence of neural and biobehavioral factors on ventricular electrical stability.... Inherent in these investigations is the development of requisite methodology that permits recording of the pharmacologic, neurophysiologic, and psychologic indices of the reaction(s)." The behavioral neuroscience that began in Donald B. Lindsley's laboratory in 1967 (Skinner, 1968, 1969) has become current state-of-the-art at the National Institutes of Health.

REFERENCES

Ashton, H., Millman, J. E., Telford, R., & Thompson, J. W. (1974). The effect of caffeine, nitrazepam, and cigarette smoking on the contingent negative variation in man. *Electroencephalography and Clinical Neurophysiology, 37*, 59-71.

Beta-blocker Heart Attack Trials Research Group. (1982). A randomized trial of propranolol in patients with acute myocardial infarctions. *Journal of the American Medical Association, 247*, 1707-1714.

Bliss, T. V. P., Goddard, G. V., & Rives, M. (1983). Reduction of long-term potentiation in the dentate gyrus of the rat following selective depletion of monoamines. *Journal of Physiology, 334*, 475-491.

Brown, D. A., & Adams, P. D. (1980). Muscarinic suppression of a novel voltage-sensitive K^+ current in a vertebrate neuron. *Nature, 283*, 673-676.

Burrell, R. J. W. (1963). The possible bearing of curse death and other factors in Bantu culture on the etiology of myocardial infarction. In T. N. James & J. W. Keys (Eds.), *The etiology of myocardial infarction* (pp. 100-101). Henry Ford Hospital International Symposium. Boston: Little, Brown.

Cannon, W. B. (1931). Again the James-Lange and the thalamic theories of emotion. *Psychological Reviews, 38*, 281-295.

Cannon, W. B. (1942). "Voodoo" death. *American Anthropologist, 44*, 169-181.

Cannon, W. B., & de la Paz, D. (1911). Emotional stimulation of adrenal secretion. *American Journal of Physiology, 28*, 64.

Cebelin, M. S., & Hirsch, C. S. (1980). Human stress cardiomyopathy: Myocardial lesions in victims of homicidal assaults without internal injuries. *Human Pathology, 11*, 123-132.

Cole, A. E., & Nicoll, R. A. (1983). Acetylcholine mediates a slow synaptic potential in hippocampal pyramidal cells. *Science, 221*, 1299-1301.

Cotman, C. W., & McGaugh, J. L. (1980). *Behavioral neuroscience*. New York: Academic Press.

Daw, N. W., Rader, R. K., Robertson, T. W., & Ariel, M. (1983). Effects of 6-hydroxydopamine on visual deprivation in the kitten striate cortex. *Journal of Neuroscience, 3*, 907-914.

Friedman, M., Manwaring, J. H., Rosenman, R. H., Donlon, G., Ortega, P., & Grube, S. M. (1973). Instantaneous and sudden deaths. *Journal of the American Medical Association, 225*, 1319-1328.

Garvey, H. L., & Melville, K. I. (1969). Cardiovascular effects of lateral hypothalamic stimulation in normal and coronary-ligated dogs. *Journal of Cardiovascular Surgery, 10*, 377-385.

Goddard, G. V. (1981). The continuing search for mechanisms. In J. A. Wada (Ed.), *Kindling 2* (pp. 1-14). New York: Raven Press.

Haas, H., & Konnerth, A. (1983). Histamine and noradrenaline decrease calcium activated conductance in hippocampal pyramidal cells. *Nature, 302*, 432-435.

Hall, R. E., Livingston, R. B., & Bloor, C. M. (1977). Orbital cortical influences on cardiovascular dynamics and myocardial structure in conscious monkeys. *Journal of Neurosurgery, 46*, 638-647.

Hall, R. E., Sybers, H. D., Greenhoot, J. H., & Bloor, C. M. (1974). Myocardial alterations after hypothalamic stimulation in the intact conscious dog. *American Heart Journal, 88*, 770-776.

Halliwell, J. V., & Adams, P. R. (1982). Voltage-clamp analysis of muscarinic excitation in hippocampal neurons. *Brain Research, 250*, 71-79.

Hashiguchi, T., Kobayashi, H., Tosaka, T., & Libet, B. (1982). Two muscarinic depolarizing mechanisms in mammalian sympathetic neurons. *Brain Research, 242*, 378-382.

Heinemann, U., & Lux, H. D. (1975). Undershoots following stimulus-induced rises of extracellular potassium concentration in cerebral cortex of cat. *Brain Research, 93*, 63-76.

Hopkins, W. J., & Johnston, D. (1984). Frequency-dependent noradrenergic modulation of long-term potentiation in the hippocampus. *Science, 226*, 350-352.

Imbert, M., Bignall, K. E., & Buser, P. (1966). Neocortical interconnections in the cat. *Journal of Neurophysiology, 29*, 382-395.

Järvinen, K. A. J. (1955). Can ward rounds be a danger to patients with myocardial infarction? *British Medical Journal, 1*, 318-320.

Johansson, G., Jonsson, L., Lannek, N., Blomgren, L., Lindberg, P., & Pouph, O. (1974). Severe stress-cardiopathy in pigs. *American Heart Journal, 87*, 451-457.

Kasamatsu, T., & Pettigrew, J. D. (1976). Depletion of brain catecholamines: Failure of ocular dominance shift after monocular occlusion in kittens. *Science, 194*, 206-208.

Krnjević, K., Pumain, R., & Renaud, L. (1971). Effects of Ba^{2+} and tetraethylammonium on cortical neurons. *Journal of Physiology, 215*, 223-245.

Kuller, L., & Lilienfeld, A. (1966). Epidemiological study of sudden and unexplained deaths due to arteriosclerotic heart disease. *Circulation, 34*, 1056.

Kuypers, H. (1977). Organization of the thalamocortical connections to the frontal lobe in the rhesus monkey. *Experimental Brain Research, 29*, 229-322.

Libet, B. (1978). Slow postsynaptic responses of sympathetic ganglion cells as models for slow potential changes in the brain. In D. A. Otto (Ed.), *Multi-disciplinary perspectives in event-related brain potential research* (pp. 12-18). Washington, DC: US Environmental Protection Agency.

Lown, B., Verrier, R. L., & Corbalan, R. (1973). Psychologic stress and threshold for repetitive ventricular response. *Science, 182*, 834-836.

Madison, D. V., & Nicoll, R. A. (1982). Noradrenaline blocks accommodation of pyramidal cell discharge in the hippocampus. *Nature, 299*, 636-638.

Manning, J. W., & Cotten, MDe. V. (1962). Mechanism of cardiac arrhythmias induced by diencephalic stimulation. *American Journal of Physiology, 203*, 1120-1124.

Marczynski, T. J. (1978). Neurochemical mechanisms in the genesis of slow potentials: Some clinical implications. In D. A. Otto (Ed.), *Multi-disciplinary perspectives in event-related brain potential research* (pp. 25-35). Washington, DC: US Environmental Protection Agency.

Moritz, A. R., & Zamcheck, N. (1946). Sudden and unexpected death of young soldiers. *Archives of Pathology, 42*, 459-494.

Nauta, W. J. H. (1964). Some afferent connections of the prefrontal cortex in the monkey. In J. W. Warren & K. Akert (Eds.), *The frontal granular cortex and behavior* (pp. 372-396). New York: McGraw-Hill.

Parkes, C. M. (1967). Bereavement. *British Medical Journal, 4*, 13.

Pavlov, I. P. (1951). The reinforcing nerve of the heart. In R. R. Ann (Ed.), *The complete collection of works of Pavlov* (Vol. 1, pp. 419-457).

Pettigrew, J. D., & Kasamatsu, T. (1978). Local perfusion of noradrenaline maintains visual cortical plasticity. *Nature, 271*, 761-763.

Rahe, R. H., Bennett, L., Romo, M., Seltanen, P., & Arthur, R. S. (1973). Subject's recent life changes and coronary heart disease in Finland. *American Journal of Psychiatry, 130*, 1222-1226.

Rees, W. D., & Lutkins, S. G. (1967). Mortality of bereavement. *British Medical Journal, 4*, 13.

Rossi, G. F., & Zanchetti, A. (1957). The brainstem reticular formation: Anatomy and physiology. *Archives Italiennes de Biologie, 95*, 199-435.

Satinsky, J., Kosowsky, B., Lown, B., & Kerzner, J. (1971). Ventricular fibrillation induced by hypothalamic stimulation during coronary occlusion. *Circulation, 44* (Suppl. II), 1160.

Schwartz, P. J., & Gerrity, R. G. (1975). Anatomical pathology of sudden unexpected cardiac death. *Circulation, 52* (Suppl. 3), 18-26.

Skinner, J. E. (1968). Regulation of electrocortical activity and behavior by the nonspecific thalamo-orbitocortical synchronizing system. *Dissertation Abstracts* 2167 B (Order No. 67-14, 281).

Skinner, J. E. (1969). Regulation of electrocortical activity and behavior by the nonspecific thalamo-orbitocortical synchronizing system. *Creative Talent Awards Program, 1966-1977 Winners' Dissertation Abstracts* (pp. 20-28). CTA Series No. 7, American Institutes for Research, Silver Springs, MD.

Skinner, J. E. (1984a). Central gating mechanisms that regulate event-related potentials and behavior. In T. Elbert, B. Rochstroh, W. Lutzenberger, & N. Birnbaumer (Eds.), *Self-regulation of the brain and behavior* (pp. 42-58). New York: Springer.

Skinner, J. E. (1984b). Psychosocial stress and sudden cardiac death: Brain mechanisms. In R. E. Beamish, P. K. Singal, & N. S. Dhalla (Eds.), *Proceedings of International Symposium: Stress and Heart Disease* (pp. 44-59). Boston: Martinus Nijhoff Publishers.

Skinner, J. E. (1985). The regulation of cardiac vulnerability by the cerebral defense system. *Journal of American College of Cardiology, 5,* 88B-94B.

Skinner, J. E., Beder, S. D., & Entman, M. L. (1983). Psychologic stress activates phosphorylase in the heart of the conscious pig without increasing heart rate and blood pressure. *Proceedings of the National Academy of Sciences, 80,* 4513-4517.

Skinner, J. E., & King, G. L. (1980a). Electrogenesis of event-related slow potentials in the cerebral cortex of conscious animals. In G. Pfurtscheller, P. Buser, F. H. Lopes da Silva, & H. Petsche (Eds.), *Rhythmic EEG activities and cortical activities* (pp. 21-32). Amsterdam: Elsevier-North Holland.

Skinner, J. E., & King, G. L. (1980b). Contribution of neuron dendrites to extracellular sustained potential shift. In H. H. Kornhuber & L. Deeke (Eds.), *Motivation, motor and sensory processes of the brain: Electrical potentials, behavior and clinical use* (pp. 89-102). Program on Brain Research. Amsterdam: Elsevier/North Holland.

Skinner, J. E., Lie, J. T., & Entman, M. L. (1975). Modification of ventricular fibrillation latency following coronary artery occlusion in the conscious pig: The effects of psychologic stress and beta-adrenergic blockade. *Circulation, 51,* 656-667.

Skinner, J. E., & Lindsley, D. B. (1971). Enhancement of visual and auditory evoked potentials during blockade of the non-specific thalamo-cortical system. *Electroencephalography and Clinical Neurophysiology, 31,* 1-6.

Skinner, J. E., & Lindsley, D. B. (1973). The nonspecific medio-thalamic frontocortical system: Its influence on electrocortical activity and behavior. In K. H. Pribram & A. R. Luria (Eds.), *Psychophysiology of the frontal lobes* (pp. 185-234). New York: Academic Press.

Skinner, J. E., Mohr, D. N., & Kellaway, P. (1975). Sleep-stage regulation of ventricular arrhythmias in the unanesthetized pig. *Circulation Research, 37,* 342-349.

Skinner, J. E., & Molnar, M. (1983). Event-related extracellular potassium ion activity changes in the frontal cortex of the conscious cat. *Journal of Neurophysiology, 49,* 204-215.

Skinner, J. E., & Reed, J. C. (1981). Blockade of a frontocortical-brainstem pathway prevents ventricular fibrillation of the ischemic heart in pigs. *American Journal of Physiology, 240,* 156-163.

Skinner, J. E., Reed, J. C., Welch, K. M. A., & Nell, J. (1978). Cutaneous shock produces correlated shifts in slow potential amplitude and cyclic 3', 5'-adenosine monophosphate level in the parietal cortex of the conscious rat. *Journal of Neurochemistry, 30*, 699-704.

Skinner, J. E., Welch, K. M. A., Reed, J. C., & Nell, J. (1978). Psychological stress reduces cyclic 3', 5'-adenosine monophosphate levels in the cerebral cortex of conscious rats, as determined by a new cryogenic method of rapid tissue fixation. *Journal of Neurochemistry, 30*, 691-698.

Skinner, J. E., & Yingling, C. D. (1976). Regulation of slow potential shifts in nucleus reticularis thalami by the mesencephalic reticular formation and the frontal cortex. *Electroencephalography and Clinical Neurophysiology, 40*, 288-296.

Skinner, J. E., & Yingling, C. D. (1977). Central gating mechanisms that regulate event-related potentials and behavior: A neural model for attention. In J. E. Desmedt (Ed.), *Progress in clinical neurophysiology* (Vol. 1, pp. 30-69). Basel: Karger.

Smith, O. A., Nathan, M. A., & Clarke, N. P. (1968). Central nervous system pathways mediating blood pressure changes. *Hypertension, 16*, 9-22.

Spiekerman, R. E., Brandenburg, J. T., Achor, R. W. P., & Edwards, J. E. (1962). Spectrum of coronary heart disease in a community of 30,000: Clinicopathologic study. *Circulation, 25*, 57.

Teece, J. J., & Cole, J. O. (1974). Amphetamine effects in man: Paradoxical drowsiness and lowered electrical brain activity (CNV). *Science, 185*, 451-453.

Tenth Report of the Director. (1982). *National Heart, Lung, and Blood Institute: Ten-year Review and Five-Year Plan* (Vol. 2, pp. 211). US Department of Health and Human Services. Public Health Service. National Institutes of Health. NIH Publication No. 84-2357.

Thompson, J. W., Newton, P., Pocock, P. V., Cooper, R.,Crow, H., McCallum, W. C., & Papakostopoulos, D. (1978). Study of pharmacology of contingent negative variation in man. In D. A. Otto (Ed.), *Multi-disciplinary perspectives in event-related brain potential research* (pp. 51-55). Washington, DC: US Environmental Protection Agency.

Thompson, R. F. (1980). In search of the engram. II. In D. McFadden (Ed.), *Neural mechanisms in behavior* (pp. 172). New York: Springer-Verlag.

Yingling, C. D., & Skinner, J. E. (1975). Regulation of unit activity in nucleus reticularis thalami by the mesencephalic reticular formation and the frontal granular cortex. *Electroencephalography and Clinical Neurophysiology, 39*, 635-642.

Yingling, C. D., & Skinner, J. E. (1976). Selective regulation of thalamic sensory relay nuclei by nucleus reticularis thalami. *Electroencephalography and Clinical Neurophysiology, 41*, 476-482.

SECTION III. NEURAL SUBSTRATES OF VISUAL DISCRIMINATION

The chapters of this section are all concerned with vision and visual behavior in either animals (Peck, [P]), or humans (Nakayama et al. [N]; and Coleman & Sydnor [C & S]). The wide range of visual responses tested in the reports of this section nicely illustrate this emphasis upon vision. Studied are visual acuity and visual pattern discrimination including two-choice visual discrimination (P); full and hemifield checkerboard pattern discrimination (C & S); and spatial frequency modulations (N). There are other common themes in the chapters of this section. Two of the chapters use average evoked-potential technique (N; C & S). And two chapters (P; C & S) are directly concerned with abnormal visual functioning, and both of these reports have important clinical implications.

The Peck investigation attempts to understand the mechanism by which strabismic misalignment of the eyes produces impairment of vision. In two experiments, cats with surgically induced rotation of one or both eyes were tested both monocularly and binocularly. The results of both experiments are consistent with the hypothesis that when the two eyes have incongruent views of the world some neural mechanism actively suppresses one of those views.

The chapter by Nakayama and colleagues describes how the technique of measuring human visual evoked potential (VEPs) may be used to isolate cortical sub-populations of cells narrowly tuned for different spatial frequencies. In a series of three experimental demonstrations, these investigators were able to selectively abolish individual peaks from a double-peak response by exploiting the sensitivity of the VEP to small changes in stimulus conditions. These investigators offer possible explanations for this paradoxical finding of "unit-like" properties in the human scalp VEPs.

In the third chapter of this section, Coleman and Sydnor use the averaged visual evoked potential technique to investigate the visual pathways of human albino subjects. Monocular and binocular visual displays were presented to one or both hemifields. They found that concisely presented hemifield stimuli revealed an abnormal projection pattern. This and other evidence discussed by the authors indicate that human albinos lack the mechanisms for binocular vision.

 DOI: 10.4324/9781003164647- 10

8 · PERCEPTUAL SUPPRESSION IN STRABISMIC CATS

C. K. Peck

University of Missouri-St. Louis

In higher mammals, including people, misalignment of the eyes during early childhood leads to severe disturbances of binocular vision (Von Noorden, 1967). Because the visual deficits persist after correction of the misalignment, it seems that there must have been changes in the pattern of connections within the nervous system. Over the past decade, a large number of animal studies have been directed toward trying to understand the mechanisms by which strabismic misalignment of the eyes produces impairment of vision. In general, it is clear that abnormal early visual experience profoundly alters not only visual behavior (Riesen, 1966) but also the physiology of cortical neurons in most species of mammals (for reviews see Barlow, 1975; Grobstein & Chow, 1975). For example, if one eye of a kitten is caused to deviate or is covered, binocular input to neurons of the visual cortex is severely and permanently reduced (Hubel & Wiesel, 1965; Wiesel & Hubel, 1963), while similar manipulations have little effect in adult cats (Hubel & Wiesel, 1970; Yinon, 1976).

It has been argued that stereoblind people also experience a loss of binocular neurons in the visual cortex. In support of this argument, Mitchell and Ware (1974), Movshon, Chambers, and Blakemore (1972), and Ware and Mitchell (1974) have shown that there is a high correlation between stereopsis and interocular transfer of visual aftereffects based on contour orientation. More generally, it has been thought that the presence of interocular transfer depends on cortical neurons with binocular input because, in both cat and monkey, neurons of the visual cortex are sensitive to stimulus orientation. Further, in normal humans, many orientation-specific aftereffects show interocular transfer, which seems to indicate that the aftereffects occur at some central neural site. However, cats with misaligned eyes show substantial interocular transfer of oriented-form discriminations (Mitchell, Giffin, Muir, Blakemore, & Van Sluyters, 1976; Peck & Crewther, 1975; Peck, Crewther, Barber, & Johannsen, 1979), which is independent of the number of binocularly driven neurons at the level of striate cortex (Blakemore, Van Sluyters, Peck, & Hein, 1975; Crewther, Crewther, Peck, & Pettigrew, 1980). These results are consistent with similar studies of both monocularly deprived kittens (Ganz, Hirsch, & Tieman, 1972) and of Siamese cats (Marzi, Simoni, & Di Stefano, 1976) which lack binocularly driven cells

DOI: 10.4324/9781003164647- 11

in striate cortex. The differences between the data on interocular transfer in humans and in animals pose problems for those who wish to create an animal model for strabismus.

An additional problem is that many patients with strabismus show an active suppression of the strabismic eye by the normal eye. The suppressed eye often shows reduced visual acuity (amblyopia), a condition which has rarely been demonstrated in animals with surgically induced deviations (Ikeda & Wright, 1976). Instead, strabismic kittens usually seem to use each eye quite well, and it has been hypothesized that they utilize a strategy of alternate fixation with each eye (e.g., Buchtel, Berlucchi, & Mascetti, 1975; Hubel & Wiesel, 1965). However, data from my laboratory indicate dramatic perceptual suppression in kittens reared with surgically induced ocular deviation, and raise the possibility that even adult cats are susceptible to perceptual disorders following surgical misalignment of the eyes.

The present experiments were guided by the intuition that since strabismic patients suffer deficits in binocular vision, deficits which are not apparent during monocular testing in animals should be apparent during binocular testing. Moreover, we were aware of the work of Buchtel et al. (1975) on monocular paralytic strabismus, and in Experiment 2 our testing procedures were designed to be similar to theirs.

EXPERIMENT I

We tested visual field perimetry in 14 cats with surgically induced rotations (see Figure 8.1) of one or both eyes (Peck, Barber, Pilsecker, & Wark, 1980). Data from 3 of these cats are shown in Figure 8.2. Two of the cats (HRC and KLV) had no visual experience prior to surgery, which was performed on the day that their eyes opened. The third cat (LAF) was allowed 6 months of normal vision and, then, well after the end of the "sensitive period" for visual deprivation (Blakemore & Van Sluyters, 1974; Hubel & Wiesel, 1970), had the same surgery. The rotational differences between the eyes were 30° for LAF, 100° for KLV, and 160° for HRC, and the degree of horizontal divergence was 10°-20°.

Perimetry testing was done after each cat had had 6-18 months of experience with the rotated eye and after extensive training on a series of visual pattern discriminations (Peck, Crewther, Barber, & Johannsen, 1979). The testing procedure, in brief, was this: the cat was hungry and was trained to fixate a piece of food while gently restrained. A second stimulus was introduced, and the presence or absence of an orienting response to the second stimulus was recorded (Peck et al., 1980; Sherman, 1973; Simoni & Sprague, 1976; Sprague & Meikle, 1965). The visual field was defined as those stimulus locations which elicited a positive response on at least 50% of the trials.

FIGURE 8.1. Cat whose left eye was intorted 100° in infancy.

Figure 8.2 illustrates the extent of the visual fields obtained for each cat during binocular vision in comparison to the monocular tests. For two of the three animals illustrated (LAF and HRC), we found deficits which were greater during binocular tests than during monocular tests. When both eyes were open, these animals did not respond to stimuli in the far periphery, i.e., stimuli which fell only on the retina of the suppressed eye. This failure to respond was unique to binocular vision, and presents a marked contrast with the results of similar tests in visually deprived cats (Sherman, 1973, 1974), in cats with surgically induced ocular convergence (Ikeda & Jacobson, 1977), and in Siamese cats with squint (Simoni & Sprague, 1976). One of the two cats that showed a loss of response to a portion of the binocular field had 6 months of visual experience prior to surgery, thus apparently indicating that some degree of plasticity persists beyond the sensitive period.

EXPERIMENT II

In 6 cats we tested a series of two-choice visual discriminations; some were learned monocularly and others were learned binocularly. We have previously shown that animals with cyclodeviations can learn such discriminations and, moreover, that when trained monocularly, they show appreciable transfer to the naive eye (Peck & Crewther, 1975; Peck, Crewther, Barber, & Johannsen, 1979; Peck & Wark, 1981). However, as shown in Figure 8.3, when similar discriminations are trained with binocular vision, less benefit accrues to the rotated eye. In fact, of the 14 cats which

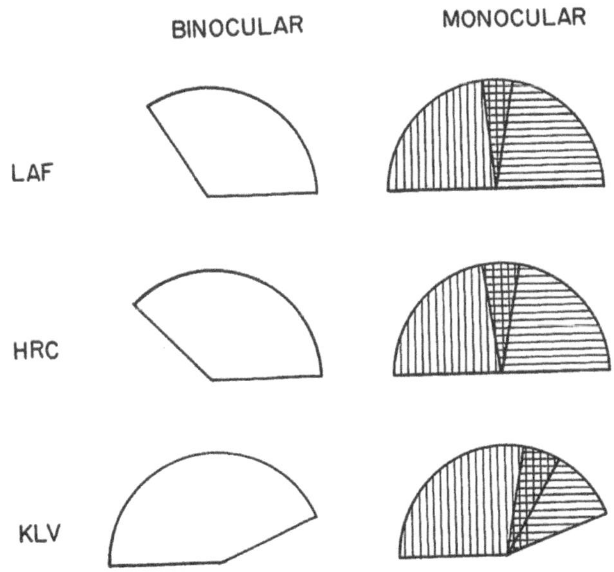

FIGURE 8.2. Schematic representation of the visual fields of three cats (LAF, HRC and KLV) with rotated eyes during monocular and binocular testing. The field represents the 180° sector ahead of the cat, bounded by an imaginary line through the outer canthi. Test stimuli were presented 35 mm from the cat; the fixation stimulus was 60 mm from the cat. Each eye was tested monocularly. The field of the right eye alone is hatched with vertical lines; that of the left, by horizontal. See text for definition of positive responses.

have been tested under both conditions (monocular and binocular learning), 6 were found to be severely handicapped in subsequent learning of the same discrimination with the rotated eye.

Figure 8.3 illustrates data for four animals with unilateral rotations after binocular learning. They performed well with the normal eye but poorly with the deviated eye. Data from two animals with binocular rotations are also illustrated. They consistently performed at or above criterion with one eye, while performing near chance with the other. In one of these cats, the suppressed eye had the greater angular displacement, but in the other, both eyes had been rotated through nearly the same angle. In general, there was little relationship between degree of rotation and percent savings after binocular learning. On the other hand, at least for cats with unilateral rotations, there was a relationship between visual acuity and percent savings after binocular learning.

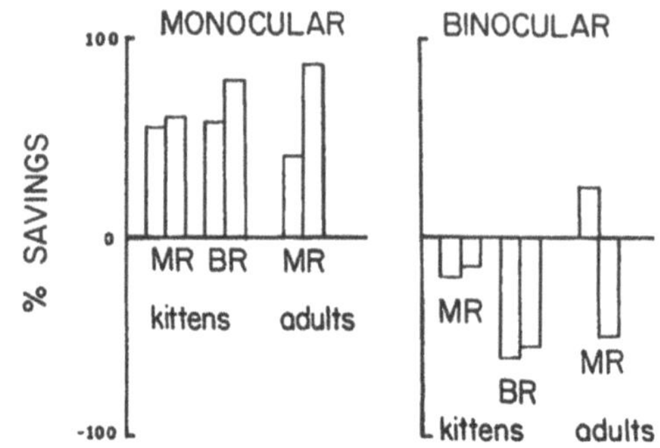

FIGURE 8.3. Results of monocular and binocular learning of two-choice visual discriminations by six cats after binocular (BR) or monocular (MR) rotations as kittens or as adults. Scores are percent savings, defined by the ratio:

$$\frac{a-b}{a+b} \times 100,$$

where *b* is the number of trials taken by the second (rotated) eye to reach criterion (90% correct on 2 successive days), and *a* is the number taken by the first eye (for monocular training) or by both eyes (open simultaneously, for binocular training). A score of 100 indicates complete transfer; zero indicates no transfer, and negative scores indicate negative transfer. Thus, paradoxically, each animal learned visual tasks more slowly after binocular training (when each eye had "seen" the stimuli) than after monocular training (when only the other eye had seen the stimuli).

Before speculating about these results, I first wish to consider some possible technical reservations about our preparation. We have attempted to assess whether, during the surgery, we might have inadvertently damaged the eye or restricted its blood supply, thus causing abnormal development of the eye itself and/or restriction of its projection to the central nervous system. In addition, because we must section the ocular muscles in order to rotate the eye, we might produce a permanent immobilization which, in itself, alters the visual system (e.g., Salinger, Garraghty, MacAvoy, & Hooker, 1980).

To answer the question about surgical damage, we prepared whole mounts of the retinas of four animals with large monocular rotations (approximately 180°) because we expected that large rotations might stress either the optic nerve or the retinal blood supply and thus maximize the probability of inadvertent damage. However, we were unable to detect any regions in which the density of ganglion cells was lower than normal. Next, we prepared the visual pathways of seven cats for autoradiography by injecting ^{3}H proline into the posterior chamber of the rotated eye. We processed sections through the dorsal lateral geniculate and superior colliculus by standard procedures, and then examined them with both light and dark field microscopy. We were unable to detect any "patchy" areas in either structure, as would be expected if there were areas of reduced function in the retina (Peck et al., 1979).

With regard to the question of ocular motility, the answer is more complicated. Ocular motility was definitely present. In young kittens, the first eye movements occurred within 4-5 days of surgery; in the adult cats, motility reappeared within a few weeks. However, the eye movements were abnormal. In general, movements of the two eyes were disparate, and out of phase by the angle of rotation. In addition, in the animals with monocular rotations, the movements of the rotated eye were often smaller than those of the normal eye, particularly among animals with the largest angular rotations. Very rarely were conjugate deviations observed in any animal.

It is possible that these atypical movements, or altered proprioceptive input from the eye muscles, do affect visual behavior. However, neither abnormal motility nor any defect within the eye itself can account for the results presented in this paper because the present experiments involve comparisons of the same animals (and the same eyes) under different conditions. In Experiment I, two animals with rotated eyes performed credibly when that eye was tested monocularly but were unresponsive to stimuli presented at some of the same locations if both eyes were open. In Experiment II, the rotated eye showed interocular transfer after monocular training, but after binocular training the rotated eye was generally unable to perform well and showed little savings in relearning.

The results of both experiments are consistent with a hypothesis that, when the two eyes have incongruent views of the world, there is a neural mechanism which actively suppresses one of those views. Whatever the mechanism, these data do not result from ocular rotation per se but from competition between two irreconcilable inputs. When only one eye is open, there is no competition and, therefore, no suppression. Although suppression may, at first, appear to be a "deficit" (because an animal fails to respond to certain stimuli), I suggest that suppression may be better viewed as an adaptive response, which functions to screen out information that may be confusing or uninterpretable.

It is tempting to speculate about the central neural mechanisms responsible for such suppressive effects but, unfortunately, there are a host of reasonable candidates. Because the perimetry data are very different from those reported in cats with lesions of either visual cortical sites (Sprague, Levy, Di Bernardino, & Berlucchi, 1977), or the superior colliculus (Sprague & Meikle, 1965), it seems unlikely that suppression can be explained by alterations at these levels of the visual system, and thus the "higher" areas, traditionally considered to be involved in sensory associations, are more plausible candidates. Therefore, the search for the neural basis of visual suppression will quite probably further our understanding of the processes of visual attention.

ACKNOWLEDGMENTS

I would like to thank Grayson Barber, Karen Kleinman, Brian Thompson and Robert Wark for their help in training the animals. Early portions of this work were supported by USPHS Grant MH-03372 to Professor R. W. Sperry and were furthered by the collaboration of Sheila Gillard Crewther. Completion of this work was supported by NS-14116 to C. K. Peck.

REFERENCES

Barlow, H. B. (1975). Visual experience and cortical development. *Nature, 258*, 199-203.

Blakemore, C., & Van Sluyters, R. C. (1974). Reversal of the physiological effects of monocular deprivation in kittens. Further evidence for a sensitive period. *Journal of Physiology (London), 237*, 195-216.

Blakemore, C., Van Sluyters, R. C., Peck, C. K., & Hein, A. (1975). Development of cat visual cortex following rotation of one eye. *Nature, 257*, 584-586.

Buchtel, H. A., Berlucchi, G., & Mascetti, G. G. (1975). Behavioral and electrophysiological analyses of strabismus in cats. In F. Vital-Durand & M. Jeannerod (Eds.), *Aspects of neural plasticity* (pp. 27-44), INSERM, 43.

Crewther, S. G., Crewther, D. P., Peck, C. K., & Pettigrew, J. D. (1980). Visual cortical comparison effects of rearing cats with monocular or binocular cyclotorsion. *Journal of Neurophysiology, 44*, 97-118.

Ganz, L., Hirsch, H. V. B., & Tieman, S. B. (1972). The nature of perceptual deficits in visually deprived cats. *Brain Research, 44*, 547-568.

Grobstein, P., & Chow, K. L. (1975). Receptive field development and individual experience. *Science, 190*, 352-358.

Hubel, D. H., & Wiesel, T. N. (1965). Binocular interaction in striate cortex of kittens reared with artificial squint. *Journal of Neurophysiology, 28*, 1041-1059.

Hubel, D. H., & Wiesel, T. N. (1970). The period of susceptibility to the physiological effects of unilateral eye closure in kittens. *Journal of Physiology (London), 206*, 419-436.

Ikeda, H., & Jacobson, S. G. (1977). Nasal field loss in cats reared with convergent squint: Behavioural studies. *Journal of Physiology (London), 270*, 367-381.

Ikeda, H., & Wright, M. J. (1976). Properties of LGN cells in kittens reared with convergent squint: A neurophysiological demonstration of amblyopia. *Experimental Brain Research, 25*, 63-77.

Marzi, C. A., Simoni, A., & Di Stefano, M. (1976). Lack of binocularly driven neurons in the Siamese cat's visual cortex does not prevent successful interocular transfer of visual form discriminations. *Brain Research, 105*, 353-357.

Mitchell, D. E., Giffin, F., Muir, D., Blakemore, C., & Van Sluyters, R. C. (1976). Behavioral compensation of cats after early rotation of one eye. *Experimental Brain Research, 25*, 109-113.

Mitchell, D. E., & Ware, C. (1974). Interocular transfer of a visual after-effect in normal and stereoblind humans. *Journal of Physiology, 236*, 707-721.

Movshon, J. A., Chambers, B. E. I., & Blakemore, C. (1972). Interocular transfer in normal humans and those who lack stereopsis. *Perception, 1*, 483-490.

Peck, C. K., Barber, G., Pilsecker, C. E., & Wark, R. C. (1980). Visual field deficits in cats reared with cyclodeviations of the eyes. *Experimental Brain Research, 41*, 61-74.

Peck, C. K., & Crewther, S. G. (1975). Perceptual effects of surgical rotation of the eye in kittens. *Brain Research, 99*, 213-219.

Peck, C. K., Crewther, S. G., Barber, G., & Johannsen, C. J. (1979). Pattern discrimination and visuomotor behavior following rotation of one or both eyes in kittens and in adult cats. *Experimental Brain Research, 34*, 401-418.

Peck, C. K., & Wark, R. C. (1981). A relationship between visual suppression and amblyopia in cats with cyclodeviations of the eyes. *Experimental Brain Research, 44*, 317-324.

Riesen, A. H. (1966). Sensory deprivation. *Progress in Physiological Psychology, 1*, 117-147.

Salinger, W. L., Garraghty, P. E., MacAvoy, M. G., & Hooker, L. F. (1980). Sensitivity of the mature lateral geniculate nucleus to components of monocular paralysis. *Brain Research, 187*, 307-320.

Sherman, S. M. (1973). Visual field defects in monocularly and binocularly deprived cats. *Brain Research, 49*, 25-45.

Sherman, S. M. (1974). Visual fields of cats with cortical and tectal lesions. *Science, 185*, 355-357.

Simoni, A., & Sprague, J. M. (1976). Perimetric analysis of binocular and monocular visual fields in Siamese cats. *Brain Research, 111*, 189-196.

Sprague, J. M., Levy, J., Di Bernardino, A., & Berlucchi, G. (1977). Visual cortical areas mediating form discrimination in the cat. *Journal of Comparative Neurology, 172*, 441-488.

Sprague, J. M., & Meikle, T. H., Jr. (1965). The role of the superior colliculus in visually-guided behavior. *Experimental Neurology, 11*, 115-146.

Von Noorden, G. K. (1967). Classification of amblyopia. *American Journal of Ophthalmology, 63*, 238-244.

Ware, C., & Mitchell, D. M. (1974). On interocular transfer of various visual after effects in normal and stereoblind observers. *Vision Research, 14*, 731-734.

Wiesel, T. N., & Hubel, D. H. (1963). Single cell responses in striate cortex of kittens deprived of vision in one eye. *Journal of Neurophysiology, 26*, 1003-1017.

Yinon, U. (1976). Age dependence of the effects of squint on cells in kittens' visual cortex. *Experimental Brain Research, 26*, 151-157.

9 · VISUAL EVOKED POTENTIALS: ISOLATION OF CORTICAL SUB-POPULATIONS NARROWLY TUNED TO SPATIAL FREQUENCY

K. Nakayama, P. Apkarian, M. Mackeben and C. W. Tyler

Smith-Kettlewell Institute of Visual Sciences

Because recordings from the human brain are usually confined to the outer perimeter of the skull, surface event related potentials (ERPs) remain as one of the few electrophysiological techniques currently available for investigation of the human brain. As this field has matured, several interpretative strategies have emerged. The most direct view is that particular voltage deflections (as extracted by time-locked averaging) reflect separate brain processes. In rare cases, this has been unusually successful. For example, it works in the interpretation of brainstem auditory evoked responses (BAER) because each short latency peak is brief and distinct (Jewett & Williston, 1971).

With most ERPs, however, this approach is too simplistic. The responses are too dispersed in time, allowing different mechanisms ample opportunity to overlap. Consequently, the identity of the individual generating components is often obliterated. In addition, the fact that one single mechanism may produce a complex polyphasic waveform adds further difficulty in the isolation of component brain processes. One possible approach to deal with this complexity is to use statistical reasoning: factor analysis, principle components analysis, etc. (Donchin, 1966). Such approaches make assumptions that are often difficult to support, however.

The techniques described above put an emphasis on *response* analysis. By itself, however, this approach is limited. A powerful complementary strategy is to manipulate the *stimulus*, or more broadly, the conditions of the experiment. This has been very successful in the area of cognitive psychophysiology, where discrete waveforms at different scalp locations can appear selectively under different cognitive demands. Most documented is the large P300 associated with uncertainty resolution, task relevance, etc. (Sutton, Braren, Zubin, & John, 1965). Another wave is the N100 frequently associated with focal attention (Hillyard, Hink, Schwent, & Picton, 1973). As yet, however, we are far from understanding such "cognitive" potentials in terms of basic neurophysiological processes.

With visual evoked potentials, however, one might expect to obtain a greater correlation with neurophysiological processes, especially with the emergence of extensive anatomical, physiological, and psychophysical data on

 DOI: 10.4324/9781003164647- 12

the visual system (Hubel & Wiesel, 1977; MacLeod, 1977). Before describing our attempt to bridge this gap between VEP and neurophysiology, using periodic rather than transient VEPs, we review some pertinent research relating psychophysics and single cell neurophysiology.

CONTRAST SENSITIVITY FUNCTION

A powerful way to describe the visual system's sensitivity to pattern stimulation is to obtain the spatial frequency sensitivity curve. Sinusoidally modulated light in the form of vertical bars is presented on a TV screen (see

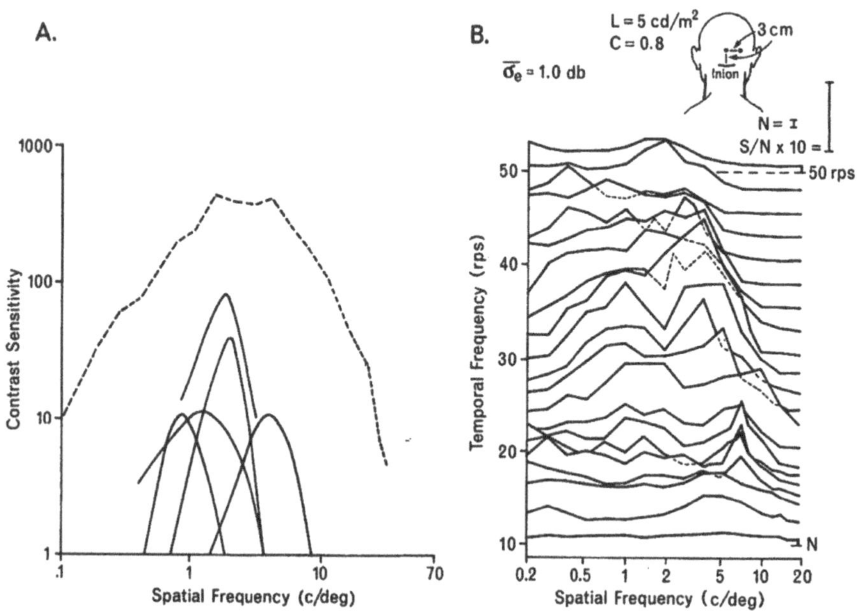

FIGURE 9.1. A. Dashed curve - spatial modulation sensitivity function of a human observer obtained psychophysically (data from Campbell and Robson, 1968). Solid curves represent typical modulation sensitivities of single cortical neurons recorded from primate visual cortex (DeValois et al., 1978). Note that the spatial frequency timing for individual units is considerably sharper than the psychophysical results. B. Human evoked potential amplitudes (signal/noise ratio) plotted as a function of spatial frequency for a wide number of reversal rates for observer CWT. Note that instead of a broad tuning function, there are numerous peaks at different spatial and temporal frequency combinations. The bar labeled S/N x 10 represents a single response 10 times the noise. Standard error of voltage signals was ±1 db. Contrast was 80%, mean luminance was 5 cd/m^2 (Tyler, Apkarian, & Nakayama, 1978).

Figure 9.2). The observer adjusts the contrast of the pattern to threshold (Campbell & Robson, 1968). Plotting the reciprocal of this threshold as a function of the number of sinusoidal cycles per degree of visual angle results in the well-known contrast sensitivity curve, relating sensitivity to spatial frequency (Figure 9.1A, top trace). This curve shows a peak sensitivity at intermediate spatial frequencies (about 4 cycles/deg) and falls off, both for high and low spatial frequencies. Thus, the visual system is not equally good at seeing all patterns, since it is less sensitive to patterns having coarse and fine gradations in luminance.

The suggested physiological substrate of this sensitivity function lies in the inhibitory and excitatory receptive field characteristics of single visual neurons, endowing such cells with contrast rather than luminance sensitivity (Hubel & Wiesel, 1968; Kuffler, 1953). High spatial frequency sinusoidal gratings (having narrow bars) will fail to stimulate the cell to the extent that the bar width becomes smaller than the receptive field center, and low spatial frequency stimuli (having wider bars) will suffer attenuation as a single bar of the grating fills the center and antagonistic surround of the receptive field.

In particular, each cortical neuron appears to have a very narrow tuning with respect to spatial frequency (DeValois, Albrecht, & Thorell, 1978; Schiller, Finlay, & Volman, 1976) and this tuning can be quantitatively predicted from the inhibitory and excitatory combinations of input from adjacent parts of the receptive field (DeValois, DeValois, & Yund, 1979). Most critical for our discussion is the narrowness of spatial frequency tuning of each individual cell in relation to the psychophysical curve. The cell's tuning is often less than one octave at half height. Thus the sensitivity of a single cell spans a much more restricted range of spatial frequencies than that of the whole visual system (see Figure 9.1A, lower solid traces).

Psychophysical evidence using selective adaptation corroborates this view (Blakemore & Campbell, 1969). After prolonged viewing, there is a decrease in contrast sensitivity for spatial frequencies centered on the adapting frequency and the bandwidth of the adaptation function is similar to that seen for single cells of monkey visual cortex. Overall contrast sensitivity of these cells is much narrower than the psychophysical function.

The apparent discrepancy can be resolved by assuming that the observer's sensitivity is due to the upper envelope of sensitivities of separate mechanisms, each sensitive to a relatively narrow range of spatial frequencies and spanning a range of frequencies. This concept is familiar to those acquainted with the narrow tuning characteristics of single auditory neurons as it relates to the broadly tuned psychophysical auditory sensitivity curve (Green, 1976).

EXISTENCE OF SPATIAL FREQUENCY PEAKS

What should we expect if we were to use these sinusoidal gratings to obtain visual evoked potentials from the human scalp? Because surface electrodes can sample from a very large population of cortical neurons, it seems at first glance that the spatial tuning curve of the human visual evoked potential (VEP) should approximate the psychophysics.

The surprise is that this is not the case. Instead of a broadly tuned function, we have shown that there are multiple tuned response peaks (Tyler, Apkarian, & Nakayama, 1978). Stimulating with counterphase modulated gratings (periodically reversing the phase and frequency filtering the occipital EEG at this frequency), we find multiple spatial frequency peaks and this spatial tuning changes as temporal frequency is altered. Thus for some rates of reversal there are two or even three distinct spatial frequency peaks, whereas at other rates, there is only one. Plotting a family of curves showing the amplitude of the visual evoked response at different spatial frequencies for different reversal rates, we can obtain an overall spatiotemporal frequency map. Figure 9.1B shows these multiple spatiotemporal peaks as they vary in amplitude across spatial and temporal frequency in a human observer. For different observers and for different electrode locations, the peaks occur at other combinations of spatial and temporal frequencies. Very narrow spatial frequency tuning can also be seen from the surface recordings in the alert trained monkey (Nakayama, Mackeben, & Sutter, 1980), often showing a single peak (Figure 9.2). It should be stressed that for a given observer (or monkey), at a given electrode location, the peaks are extremely consistent and reproducible for periods exceeding one year.

Having established that the existence of very narrow tuning is quite reproducible, and can also occur across species, we are faced with the issue as to whether these peaks represent electrical signs of separate brain processes. To start, it is best to return to the idea of "components" mentioned earlier. Since voltages superimpose passively in the brain, electrical signs of separate components can summate, often obscuring the characteristics of individual components. Thus, it is important to develop a technique that will disentangle the individual components.

With the transient evoked potential, it is well recognized that various peaks can add and subtract from each other to give a composite which may have little resemblance to individual mechanisms. With steady state potentials an analogous problem is present: multiple component signals at the same temporal frequency can lead to bigger or smaller composite signals depending on the relative temporal phase between the component signals. To the extent that component signals are out of phase (the maximum being 180 degrees) they will subtract, otherwise they will add. As such, it is conceivable that the trough between some of the peaks seen in Figure 9.1 is

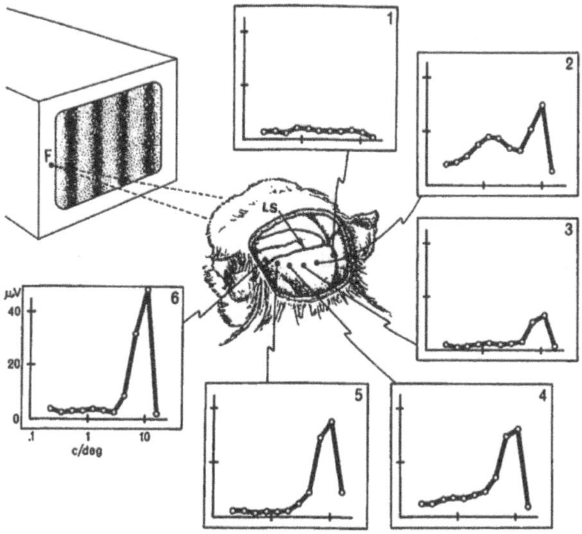

FIGURE 9.2. VEP amplitude (in microvolts) plotted as a function of spatial frequency (cycles/deg) for six different electrodes placed over the left occipital region of the alert monkey brain. Recording frequency = 16 Hz. Each curve consists of the average of four sets (24 values per point) recorded on different days. Upper left of figure shows the stimulus field containing the vertical sinusoidal gratings and the fixation point (f). Note that the same type of gratings were also used for the human experiments described herein. Scaling of the axes in each of the curves is identical (Nakayama, Mackeben, & Sutter, 1980).

due to this type of response cancellation.

This "cancellation" explanation is not applicable to most of our experimental findings, however. First we show that for many reversal rates, the recorded phase of the sinusoidal response at one spatial frequency peak is essentially identical to the phase at another peak. In other words they are not 180 degrees out of phase and could not cancel electrically (Tyler et al., 1978). In other cases there is only one peak present in the spatial frequency plot (see Figures 9.2 and 9.3), removing any possibility of cancellation by a second mechanism.

So far we have dealt with the nature of the response, indicating that it is compatible with multiple mechanisms, each having narrow spatial frequency tuning. In the remainder of this paper we further establish the existence of these peaks at a neurophysiological level by a complementary approach, further manipulating the set of stimulus conditions presented. We vary stimuli, using a logic very similar to that used in psychophysics and in the examination of single visual neurons.

ISOLATION OF SPATIAL FREQUENCY PEAKS

METHOD

For all experiments we presented vertical sinusoidal gratings on an oscilloscope screen. Mean luminance and contrast were held constant. Spatial frequency and reversal rate of the gratings were under experimental control. Electrical activity recorded from the human scalp was sampled with bipolar electrodes, one on the midline 3 cm above the inion and one lateral, 3 cm to the right of the midline placement. The response was averaged for 10 seconds using a narrow bandfilter (less than 0.3 Hz bandwidth) exactly centered on the grating reversal frequency. The filtered response was full wave rectified and integrated, providing a single numerical measure of evoked potential amplitude. For details of the method see Tyler et al. (1978).

EXPERIMENT I: BINOCULAR vs. MONOCULAR RESPONSES.

As a first example, we show how a comparison of monocular and binocular responses can be used to fractionate a double peaked spatial frequency tuning function. In observer DS, we obtain a binocular response as a function of spatial frequency (Fig. 9.3A, upper curve) showing a peak at approximately 2 cycles/deg and one around 5 cycles/deg. By confining the stimulus to only one eye (lower curve), we find that the double peaked response has been transformed into one having only the single high frequency peak. Thus, there appears to be a selective facilitation of the binocular response to the low frequency gratings relative to the monocular response. In this spatial frequency range, the binocular response is much greater than twice the monocular response. In contrast, the high spatial frequency response shows little indication of binocular facilitation, with the binocular response even less than twice the monocular response. (See Apkarian, Nakayama, & Tyler, 1981; Nakayama, Apkarian, & Tyler, 1982, for further details.)

EXPERIMENT II: SEGREGATION BY RETINAL AREA.

To further isolate these two spatial frequency peaks in the same observer, we also restricted the stimulus to discrete retinal areas. In the middle trace of Figure 9.3B we confined our stimulus grating to a small annular region (2 degrees inner diameter and 4 degrees outer diameter) surrounding the fovea. Note that in this paramacular region we obtain only the high frequency response to fine gratings. Restricting the stimulation, however, to a more peripheral annulus (4 degrees inner diameter, 8 degrees outer diameter) leads to the appearance of the response to coarse gratings

(Figure 9.3B, top middle trace). Thus, different retinal areas appear to make different contributions to the peaks in the evoked response recorded at this reversal rate and this is in accord with the well-known fall-off in visual resolution with increasing retinal eccentricity. For more details see Tyler and Apkarian (1982).

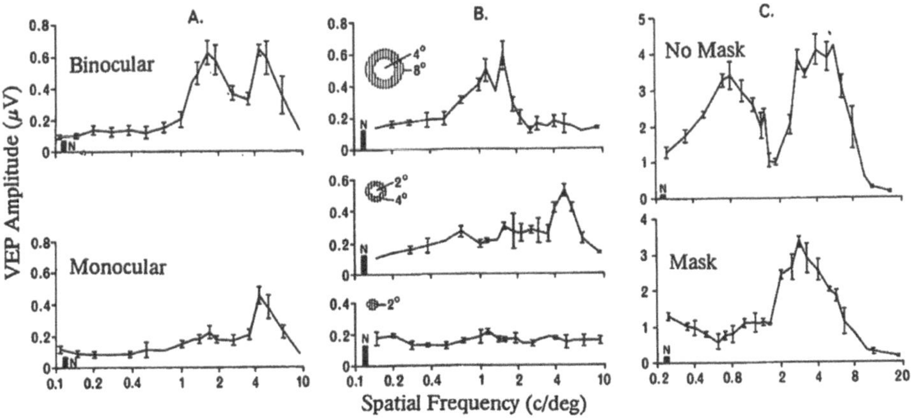

FIGURE 9.3. Isolation of spatial frequency peaks by alteration of stimulus conditions. A. Comparison of monocular and binocular responses in observer DS. Reversal rate was 30 rps, contrast was 68%. Upper trace is the spatial frequency curve for binocular viewing. Bottom trace is a similar tuning curve for monocular viewing. B. Comparison of responses from differing retinal areas. Using the same electrode location and reversal rate in observer DS (see A above), we restricted the stimulation to a central 2° diameter foveal disk (bottom trace), a parafoveal annular region, 2-4° (middle trace), and a peripheral annular region, 4-8° (upper trace). C. Effect of an occluding mask consisting of horizontal bars. Top trace is the spatial frequency response of observer DL to a reversal rate of 20 rps under binocular conditions. Lower curve is for the same, except that horizontal occluding bars were placed in front of the vertical stimulus grating. Masking bars had a spatial frequency of 1 cycle/deg.

EXPERIMENT III: MATCHING THE SHAPE OF CORTICAL RECEPTIVE FIELDS.

As a final example of a selective fractionalization of a multiple peaked response, consider the response of observer DL as a function of spatial frequency at 20 reversals/sec (Figure 9.3C, top trace). It shows a peak at 0.8 and 5 cycles/deg, corresponding to bar widths of 38 and 6 minutes of arc, respectively. It is well known that the receptive fields of cortical neurons are elongated and the receptive field length is roughly three times the receptive field width (Hubel & Wiesel, 1962). To isolate those spatial frequency mechanisms selective for narrow bars we chose a stimulus which might be considered to degrade selectively the response to wide bars (low spatial frequency gratings). This was accomplished by placing a square wave occluding mask (1 cycle/deg) horizontally across the vertical sine wave grating. With such a mask, one sees the vertical sine wave grating between the horizontal bars, so that each vertical bar subtends 30 minutes of visual angle. Under these circumstances, high spatial frequency bars will be long relative to their width, staying above the 3:1 ratio mentioned above. For low frequency bars, however, the mask will truncate the relative length of the wide bars and thus diminish its ability to stimulate elongated receptive fields.

The effect of this mask on the responses at different spatial frequencies was dramatic and in the predicted direction. The low frequency peak was completely abolished, leaving only the high frequency peak with somewhat reduced amplitude, as might be expected merely from the reduction in area of the flickering stimulus (Figure 9.3C, bottom trace).

DISCUSSION

In all of the above cases, we have been able to selectively abolish individual peaks from a double peaked response by exploiting the extreme sensitivity of the VEP to small changes in stimulus conditions. As such, we have additional information to suggest that the spatial frequency peaks represent a distinct subset of the total cortical neuronal population.

This brings us to an obvious yet puzzling question. Why should a scalp potential, supposedly recording from many different classes of nerve cells, have properties more similar to single neurons?

To deal with the paradox of finding "unit-like" properties in the scalp VEP, we offer two hypotheses: one dealing with the stimulus specificity of cortical neurons under repetitive stimulation and the other dealing with the extracellular voltage fields surrounding sets of nerve cells.

Our first suggestion is that some neurons which share similar spatial receptive field characteristics also share some accidental temporal characteristics. Thus a given class of cells sensitive to a particular spatial frequency might share the same temporal characteristics: time constants, delay times, etc. If the microcircuitry of this set of spatially tuned neurons is sufficiently uniform, some temporal frequencies may be unusually effective in driving homogeneous sub-populations of neurons. As such, these cells would show "resonant-like" characteristics and repetitive stimulation at one temporal frequency could activate only those cells sensitive to particular temporal, and hence spatial frequencies. (The term "resonant-like" is not intended to carry any implication that the system is linear.) As yet, a comparable selectivity in temporal frequency has not been seen for single cortical neurons (Movshon, Thompson, & Tolhurst, 1978).

Our second suggestion concerns the geometry of generating sources. Whether or not a large potential is seen at relatively distant electrodes depends on the symmetry and distribution of synaptic currents (Lorente de No, 1947). Thus with "closed fields," there is little current flow outside of the immediate vicinity of the neuron and extracellular potential fluctuations at any distance are minimal (Baker & Precht, 1972). In contrast, widespread currents surround the synaptic activation of the classic dipole or "open field" which can be seen as potential fluctuations over a large volume of brain tissue (Freeman, 1975). To the extent that some sets of visual neurons may have self-canceling extracellular current fields, the activity of these cells will be essentially invisible to surface recording electrodes. As such, the geometry of synaptic currents provides an additional reason for the selective nature of VEP recording.

So far the relation between the functional characteristics of cortical neurons and their geometry remains to be elucidated. New techniques are emerging, however. Single cell recordings, for example, can be correlated with morphology using HRP techniques (Gilbert & Wiesel, 1979) and the microscopic location of synaptic terminals associated with very specific forms of visual stimulation can be obtained by de-oxyglucose methods (Sharp & Kilduff, 1981).

The preceding results raise the hope that with an appropriate animal model (e.g., the monkey, see Nakayama, 1982), particular components of the VEP might be linked to physiologically and anatomically determined classes of neurons in monkey visual cortex. To the degree to which the results could be extrapolated to humans, the VEP could become a surprisingly powerful method to reveal the activity of small neuronal populations in the human brain.

ACKNOWLEDGMENTS

We acknowledge the support of Grants EY-01582, EY-01186, and EY-02124 from the National Institute of Health and grants from the Smith-Kettlewell Eye Research Foundation.

REFERENCES

Apkarian, P., Nakayama, K., & Tyler, C. W. (1981). Binocularity in the human visual evoked potential: Facilitation, summation, and suppression. *Electroencephalography and Clinical Neurophysiology, 51,* 32-48.

Baker, R., & Precht, W. (1972). Electrophysiological properties of trochlear motoneurons as revealed by IVth nerve stimulation. *Experimental Brain Research, 14*, 127-157.

Blakemore, C., & Campbell, F. W. (1969). On the existence of neurones in the human visual system selectively sensitive to the orientation and size of retinal images. *Journal of Physiology (London), 203*, 237-260.

Campbell, F. W., & Robson, J. G. (1968). Application of Fourier analysis to the visibility of gratings. *Journal of Physiology (London), 197*, 551-566.

DeValois, R., Albrecht, D. G., & Thorell, L. G. (1978). Cortical cells: Bar and edge detectors, or spatial frequency filters? In S. Cool (Ed.), *Frontiers of visual science* (pp. 544-556). Berlin: Springer-Verlag.

DeValois, K. K., DeValois R., & Yund, D. (1979). Responses of striate cortex cells to grating and checkerboard patterns. *Journal of Physiology (London), 291*, 483-505.

Donchin, E. (1966). A multivariate approach to the analysis of average evoked potentials. *IEEE Transactions on Bio-medical Engineering, BME-13*, 131-139.

Freeman, W. (1975). *Mass action in the nervous system.* New York: Academic Press.

Gilbert, C., & Wiesel, T. N. (1979). Morphology and intercortical projections of functionally characterized neurones in the cat visual cortex. *Nature, 280*, 120-125.

Green, D. M. (1976). *An introduction to hearing.* Hillsdale, NJ: Lawrence Erlbaum Associates.

Hillyard, S. A., Hink, R. F., Schwent, V. L., & Picton, T. W. (1973). Electrical signs of selective attention in the human brain. *Science, 182*, 171-180.

Hubel, D. H., & Wiesel, T. N. (1962). Receptive fields, binocular interaction and functional architecture in the cat's visual cortex. *Journal of Physiology, 160*, 106-154.

Hubel, D. H., & Wiesel, T. N. (1968). Receptive fields and functional architecture of monkey striate cortex. *Journal of Physiology (London), 195*, 215-243.

Hubel, D. H., & Wiesel, T. N. (1977). Ferrier Lecture: Functional architecture of macaque visual cortex. *Proceedings of the Royal Society of London, Series B, 198*, 1-59.

Jewett, D., & Williston, J. S. (1971). Auditory evoked far-fields averaged from the scalp of human. *Brain, 94*, 681-696.

Kuffler, S. W. (1953). Discharge patterns and functional organization of mammalian retina. *Journal of Neurophysiology, 16*, 37-68.

Lorente de No, R. (1947). *A study of nerve physiology (Part II), 132.* New York: Rockefeller Institute of Medical Research.

MacLeod, D. I. A. (1977). Visual sensitivity. *Annual Review of Psychology, 29*, 613-645.

Movshon, J. A., Thompson, I. D., & Tolhurst, D. J. (1978). Receptive field organization of complex cells in the cat's striate cortex. *Journal of Physiology (London), 283*, 101-120.

Nakayama, K. (1982). The relationship of VEP to cortical physiology. In I. Bodis-Wollner (Ed.), *Evoked potentials* (pp. 21-36). New York: New York Academy of Sciences.

Nakayama, K., Apkarian, P., & Tyler, C. W. (1982). VEP evidence for spatial frequency limitation in binocular neurons. In I. Bodis-Wollner (Ed.), *Evoked potentials* (pp. 610-614). New York: New York Academy of Sciences.

Nakayama, K., Mackeben, M., & Sutter, E. E. (1980). Narrow spatial and temporal frequency tuning in the alert monkey VEP. *Brain Research, 193*, 263-267.

Schiller, P. H., Finlay, B. L., & Volman, S. F. (1976). Quantitative studies of single cell properties in monkey striate cortex. III. Spatial frequency. *Journal of Neurophysiology, 39*, 1334-1351.

Sharp, F. R., & Kilduff, F. (1981). The 2-deoxyglucose neuroanatomical mapping technique. *Trends in Neurosciences, 4*, 144-148.

Sutton, S., Braren, J., Zubin, J., & John, E. R. (1965). Evoked potential correlates of stimulus uncertainty. *Science, 150*, 1187-1188.

Tyler, C. W., & Apkarian, P. (1982). Properties of localized pattern evoked potentials. In I. Bodis-Wollner (Ed.), *Evoked potentials* (pp. 662-670). New York: New York Academy of Sciences.

Tyler, C. W., Apkarian, P., & Nakayama, K. (1978). Multiple spatial frequency tuning of electrical responses from the human visual cortex. *Experimental Brain Research, 33*, 535-550.

10 · THE HUMAN ALBINO: AVERAGED VISUAL EVOKED POTENTIAL STUDY OF ABNORMAL VISUAL PATHWAYS

J. R. Coleman and C. F. Sydnor

University of South Carolina and *Duke University*

Study of central visual organization of albino mammals has captured considerable interest because of systematic alterations observed in visual pathways. Unlike the decussation pattern typically observed at the optic chiasm, albino mammals have aberrant projections arising from the temporal retina. Abnormal organization of this type has been demonstrated by neuroanatomical investigations of albinos of numerous species including rat (Creel & Giolli, 1976; Lund, Lund, & Wise, 1974), rabbit (Giolli & Guthrie, 1969; Sanderson, 1975), and cat (Guillery, 1969; Guillery & Casagrande, 1975; Kalil, Jhaveri, & Richards, 1971). Characteristically, fibers arising from the temporal retina located in the region near the zero vertical meridian in albinos decussate at the chiasm instead of remaining ipsilateral as in subjects with normal eye pigment. Neurophysiological investigations have confirmed such findings and have provided further insight into abnormal organization at the cortical level in albinos (Cooper & Blasdel, 1980; Creel, Dustman, & Beck, 1970; Hubel & Wiesel, 1971; Kaas & Guillery, 1973).

Investigations into the nature of central visual organization of the human albino have revealed anomalies similar to those observed in albino mammals. Six forms of oculocutaneous albinism have been described which are characterized by decreased pigmentation of eyes, hair, and skin, as well as decreased visual acuity (Witkop, White, & King, 1974). The oculocutaneous human albino lacks a fovea, and the central cones have been reported to resemble, in structure and distribution, those of the normal parafoveal areas (Fulton, Albert, & Craft, 1978). Histological examination of the lateral geniculate nucleus of a human albino has revealed abnormal fusions between different layers which occur in both parvocellular and magnocellular laminae (Guillery, Okoro, & Witkop, 1975); this pattern is characteristic of cell groups that receive an abnormally crossed input (Gross & Hickey, 1980; Guillery & Kaas, 1971). Finally, recordings of visual evoked potentials in human albinos have disclosed abnormal patterns (Coleman, Sydnor, Wolbarsht, & Bessler, 1979; Creel, O'Donnell, & Witkop, 1978; Creel, Witkop, & King, 1974). Monocular stimulation produces asymmetrical hemispheric responses; components before 100 msec are reduced or absent in the hemisphere receiving nondecussated input.

DOI: 10.4324/9781003164647- 13

10. HUMAN ALBINO: ABNORMAL VISUAL PATHWAYS

In order to further define the nature of visual organization in the human albino, the averaged evoked potential method was applied using monocular and binocular visual displays presented to one or both hemifields. It was hypothesized that concisely presented hemifield stimuli would reveal an abnormal retinal projection pattern; and that if presented more remotely to temporal retina, the patterns could demarcate a normal projection zone. A previous report describes other details in our investigation of pigmented and albino human subjects (Coleman et al., 1979).

METHODS

Eight oculocutaneous albinos, one ocular albino, and ten normally pigmented adults participated as subjects in this study. Albino subjects had monocular visual acuities varying from 20/40 to 20/200 and full monocular visual fields. Flashed stimuli presented to all subjects consisted of square checkerboard patterns subtending a visual angle of 10° presented binocularly and monocularly to the nasal-temporal retina, and 5° (horizontal field) X 10° hemipatterns presented to the nasal or temporal retina of each eye, while the other eye was covered with a light-tight eye patch. From the fixation point hemipattern stimuli were presented to the temporal retina at 0°, 15°, 30°, and 45°. Individual checks of 14-min arc were used, and dark-adapted subjects fixated on a 10-min dim red light. Defocused flash stimuli were also used and responses compared to those from patterned stimuli. The flash unit of a Grass PS22 photic stimulator mounted in a projector produced stimuli (10^6 candlepower peak intensity) which passed through a checkerboard transparency to illuminate a plastic screen. Using a chin-rest, each subject was adjusted so that the eyes were 213 cm from the screen.

Monopolar scalp recordings made with silver disc electrodes located 3 cm above the inion and 2.5 cm from midline using an earlobe reference were amplified with Grass P511 amplifiers and averaged on a BioMac computer (640-msec storage). Repeated runs of 64 poststimulus samples each observed at 1000-msec intervals were photographed with a Polaroid camera mounted on the averager.

RESULTS

Results from monocular full pattern stimulation of nasal-temporal retina demonstrated that albino subjects show hemispheric asymmetries. No hemispheric asymmetries were detected following binocular stimulation. Reduced amplitude, absence or reversal of certain response components recorded in the hemisphere ipsilateral to the eye of stimulation, was observed in seven of nine albinos. In particular, early latency components (95-100 msec) seemed to best reflect hemispheric differences. Examples of contrasting differences between O_1 and O_2 are shown in Figure 10.1. Albino subject SS has prominent NPNPN deflections at O_2 to stimulation of the left eye (OS), and at O_1 to stimulation of the right eye (OD). In contrast, the hemisphere ipsilateral to stimulation in SS shows an absence of the first NP components observed in the opposite hemisphere and actually reversal of the positive component. Subsequent PNP components are also reduced. A second albino subject (NA) also showed reduced amplitude or absence of early latency components in the hemisphere receiving nondecussating fibers, although no clear component reversals were observed. Peak amplitude assessment of early latency components observed across subjects by analysis of variance revealed asymmetry in albino subjects ($p<.001$); differences in normally pigmented subjects were not significant.

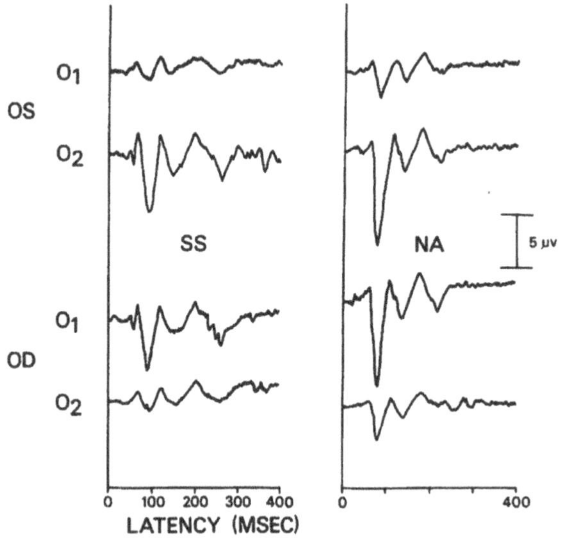

FIGURE 10.1. Visual evoked potentials to 10° patterned stimuli presented to the left (OS) and right (OD) eyes of two albino subjects. Each tracing represents 64 sweeps; negativity at scalp produces upward deflection. In all examples, the most prominent deflections appear over the hemisphere contralateral to stimulation.

Recordings involving monocular hemifield stimulation revealed systematic response patterns. Stimulation of the nasal retina produced larger amplitude early latency components in the hemisphere opposite the side of stimulation (see Figure 10.2; $p<.001$); these findings are similar to those observed in our normal subjects. Contralateral records in albinos showed response components which corresponded to those observed under nasal-temporal retina stimulation. Patterned stimulation of the temporal retina produced dramatic differences between albino and normal subjects. As

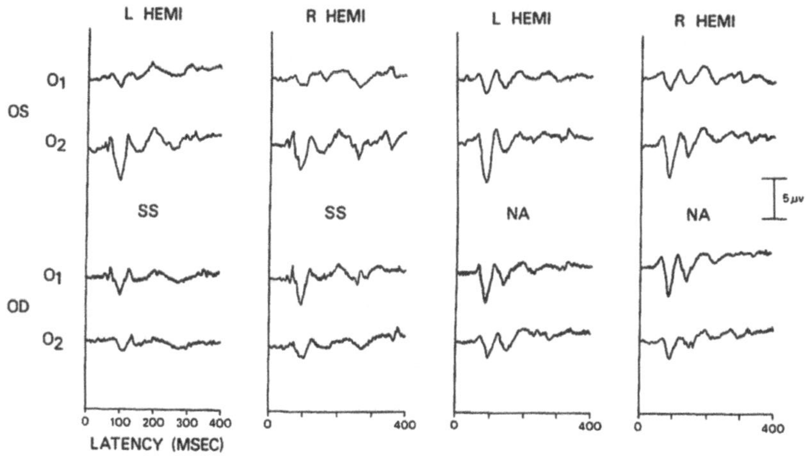

FIGURE 10.2. Visual evoked potentials to 5° patterned hemifield stimuli presented to the left and right eyes of two albino subjects. Subjects are fixated at 10° from the edge of the pattern; each tracing represents 64 sweeps and negative is up. Highest amplitude deflections appear over the contralateral hemisphere regardless of the hemifield of stimulation.

illustrated in Figure 10.2, averaged recordings of albinos showed contralateral responses to temporal retina stimulation which were comparable in latency and amplitude to tracings from nasal stimulation of the contralateral eye and larger than those recorded ipsilaterally ($p<.001$). In normally pigmented subjects temporal retina stimulation yielded larger amplitude components ipsilateral to the eye of stimulation rather than contralateral ($p<.001$).

Finally, presentation of hemipattern stimuli to restricted central and peripheral sectors of temporal retina provided further information concerning organization of visual input to the brain in albino subjects. When the stimulus was directed to temporal retina beginning at 0° or 5° the typical type of response reported above was observed; that is, larger amplitude early latency components appeared at the contralateral hemisphere. Presenting the stimulus at 15° from the center of the retina evoked a qualitatively new pattern. Early latency components of the contralateral hemisphere were reduced in amplitude, while new and higher amplitude components now appeared at the hemisphere ipsilateral to the eye of stimulation. The ipsilateral response was prominent following 30° stimulation; a further diminution of contralateral response amplitude was observed. Stimulation at 45° resulted in greater reduction in amplitude of components, especially those recorded contralaterally. In normally pigmented subjects presentation of stimuli to progressively more peripheral parts of the temporal retina resulted in a substantial, but irreversible, decline in the amplitude of response components with no shift of prominent responses to the opposite hemisphere. Absence of a more dramatic decline in response amplitude in albinos following more peripheral presentation may in part reflect eye movements due to pendular nystagmus.

DISCUSSION

Our results further establish the existence of anomalous retinal projections in the human albino. The predominant visual evoked cortical response to nasal-temporal, nasal, or temporal stimulation is in the hemisphere contralateral to the eye of stimulation. However, illumination of more peripheral temporal retina results in normal ipsilateral responses. These results confirm and extend the original recording observations on human albinos by Creel et al. (1974). That the defect observed is not a function of certain other abnormalities often associated with albinism such as strabismus is supported by electrophysiological findings (McCormack, 1975). In addition, this application of the averaged visual evoked potential method for investigation of human albinos underscores the expanded usage of this technique for studying visual disorders (Celesia & Daly, 1977; Feinsod, Hoyt, Wilson, & Spire, 1976; Vaughan & Katzman, 1964).

Visual evoked potentials from hemifield stimulation have been used to study correlates of laterality of function (e.g., Buchsbaum & Fedio, 1970; Eason, Groves, White, & Oden, 1967; Vella, Butler, & Glass, 1972) or simply to investigate characteristic effects of field stimulation (Barrett, Blumhardt, Halliday, Halliday, & Kriss, 1976; Cobb & Morton, 1970; Jeffreys & Axford, 1972; Shagass, Amadeo, & Roemer, 1976). Visual evoked potentials in the present study showed some interindividual differences in latency and amplitude among both albino and normal subjects; this appears characteristic of other studies (e.g., Jeffreys & Axford, 1972) in which locus of electrodes is reported to be an important consideration. Our results in normally pigmented subjects conform to previous findings of predominant components recorded in the hemisphere contralateral to stimulation (Shagass et al., 1976). The importance of input from the contralateral eye for response integrity is illustrated by recording observations from a split chiasm patient (Lehmann, Kavanagh, & Fender, 1969). Further evidence from patients with visual system disorders involving field defects also supports this concept (Celesia & Daly, 1977; Vaughan & Katzman, 1964; Wildberger, Van Lith, Wijngaarde, & Mak, 1976). Our results in human albinos which show predominance of responses to contralateral eye stimulation regardless of the hemifield of presentation reflect a form of abnormal input characterized by crossing of fibers originating from a portion of the temporal retina.

Our data show that illumination of the albino's temporal retina at successively more peripheral eccentricities from the fixation point with a 5° stimulus eventually yields enhanced responses in the ipsilateral hemisphere. This is in contrast to our normally pigmented subjects in which component amplitudes decrease with eccentricity, although response latencies change; a decrement in response with eccentricity has previously been reported (Harter, 1970; Rietveld, Tordoir, Hagenouw, Lubbers, & Spoor, 1967). Ipsilateral responses to stimulation at 15° eccentricity and beyond may reflect the normal nondecussating pathway. Because the contralateral response is greatly reduced following stimulation at 30° stimulation, it seems likely that the decussation point in human albinos is closer to 15° than 30° eccentricity. Some evidence suggests that human albinos show variability in decussation patterns (Creel, Spekreijse, & Reits, 1981). The presence of pendular nystagmus in these subjects makes precise definition of the decussation point in human albinos more difficult to determine. The Siamese cat, which also has the defect of a relative reduction of nondecussated input, has an abnormal temporal strip extending from the vertical meridian to about 20° eccentricity (Guillery & Kaas, 1971; Hubel & Wiesel, 1971; Kaas & Guillery, 1973; Shatz, 1977). The defect observed in humans appears related to genetic mechanisms producing albinism in all mammals, which accounts for the similarity in organization of anomalous visual pathways observed in widely separated species.

If the input to striate cortex is abnormal in albinos, the question remains as to how the visual system becomes reorganized. Single unit recordings of Siamese cats show that almost all neurons of areas 17 and 18 are monocularly driven (Hubel & Wiesel, 1971; Kaas & Guillery, 1973). Perimetric analysis of human albinos in our study revealed normal monocular visual fields (Coleman et al., 1979). However, when these subjects are tested with random dot stereograms none are able to demonstrate stereopsis (Coleman et al., 1979; Creel et al., 1981). Therefore, human albinos lack mechanisms for binocular vision. It may be that the striate cortex of each hemisphere in human albinos has separate representations of both the ipsilateral and abnormal contralateral visual fields such as observed in "Boston" Siamese cats (Hubel & Wiesel, 1971; Shatz, 1977). The zero meridian may not be located at the area 17-18 border, but instead be intercalated between the normal and abnormal field representations. Since albino subjects show asymmetry of evoked potentials following monocular nasal-temporal stimulation and predominant components from contralateral rather than ipsilateral stimulation, the abnormal crossed representation seems to be expressed at the cortical level. Details of the cortical topography of this representation in human albinos await further research.

REFERENCES

Barrett, G., Blumhardt, L. D., Halliday, A. M., Halliday, E., & Kriss, A. (1976). A paradox in the lateralization of the visual evoked response. *Nature, 261*, 253-255.

Buchsbaum, M., & Fedio, P. (1970). Hemispheric differences in evoked potentials to verbal and nonverbal stimuli in the left and right visual fields. *Physiology and Behavior, 5*, 207-210.

Celesia, G. C., & Daly, R. F. (1977). Visual electroencephalographic computer analysis (VECA): A new electrophysiological test for the diagnosis of optic nerve lesions. *Neurology, 27*, 637-641.

Cobb, W. A., & Morton, H. B. (1970). Evoked potentials from the human scalp to visual half-field stimulation. *Journal of Physiology (London), 208*, 39P-40P.

Coleman, J., Sydnor, C. F., Wolbarsht, M. L., & Bessler, M. (1979). Abnormal visual pathways in human albinos studied with visually evoked potentials. *Experimental Neurology, 65*, 667-679.

Cooper, M. L., & Blasdel, G. G. (1980). Regional variation in the representation of visual field in the visual cortex of the Siamese cat. *Journal of Comparative Neurology, 193*, 237-253.

Creel, D., Dustman, R. E., & Beck, E. C. (1970). Differences in visually evoked responses in albino versus hooded rats. *Experimental Neurology, 29*, 298-309.

Creel, D., & Giolli, R. A. (1976). Retinogeniculate projections in albino and ocularly hypopigmented rats. *Journal of Comparative Neurology, 166*, 445-456.

Creel, D., O'Donnell, F. E., Jr., & Witkop, C. J., Jr. (1978). Visual system anomalies in human ocular albinos. *Science, 201*, 931-933.

Creel, D., Spekreijse, H., & Reits, D. (1981). Evoked potentials in albinos: Efficacy of pattern stimuli in detecting misrouted optic fibers. *Electroencephalography and Clinical Neurophysiology, 52*, 595-603.

Creel, D., Witkop, C. J., Jr., & King, R. A. (1974). Asymmetric visually evoked potentials in human albinos: Evidence for visual system anomalies. *Investigative Ophthalmology, 13*, 430-440.

Eason, R. G., Groves, P., White, C. T., & Oden, D. (1967). Evoked cortical potentials: Relation to visual field and handedness. *Science, 156*, 1643-1646.

Feinsod, M., Hoyt, W. F., Wilson, W. B., & Spire, J. (1976). Visually evoked response: Use in neurologic evaluation of posttraumatic subjective visual complaints. *Archives of Ophthalmology, 94*, 237-240.

Fulton, A. B., Albert, D. M., & Craft, J. L. (1978). Human albinism: Light and electron microscopy study. *Archives of Ophthalmology, 96*, 305-310.

Giolli, R. A., & Guthrie, M. D. (1969). The primary optic projections in the rabbit. An experimental degeneration study. *Journal of Comparative Neurology, 136*, 99-126.

Gross, K. J., & Hickey, T. L. (1980). Abnormal laminae patterns in the lateral geniculate nucleus of an albino monkey. *Brain Research, 190*, 231-237.

Guillery, R. W. (1969). An abnormal retinogeniculate projection in Siamese cats. *Brain Research, 14*, 739-741.

Guillery, R. W., & Casagrande, V. A. (1975). Adaptive synaptic connections formed in the visual pathways in response to congenitally abnormal inputs. *Cold Springs Harbor Symposium on Quantitative Biology, 40*, 611-617.

Guillery, R. W., & Kaas, J. H. (1971). A study of normal and congenitally abnormal retinogeniculate projections in cats. *Journal of Comparative Neurology, 143*, 73-100.

Guillery, R. W., Okoro, A. N., & Witkop, C. J., Jr. (1975). Abnormal visual pathways in the brain of a human albino. *Brain Research, 96*, 373-377.

Harter, M. R. (1970). Evoked cortical responses to checkerboard patterns: Effect of check-size as a function of retinal eccentricity. *Vision Research, 10*, 1365-1376.

Hubel, D. H., & Wiesel, T. N. (1971). Aberrant visual projections in the Siamese cat. *Journal of Physiology (London), 218*, 33-62.

Jeffreys, D. A., & Axford, J. G. (1972). Source locations of pattern-specific components of human visual evoked potentials. I. Component of striate cortical origin. *Experimental Brain Research, 16*, 1-21.

Kaas, J. H., & Guillery, R. W. (1973). The transfer of abnormal visual field representations from the dorsal lateral geniculate nucleus to the visual cortex in Siamese cats. *Brain Research, 59*, 61-65.

Kalil, R., Jhaveri, S., & Richards, W. R. (1971). Anomalous retinal pathways in the Siamese cat: An inadequate substrate for normal binocular vision. *Science, 174*, 302-305.

Lehmann, D., Kavanagh, R. N., & Fender, D. H. (1969). Field studies of averaged visually evoked EEG potentials in a patient with a split chiasm. *Electroencephalography and Clinical Neurophysiology, 26*, 193-199.

Lund, R. D., Lund, J. S., & Wise, R. P. (1974). The organization of the retinal projection to the dorsal lateral geniculate nucleus in pigmented and albino rats. *Journal of Comparative Neurology, 158*, 383-404.

McCormack, G. L. (1975). Electrophysiologic evidence for normal optic nerve fiber projections in normally pigmented squinters. *Investigative Ophthalmology, 14*, 932-935.

Rietveld, W. J., Tordoir, W. E. M., Hagenouw, J. R. B., Lubbers, J. A., & Spoor, T. A. C. (1967). Visual evoked responses to blank and to checkerboard patterned flashes. *Acta Physiologica et Pharmacologica Neerlandica, 14*, 259-285.

Sanderson, K. J. (1975). Retinogeniculate projections in the rabbits of the albino allelomorphic series. *Journal of Comparative Neurology, 159*, 15-28.

Shagass, C., Amadeo, M., & Roemer, R. A. (1976). Spatial distribution of potentials evoked by half-field pattern-reversal and pattern-onset stimuli. *Electroencephalography and Clinical Neurophysiology, 41*, 609-622.

Shatz, C. (1977). A comparison of visual pathways in Boston and Midwestern Siamese cats. *Journal of Comparative Neurology, 171*, 205-228.

Vaughan, H. G., Jr., & Katzman, R. (1964). Evoked response in visual disorders. *Annals of the New York Academy of Sciences, 112*, 305-319.

Vella, E. J., Butler, S. R., & Glass, A. (1972). Electrical correlate of right hemisphere function. *Nature New Biology, 236*, 125-126.

Wildberger, H. G. H., Van Lith, G. H. M., Wijngaarde, R., & Mak, G. T. M. (1976). Visually evoked cortical potentials in the evaluation of homonymous and bitemporal visual field defects. *British Journal of Ophthalmology, 60*, 273-278.

Witkop, C. J., Jr., White, J. G., & King, R. A. (1974). Oculocutaneous albinism. In W. L. Nyhan (Ed.), *Heritable disorders of amino acid metabolism* (p. 171). New York: Wiley.

SECTION IV. NEURAL SUBSTRATES OF ATTENTION

William James in his *Principles of Psychology* was one of the first modern psychologists to write extensively about the topic of attention. Some 80 years later, E. G. Boring (see Boring's chapter in Mostofsky, 1970) was able to name 10 different uses of the term "attention" before 1930. Posner and Boies (1971), recognizing this diversity of meanings for the term attention, suggested that all types of attention could be classified into three "components of attention," namely selective attention, alertness, and attentional capacity. Each of these components is concerned with different questions about attention. Selective attention is concerned with what is attended to, and why. Alertness is concerned with the level of attention--how attentive is an individual and what factors, external and internal, modify this alertness. Attentional capacity has to do with limits to the amount of information that an individual can attend to and factors that influence them.

Posner and Boies (1971) did not just discuss the three components of attention in an abstract, theoretical way; they actually did experiments to investigate each component, using different experimental manipulations to measure each. Although it seems likely that the three "components of attention" are interrelated, they are still sufficiently distinct to allow this grouping of components to be used preliminarily to help clarify the confusion over the many uses of the term attention.

Sustained attention (vigilance) in Posner and Boies' (1971) "components" conceptualization is alertness, and Mirsky and Ray's chapter is about vigilance or alertness of humans and animals as measured by the continuous performance test (CPT). They report on the relationship between petit mal epilepsy as manifested by spike-wave activity of human subjects and performance on the CPT. The results of several studies show the relationship to be complex. For example, one patient responded correctly to critical letters that appeared before or after a burst of spike-wave activity associated with a seizure but failed to respond to critical letters during his seizure burst. However, some patients were able to perform the CPT correctly *during* the time that they displayed spike-wave electrical activity. Other patients showed electroretinogram B-wave reduction during the seizure bursts.

Mirsky and Ray trained monkeys to perform a simian version of the sustained attention task used in humans. Three experimental techniques were utilized: (1) subcortical lesions produced by aluminum cream implants; (2) stimulation of brain structures intercurrent with performance; and (3) recording of cellular activity intercurrent with performance. The results from these experimental techniques suggest that the same rostral brain stem areas

 DOI: 10.4324/9781003164647- 14

that the classical studies of Lindsley, Moruzzi, French and others have implicated in consciousness and in tonic arousal and EEG effects is also involved in a spectacular way in moment-to-moment phasic fluctuations of attention. Further recordings of cellular units have displayed two general kinds of performance-related firing patterns. One class of units exhibited a go only or type 1 pattern; the firing rates changed significantly only during the go trials. A second population of cells, referred to as type 2, displayed similar firing rate change during both the go and no-go trials.

Beatty's chapter investigates whether activation responses accompanying the processing of sensory stimuli can be influenced by the control of selective attention. Specifically, several investigations have shown that the degree of pupillary dilation observed during the execution of different functions is directly proportional to the demands that these functions place on the cognitive system. Eight young adult subjects were asked to respond to two auditory tone bursts of high and low frequencies with instructions to attend to either the high or the low frequency tones. Comparisons were made for all subjects under all conditions, i.e., a repeated measures design. It was found that pupillary dilation appeared for the stimuli when a channel was attended to but not otherwise. Finally, the study demonstrates the value of employing relevant physiological measures (in this case, pupillary dilation) in the study of cognitive processing (selective attention and attentional capacity).

Näätänen's chapter is a detailed, scholarly review of event-related potential (ERP) measures of selective attention taken in conjunction with regional cerebral blood-flow (rCBF) measures. The two techniques, ERPs and rCBF, complement each other nicely. The ERPs provide excellent temporal resolution data regarding the phasic aspects of selective attention while rCBF studies provide excellent spatial resolution information regarding the more tonic aspects of selective attention. The selective-attention related rCBF and ERP patterns of response seem to correspond most closely to the frontal regions of the brain. The first major, negative component of the ERP, the N1 response, is the component that corresponds most closely to selective attention.

The chapter by Schwartz and co-workers investigates the relationship between unilateral neglect and sensory extinction to determine if sensory extinction is simply a mild manifestation of unilateral neglect. One hundred and twenty-nine right-handed patients who showed radiographically verified hemisphere lesions and also displayed either neglect or extinction, or both, were the subjects. The patients were compared with a group of 83 non-brain-damaged control subjects on the measure of extinction and unilateral neglect. The results support the conclusion that sensory extinction is qualitatively different from unilateral neglect. It is suggested that at least two different mechanisms are needed to described these neurological impairments. Further testing is needed to determined exactly how these disorders relate to attention and arousal.

In the Schlag-Ray and Schlag chapter, the visual attention studied is that of visual fixation, i.e., the ability to direct one's gaze to a point in space and fixate that point, no matter what conditions of visual stimulation prevail. Curiously, neurons firing before and during eye movements were found in the thalamic internal medullary lamina (IML) of the cat. For several reasons these neurons may be involved in the guidance of gaze. The firing characteristics of approximately 300 IML units were recorded when the anticipatory fixation was on the correct site, which was reinforced by a light signal invariably followed by milk delivery. Four major properties of these IML cells are described. It is concluded that the same cells receive both information of retinal origin and information on the position of the eyes. These results show the importance of IML neurons in the study of visual attention and fixation.

REFERENCES

Boring, E. G. (1970). Attention: Research and beliefs concerning the conception in scientific psychology before 1930. In D. I Mostofsky (Ed.), *Attention: Contemporary theory and analysis* (pp. 5-8). New York: Appleton-Century-Crofts.

Posner, M., & Boies, S. J. (1971). Components of attention. *Psychological Review, 78,* 391-408.

11 · STUDIES IN THE NEUROPSYCHOLOGY OF ATTENTION IMPAIRMENT: HUMAN SYMPTOMS AND ANIMAL MODELS

A. F. Mirsky and C. Ray

National Institute of Mental Health
and
Boston University School of Medicine

It is a delight to participate in this *Festschrift* that honors Professor Donald Lindsley. Neither of the present authors had the good fortune to have been either a student or colleague of Don's; he left New England a number of years before either of us entered graduate school. Contact with him was entirely through his published writings, at least until 1967. That year, one of us (AFM) had the occasion to meet Don in Teddington, England; we were both participants in a conference on attention in neurophysiology. It was a very gratifying experience to meet him in person and to realize that this giant in the field, the author of "Psychological Phenomena and the Electroencephalogram" (Lindsley, 1952); of "Behavioral and EEG Changes Following Chronic Brainstem Lesions in Cats" (Lindsley, Schreiner, Knowles, & Magoun, 1950); of "Effects Upon the EEG of Acute Injury to the Brainstem Activating System" (Lindsley, Bowden, & Magoun, 1949) and of the monumental review in the *Handbook of Physiology* titled "Attention, Consciousness, Sleep and Wakefulness" (Lindsley, 1960), that this titan among neurophysiological psychologists was a kind and friendly person. My (AFM's) grandmother used to say that all of the really great people in the world were really nice, if you could just have a chance to meet them. She was right, as usual, and would have enjoyed meeting Don.

However, we do not wish to devote all of this presentation to a mere recitation of admiration and warm feelings for Donald B. Lindsley. We wish instead to report about a research program that has been going on since about the time of publication of "Psychological Phenomena and the Electroencephalogram." This work has been influenced, both directly and indirectly by the work of Don and his students, and it represents, we hope, the kind of blend of clinical and experimental interests that has been characteristic of the Lindsley approach. In addition, we believe, it suggests a role for parts of our beloved brainstem reticular formation that was perhaps not fully anticipated.

The narrative begins with an effort to understand altered attentiveness seen in patients with petit mal ("centrencephalic") epilepsy at the time of

DOI: 10.4324/9781003164647- 15

their characteristic seizure discharge. Figure 11.1 shows a sample tracing from such a patient who was being administered a test of sustained visual attention. This continuous performance test or CPT (Rosvold, Mirsky, Sarason, Bransome, & Beck, 1956) required the patient to watch a visual display and to respond by pressing a response key whenever a critical letter, "X," appeared. He was to withhold responses to all other letters. In the

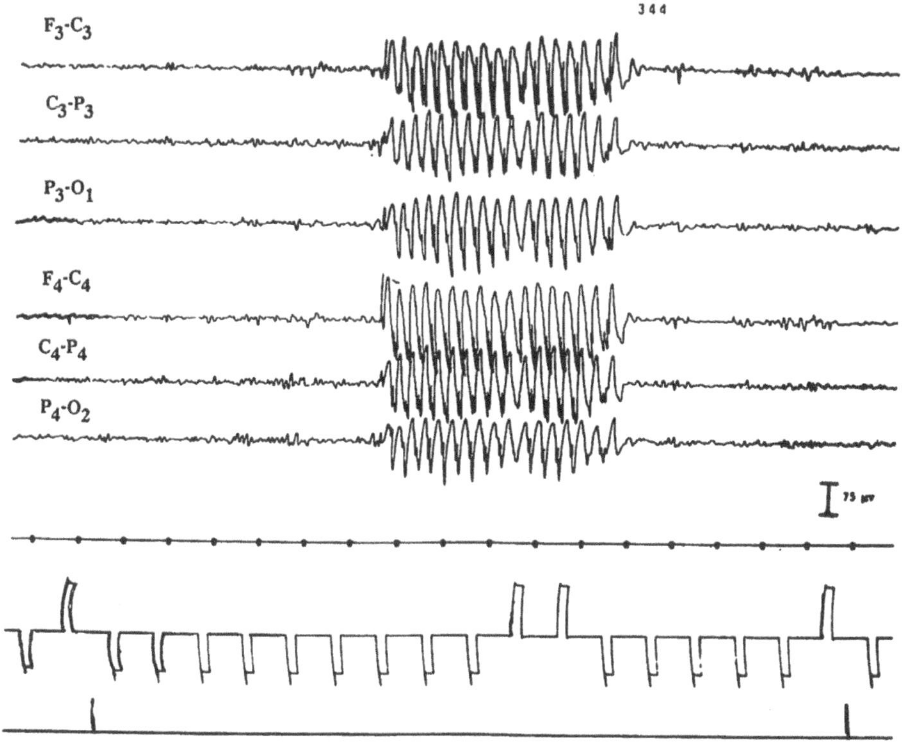

FIGURE 11.1. The relation between a burst of spike and wave activity and CPT performance in a patient suffering from petit mal epilepsy. The top six channels in the tracing represent a standard anteroposterior EEG run, with electrode placements determined by the "10-20" system. The seventh channel below this is a 1-sec time mark. Below this in channel 8 are represented the stimuli (duration = 0.2 sec) shown to the patient; those requiring a response (the letter "X") are seen as deflections above the baseline; other letter stimuli appear as deflections below the baseline. The patient's response appears on channel 9 as an upward deflection. In this sample, the patient responded correctly to X's presented before and after the spike and wave burst but failed to respond to the two occurring within the burst (from Mirsky & Tecce, 1966).

tracing shown here the patient responded correctly to critical letters which appeared before or after the burst of spike-wave activity but failed to respond to critical letters during the burst. This is a behavioral measure of an "absence" attack; as is fairly typical for such attacks the patient showed no outward physical manifestation of the seizure discharge except the blank stare or fixation of gaze. After considerable behavioral study of this phenomenon, we became reasonably convinced that this instance of impaired attention or reduced consciousness was due at least in part to reduced or altered sensory input or, possibly, to defective integration during spike-wave bursts (Mirsky & Van Buren, 1965).

This impression from behavioral studies has received some support from electrophysiological investigations we have performed. This work suggests that visual evoked potentials (VEP) are reduced markedly in size during spike-wave activity (Figure 11.2) and that the reduction in VEP amplitude actually begins before the spike-wave burst activity can be recorded from the scalp (Orren, 1974, 1978). This can be seen clearly in Figure 11.3, which illustrates the data obtained from a "sliding average" technique used by our colleague Merle Orren. The fact that the altered sensory input (as reflected by VEPs) precedes the spike-wave burst by approximately one second is in agreement with the results of several behavioral studies. These suggest that

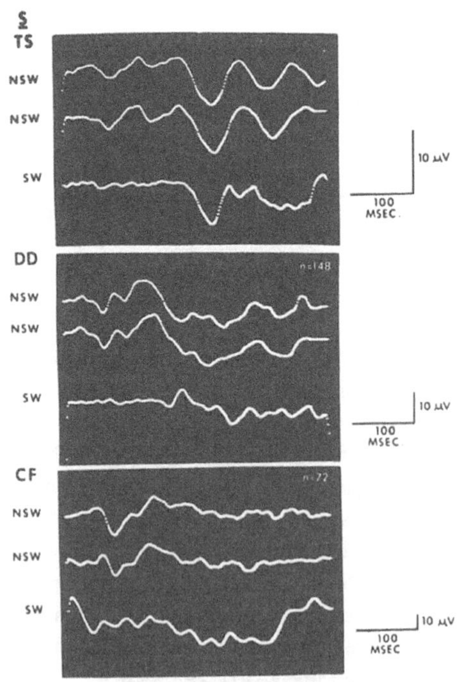

FIGURE 11.2. Average evoked potentials during non-spike-wave (NSW) and spike-wave (SW) activity; subjects TS, DD, and CF. The averages were obtained from parietal-occipital recordings (P_3-O_1 or P_4-O_2). An upward deflection indicates positivity at the occipital relative to the parietal electrode (from Orren, 1974).

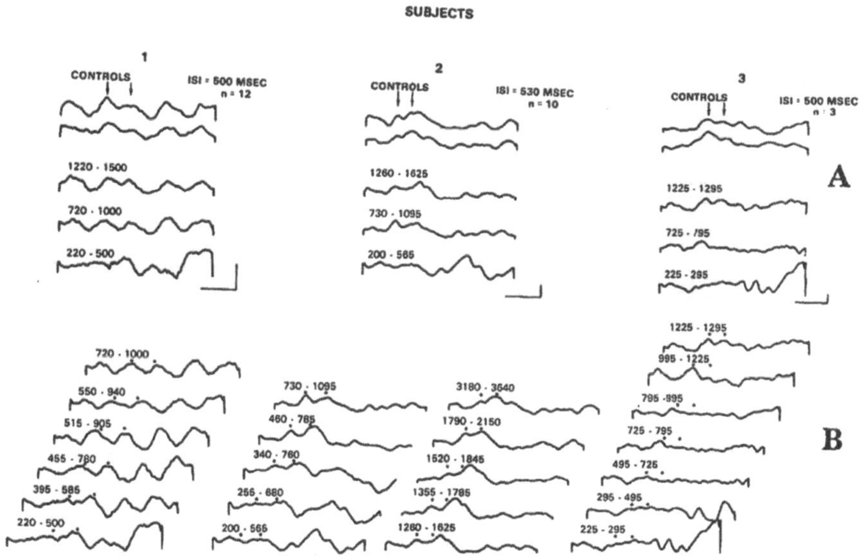

FIGURE 11.3. Changes in the average VEP of three subjects prior to the onset of spike and wave bursts. Single responses were averaged as a function of flash-to-burst interval, indicated, in msec, above each curve. Interstimulus interval (ISI) and VEP sample size are shown for each subject. A. Average VEPs (P-O recording) during the fourth pre-burst second (controls) and during three successive intervals preceding burst onset. Arrows point to components common to both control responses. B. The altered average is displaced through the interval of change to show transitional waveforms. Dots are at fixed latencies corresponding to reliable components in the control curves. Calibrations: 10 μV and 100 msec. Relative positivity at occipital electrode indicated by an upward deflection (Orren, 1978).

the patients' tendency to exhibit attention loss may precede the spike-wave burst by the order of about one second (Geller & Geller, 1970; Mirsky & Van Buren, 1965). The fact that the behavioral symptom of omission errors or failure to respond may precede or be slightly out of phase from the EEG symptom is particularly significant; we shall return to it in a moment.

To buttress the point about significant reduction of visual sensory input during spike-wave seizures, Figure 11.4 illustrates the results of a study of the electroretinogram (ERG) both prior to and during bursts of spike-wave activity (Ricks & Mirsky, 1977). This illustrates what we believe to be happening, i.e., reduced sensitivity or receptivity at the level of the receptor during spike-wave bursts. The explanation of how the retina gets turned down during this seizure activity is problematical; however, it may reflect the influence of centrifugal afferents in the optic nerve which exert inhibitory effects on the retina during spike-wave bursts (Mirsky, Bloch, Tecce, Lessell, & Marcus, 1973). From some studies we have done involving animal models

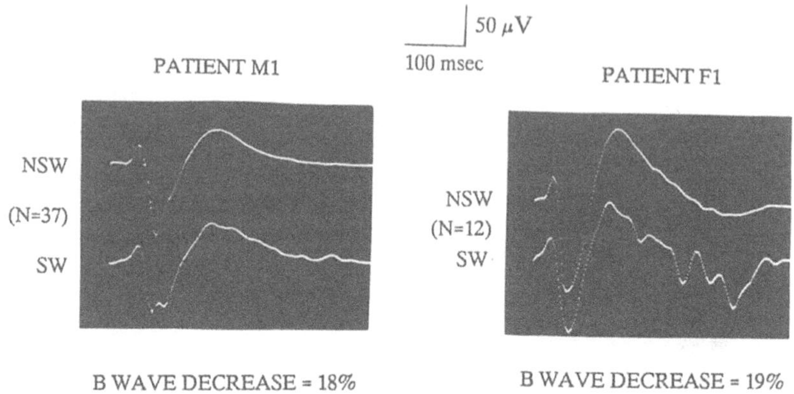

FIGURE 11.4. Electroretinograms (ERG) recorded during non-spike-wave and spike-wave EEG activity in patients M1 and F1. The measured decrease in B-wave was about 18% for M1 and 19% for F1. The decrease was statistically significant in each case. Intensity = "8" on a Grass PS-2 stimulator located 20 cm from the eye. The pupil was not dilated (unpublished observations from Ricks and Mirsky, 1977).

of spike-wave EEG activity and its effects on visual system functioning, we believe that the reduced VEP cannot be explained entirely on the basis of the retinal effect; rather it appears that there is a general inhibitory effect which may be exerted more or less independently at retinal and cortical levels.

Let us return to the issue of the time relation between behavioral and EEG changes in petit mal seizures. The fact that the behavioral sign of the seizure can precede the electrographic sign is subject to a number of interpretations: (1) the EEG change may be occurring relatively late as a reflection of a "centrencephalic" process which is propagated to cortex, but which affects subcortical "attention" systems first; (2) the EEG and behavioral changes may reflect the action of closely related but independent or separable systems. Most of the evidence, which we will not review here, supports the latter view; some of this evidence, however, is presented in Figure 11.5. This is a dramatic instance of a patient continuing to perform correctly on our visual attention task despite the presence of bilaterally symmetrical spike-wave discharge activity. This is a relatively unusual but not rare occurrence. We note here that Don Lindsley wrestled with this problem of the variable behavioral effects of seizures, or at least alluded to it (Lindsley, 1960, p. 1583).

If we assume that the occasional dissociation of the EEG and the behavioral systems in petit mal reflect separate neural mechanisms, then we can design our research appropriately. One strategy that may be followed is to investigate the behavioral, rather than the electrographic features of the disease. This may be pursued in both human subjects and animal models.

Thus, rather than focusing on the production of spike-wave activity (by electrical brain stimulation, or by local or systemic administration of drugs) we can concentrate on what is the important clinical feature of this illness -- the staring spell, absence attack, or as we have defined it, omission errors on our visual attention task.

As is well known, it is possible to train monkeys to perform marvelously complex visual tasks; we trained them to perform a simian version of the attention task we used in man: press a response key for a red stimulus light, and withhold pressing for a green or blue stimulus light. This successive go, no-go visual discrimination task served as our operational measure of attention, and some preliminary work with drugs and sleep deprivation demonstrated that it behaved much like the human version (Mirsky & Bloch, 1967).

The use of the animal model of this sustained attention function has involved three types of experimental techniques: (a) subcortical aluminum cream lesions; (b) stimulation of brain structures intercurrent with performance; and (c) recording of cellular activity intercurrent with performance. The question of where to intervene in the brain in these experiments was not difficult to resolve. Guidance is provided by the centrencephalic school of Penfield and Jasper (1954); the "centrencephalon" is a kind of Canadian version of the ascending reticular activating system.

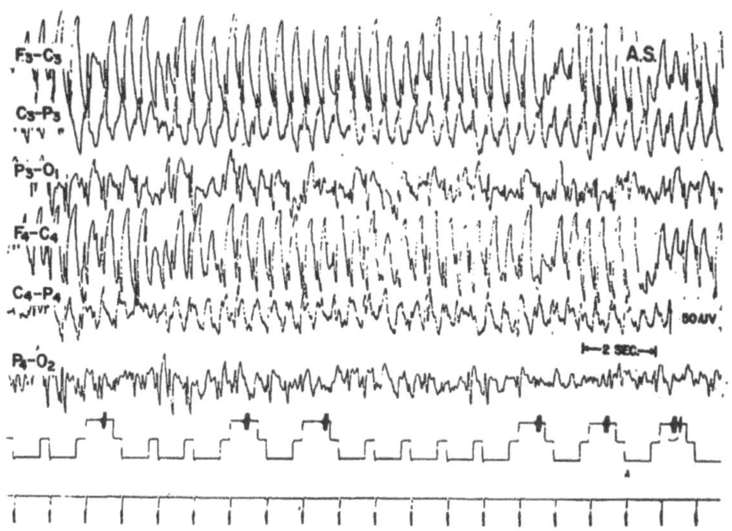

FIGURE 11.5. Tracing from a patient with petit mal epilepsy who is able to perform correctly on a visual attention task during spike-wave activity. Critical stimuli (letter "X") appear as two-stage deflections on next-to-bottom channel. Responses appear as heavy marks superimposed on this channel.

The work of these investigators suggested that manipulation of both the thalamic and mesopontine regions of the centrencephalic system would probably be most fruitful in terms of some of the earlier studies done both in Montreal and Los Angeles. We refer to some of the highlights of these investigations. Both the aluminum cream implant studies and the brain stimulation studies have suggested that the brainstem rather than the thalamic portions of the activating system are important for the maintenance of this sustained visual attention task. The next two figures are representative coronal sections showing the maximal damage in thalamus (Figure 11.6) and brainstem (Figure 11.7) we achieved with this technique. Figure 11.8 illustrates the behavioral effects we obtained (Mirsky & Oshima, 1973). No impairment was seen in any thalamus animal; on the other hand, every brainstem preparation showed at least temporary impairment of the visual attention task. In two animals it was permanent. Animal B-2, who died in status epilepticus, showed generalized seizures presumably originating in the brainstem. This is illustrated in Figure 11.9. One structure which was damaged in all four brainstem preparations was the mesencephalic or pontine reticular formation.

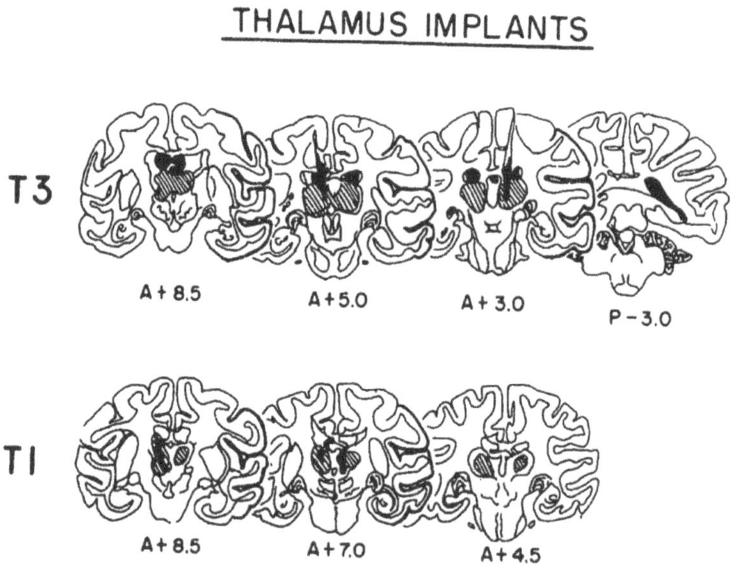

FIGURE 11.6. Coronal sections through brains of two animals with aluminum cream implants in medial thalamus. Black represents visual aluminum cream, cross-hatched area represents region of complete cell loss (Mirsky & Oshima, 1973).

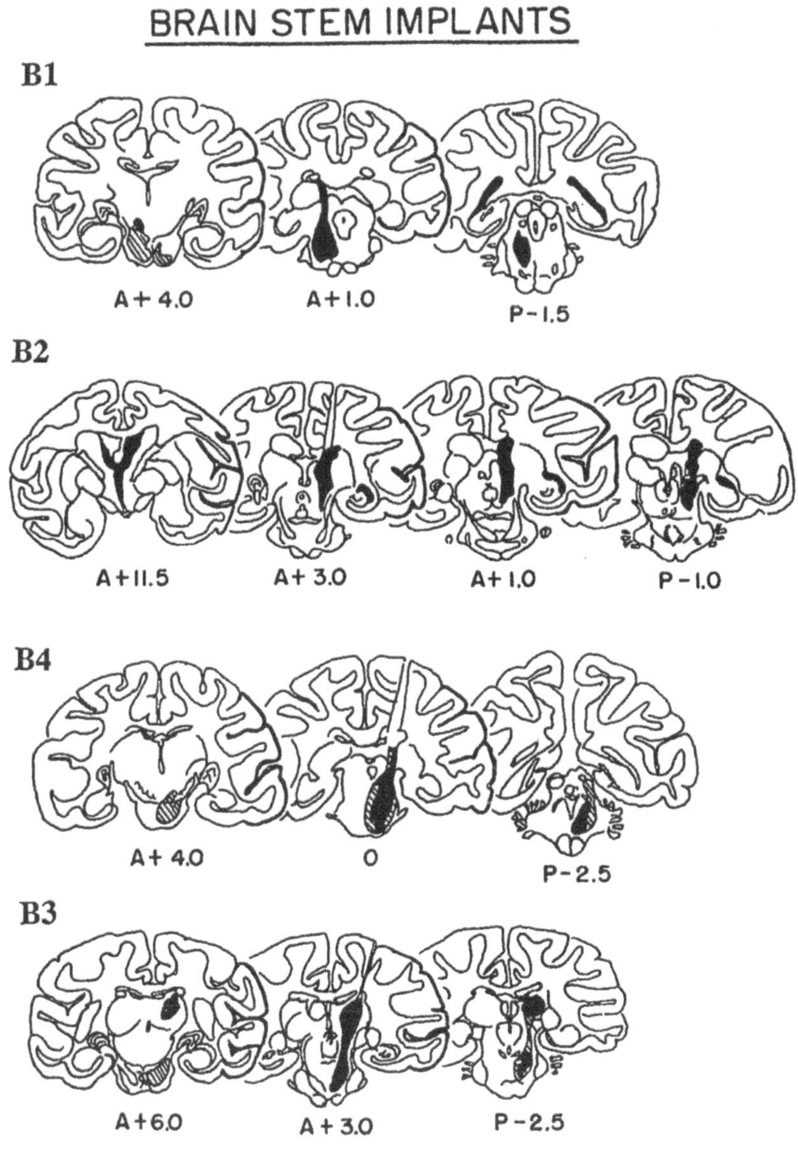

FIGURE 11.7. Coronal sections through brainstem aluminum cream implants. See legend for Figure 11.6 (Mirsky & Oshima, 1973).

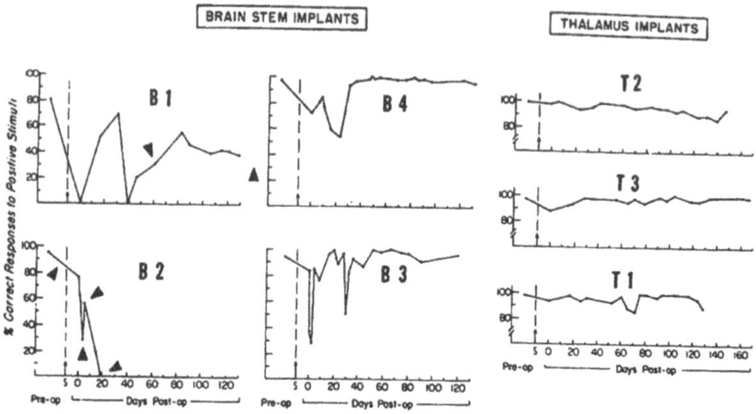

FIGURE 11.8. Percentage of correct responses to positive stimuli for animals with brainstem (B 1-4) and thalamic (T 1-3) implants. The vertical dashed line represents the time of surgical implant ("S"). Numbers along the abscissa refer to days after the resumption of postoperative testing. Solid triangles indicate days when EEG samples were obtained in B1 and B2 (from Mirsky & Oshima, 1973).

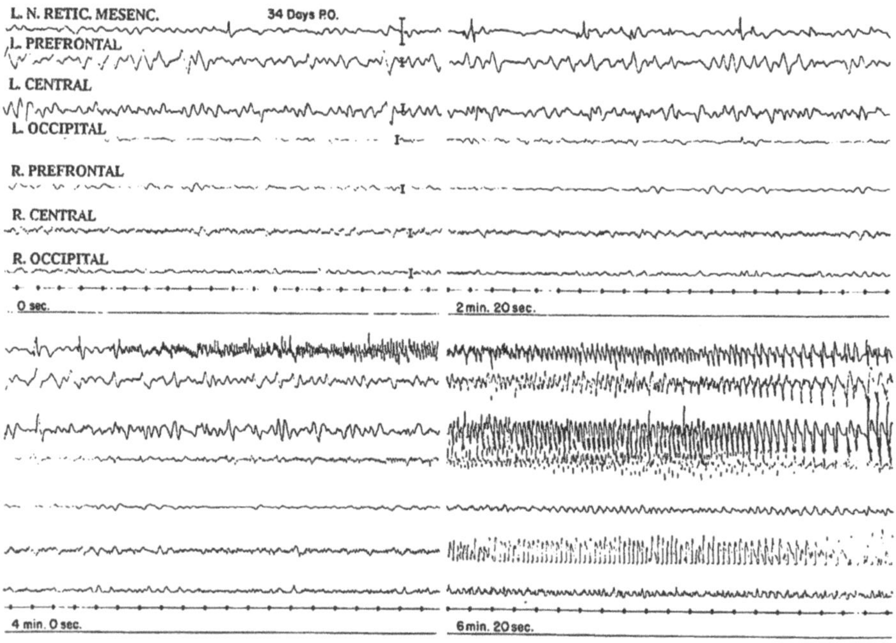

FIGURE 11.9. Development of seizure in monkey B2 showing origin in reticular formation of mesencephalon. Note spike in top channel, gradually increasing in frequency until seizure becomes generalized, in bottom right tracing.

However, because the lesions were, in general, so large we elected in our next study to use a more refined technique, that of brain stimulation. The animals in the next phase of the work were fitted with guides that enabled us to stimulate via movable electrodes many points within the brain. This we did, while animals were performing the go, no-go attention task, or other tasks. The outcome of the stimulation study is summarized in the next few figures. Figure 11.10 presents three sagittal sections summarizing hundreds of individual stimulation experiments in five animals. We direct your attention to the omission error points and to their clustering in an area representing the junction of the mesencephalic and pontine reticular formations. Omission error points were also encountered in the pretectal area and in the red nucleus but were rare in any areas of the thalamus or

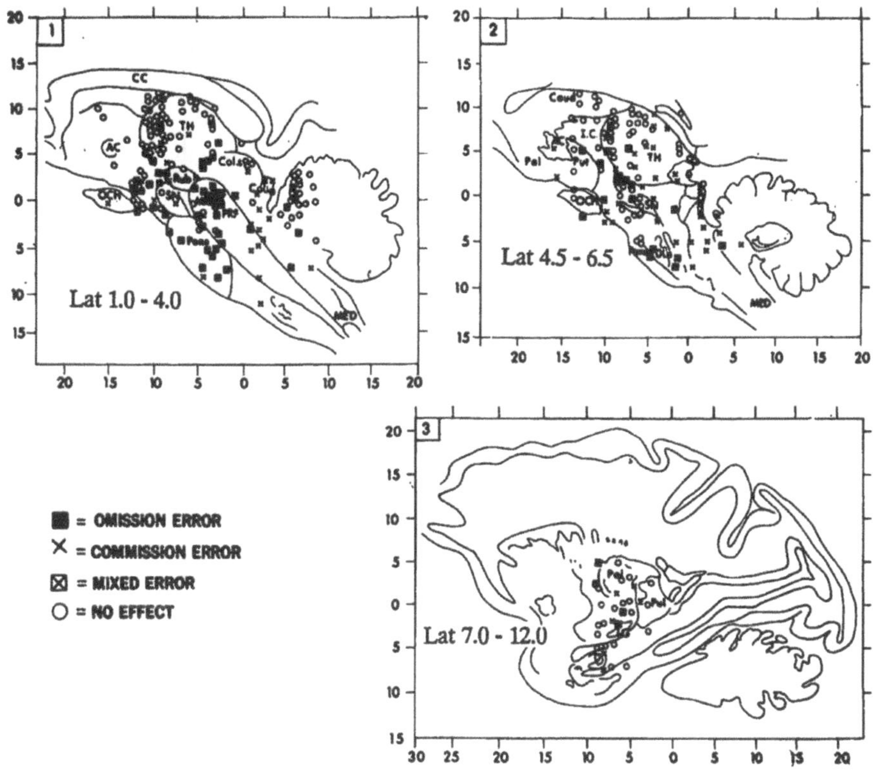

FIGURE 11.10. Sagittal sections through monkey brain showing loci of electrical stimulation effects on attentive performance. Omission error points are seen to cluster in the area of the mesencephalic and pontine reticular formation (modified from Pragay, Mirsky, Fullerton, Oshima, & Arnold, 1975).

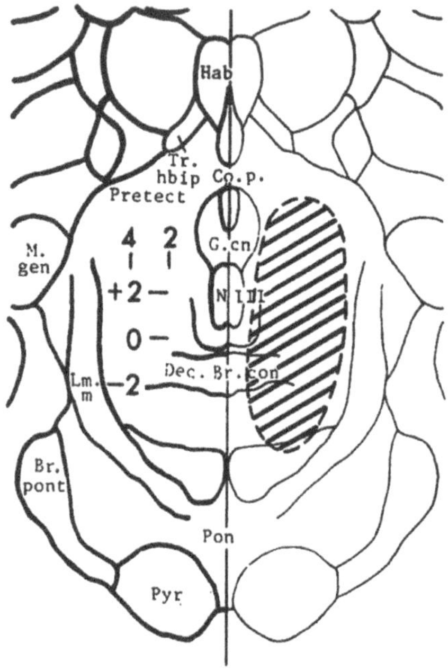

AP + 5.5 through + 1.5

FIGURE 11.11. Area in brainstem in which significant effects resembling petit mal epilepsy are produced by stimulation in the monkey brain.

those parts of the basal ganglia we explored. Moreover, several other interesting phenomena were found, which are alluded to in Figure 11.11. Stimulation in the area represented by the cross-hatching tended to produce not only errors of omission but arrest of gaze (Figure 11.12) and reduced visual evoked potentials. The latter is represented in Figure 11.13. Finally, Figure 11.14 summarizes the several effects of most interest for this discussion: errors of omission, arrest of gaze, and reduced VEPs. These three phenomena or effects are all seen in the petit mal seizure and collectively raise the possibility that this region of the brain may be the primary locus of the pathophysiology of this illness.

This supports the findings of several other workers (e.g., Weir, 1964) who found this meso-pontine junction to be particularly effective in eliciting spike-wave discharges in the cat. It must be emphasized further, however, that this suggests that the same rostral brainstem area which the classical studies of Lindsley, Magoun, Moruzzi, French and others have implicated in consciousness and in the *tonic* arousal and EEG effects is also involved in a spectacular way in moment-to-moment *phasic* fluctuations of attention. This view of the pathophysiology of petit mal or absence epilepsy is not universally

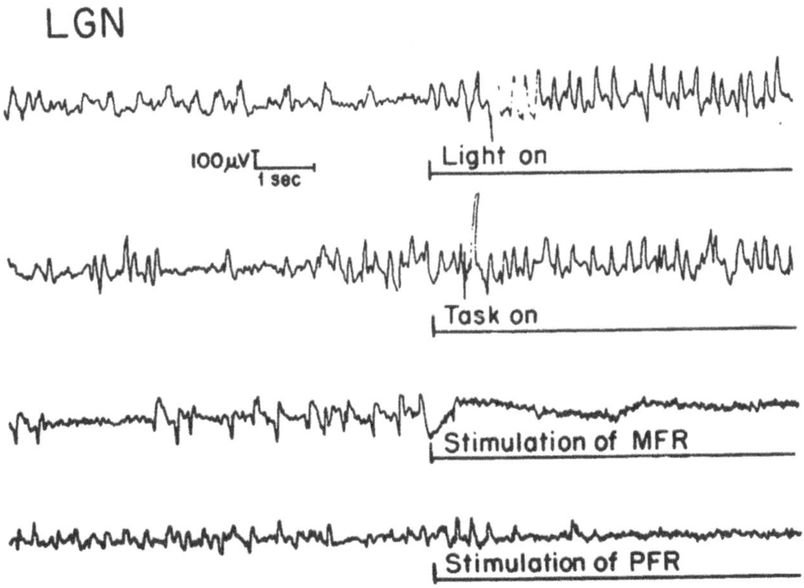

FIGURE 11.12. Eye movement spike potentials recorded under various experimental conditions in lateral geniculate nucleus (LGN) of macaque. Top two traces show increase in spikes with visual stimulation. Bottom two traces show reduced spiking with electrical stimulation of mesencephalic (MFR) and pontine (PFR) reticular formation (from Mirsky, Pragay, & Harris, 1977).

accepted. Gloor (1978), on the basis of a feline preparation in which seizure activity is induced by injection of penicillin, has suggested that the cortex may be the critical locus of the behavioral effects seen in absence epilepsy. He has coined the term "*cortico*-reticular epilepsy" (emphasis added), which indicates his conception of the relatively subsidiary role of brainstem structures in the manifestations of the disease (Gloor, 1969, 1978). However, the penicillin model of absence epilepsy is not universally accepted, since most of the seizure manifestations appear to be of the generalized tonic-clonic variety, rather than the absence type. The difference between Gloor's view and the one expressed here may be a matter of emphasis.

Most recently we have recorded from single neurons in the rostral brainstem during task performance in intact animals using Evarts' (1968) chronic single unit recording technique. This research is directed at determining whether the firing patterns of cells in the region shown in Figure 11.15 are related to attentive-discriminative activity, and what the nature of that relation might be.

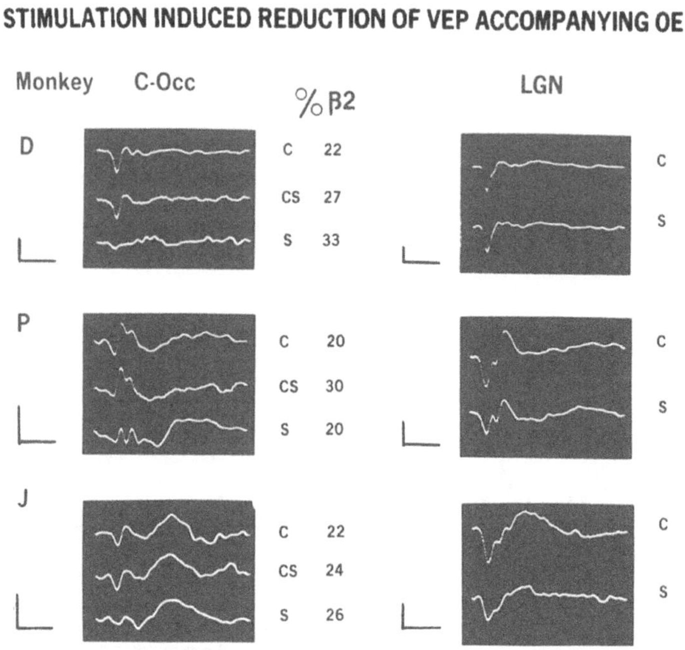

FIGURE 11.13. Reduction in visual evoked potentials from stimulation in mesencephalic or pontine reticular formation (cross-hatched area in center panel of Figure 11.9) in monkeys D, P, and J. Tracings at the left are recorded from central-occipital cortex; those at the right are from lateral geniculate nucleus (LGN). Within each tracing set the topmost curve ("C") represents the control response to 2 Hz photic stimulation; "S" at the bottom, represents the response to photic stimulation during reticular formation stimulation; "CS" indicates the response seen during stimulation of a control structure, usually located several mm from the behaviorally effective point. The numerical values to the right of the "C," "CS," or "S" designations indicate the percentage of $\beta 2$ activity (27-40 Hz) determined from a baseline-crossing frequency analysis of the EEG at the time of the combined photic and reticular stimulation. The reduction in cortical VEPs during stimulation of behaviorally effective points is clearly seen. The effects are much less striking at LGN, and neither change is clearly related to any change in "activation" as measured by the amount of activity in the EEG. Each sweep is an average of 50 to 60 photic stimuli; horizontal calibration is 100 msec, vertical calibration is 50 μV for D and 100 μV for P and J (from Mirsky, Pragay, & Harris, 1977).

Figure 11.16 shows the version of the attention task we used in these investigations. This task requires the animal to discriminate between a red (go) light and a green (no-go) light. The monkey initiates a trial by pressing and holding the button for two seconds. At the end of this interval the upper stimulus button is illuminated in a semirandom order with either a red or a

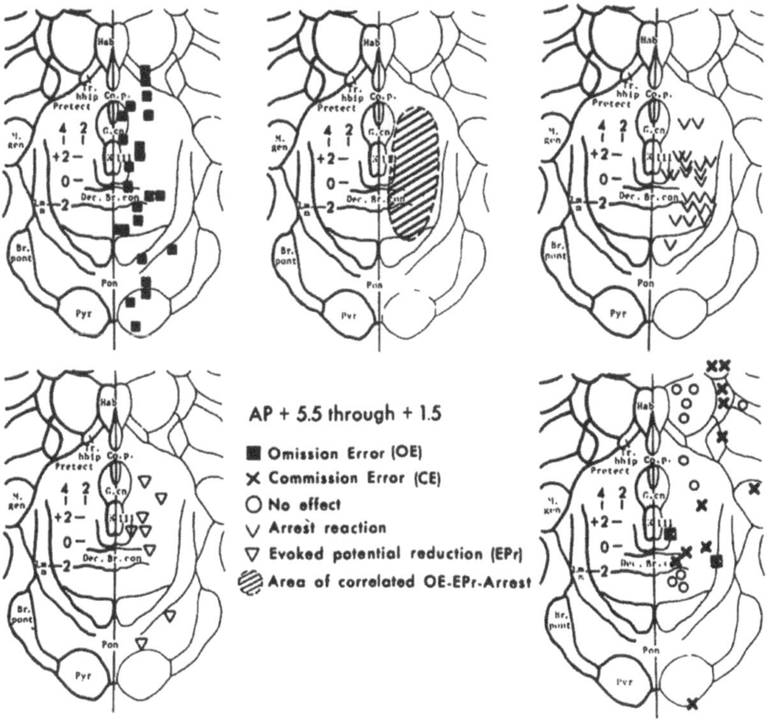

FIGURE 11.14. Area of brainstem in macaque yielding stimulation-induced behavioral and evoked potential effects similar to those seen in human petit mal epilepsy. Upper left = omission error effects on CPT; lower left = reduced visual evoked potentials; upper right = "arrest" reaction. Central panel shows area of overlap of these effects (from Mirsky, Pragay, & Harris, 1977).

green light. If the light is red, the monkey is required to lift its hand off the hold button and hit the stimulus button (go response). Correct responses are reinforced with drops of flavored water. When the light is green, the animal is rewarded for continuing to hold the bottom button (no-go "response") for one second until the green light is turned off.

The units we observed have displayed two general kinds of performance-related firing patterns (Pragay, Mirsky, Ray, Turner, & Mirsky, 1978). One class of unit exhibited a go only or type 1 pattern; the firing rates of these units changed significantly only during the go trials. A separate population of cells referred to as type 2 displayed similar firing rate changes during both the go and the no-go trials. No-go only patterns were rare. Examples of the most common type 1 and type 2 patterns are illustrated in Figure 11.17. It is notable that the median response latencies of the type 1 and type 2 populations differ significantly ($p<.0001$). As shown in Figure 11.18 the

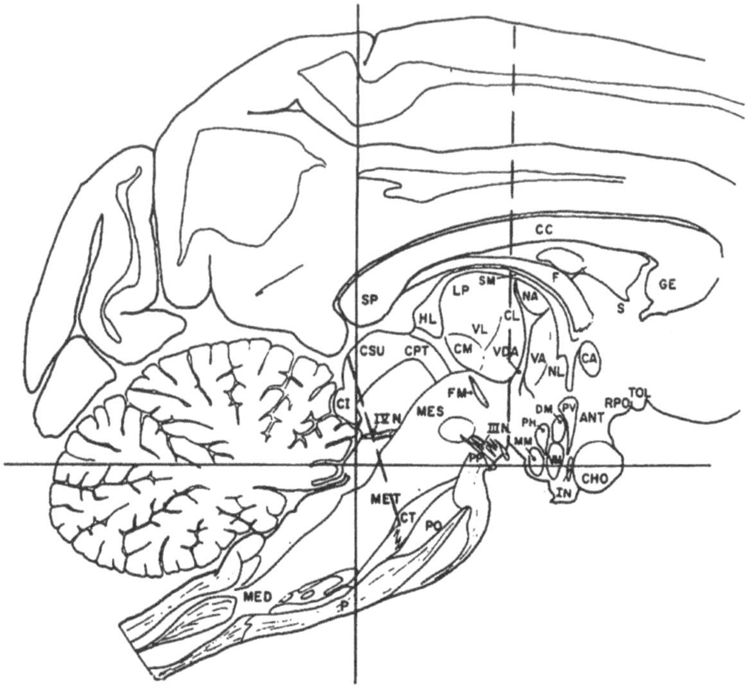

FIGURE 11.15. Sagittal view of area of exploration. The dashed lines represent its anterior (A8.5) and posterior boundaries (P2.0). The following anatomical abbreviations were used: ANT, anterior hypothalamic area; CA, anterior commissure; CC, corpus callosum; CHO, optic chiasm; CI, inferior colliculus; CL, central lateral nucleus; CM, centre median nucleus; CPT, posterior commissure; CSU, superior colliculus; CT, trapezoid body; DM, dorsomedial hypothalamic nucleus; F, fornix; FM, habenulopeduncular tract; GE, genu of the corpus callosum; HL, habenular nuclei; LP, lateral posterior nucleus; MED, medulla; MES, midbrain; MET, pons; MM, mammillary body; NA, anterior nucleus; NBO, supraoptic nucleus; NL, reticular nucleus; P, pyramid; PH, posterior hypothalamic area; PO, pons; PP, cerebral peduncle; PV, paraventricular nucleus; RPO, preoptic area; S, septal region; SM, medullary stria; SP, splenium of the corpus callosum; TOL, lateral olfactory tract; VA, ventral anterior nucleus; VDS, mammillothalamic tract; VL, ventral lateral nucleus; VM, ventromedial nucleus; I N, optic nerve, III N, oculomotor nerve; IV N, trochlear nerve. Adapted from Russell (1961).

majority of the task-related responses displayed by the go, no-go units have an earlier onset than those of the go only or type 1 units. The latency difference, together with the differential involvement in go and no-go trials, suggests that these two types of cells occupy different positions in the functional chain leading from stimulus reception to overt motor response.

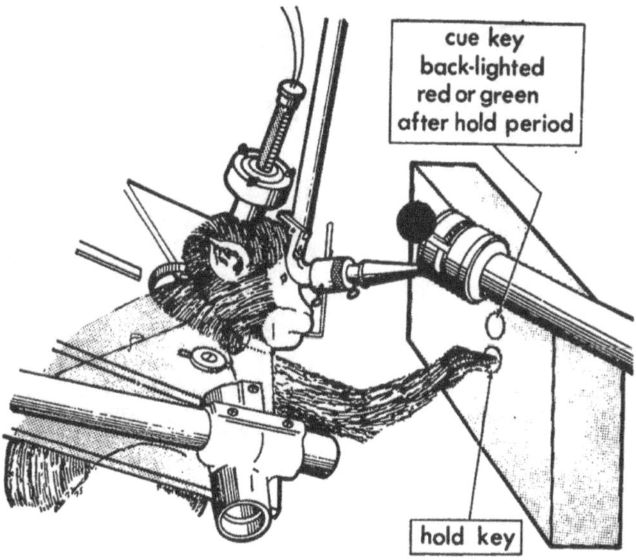

FIGURE 11.16. Illustration of the modified version of the monkey apparatus used in single-unit studies. The animal is trained in stages to keep the hold key depressed for two seconds. The cue key is then illuminated with either a red or green light. If red, the animal must lift off the hold key and press or hit the cue key within one second to turn off the red light and to receive a juice reward; if green, the animal must maintain pressure on the hold for one additional second. At the end of this time, juice is delivered via a tube near the animal's mouth. The electrode microdrive apparatus is illustrated, mounted on the animal's head during the testing session (from Otero, 1975).

The majority of the type 1 responses appeared to be associated with the flux of sensorimotor activity involved in the execution of the go response. Task-related units from specialized motor/somatosensory structures, e.g., red nucleus, sensorimotor cortex, and ventrolateral thalamus, were almost exclusively type 1. Their "sensorimotor" character could often be verified by *kitzelig*, i.e., observing unit responses in relation to spontaneous or passive movements.

An electrooculographic analysis of saccadic eye movements in three animals revealed that there is an eye movement set associated with the execution of the go response which must be considered when interpreting a type 1 response (Ray, 1978). The components of this set consisted of a fixation of gaze during the go response followed less regularly by a blink as the monkey pressed the stimulus button. Regular occurrences of fixation-blinks during go trials could produce associated changes in eye movement-

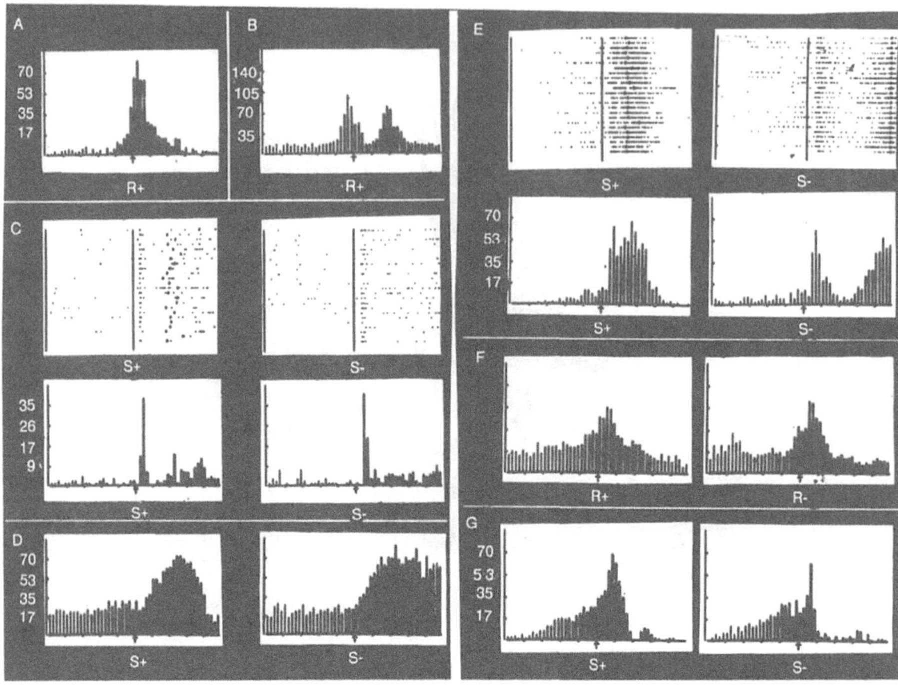

FIGURE 11.17. A and B. Response histograms of type Ia (left) and Ib (right) units from post-central gyrus, hand area. Ordinate represents firing frequencies per second; abscissa represents frequencies cumulated in 40 msec-bin intervals 1 sec pre- and 1 sec post-lift response (indicated by arrow). Type 1a unit shows one burst of increased firing frequency around the lift response. Type 1b unit shows two bursts: one around the lift response and a second one around the replacement of the paw on the hold button; this occurs about 450 msec later than the lift. C. Top row: stimulus "rasters" or alignment of cellular firing in go (S+) trials (left) and no-go (S-) trials (right) with respect to the stimuli in the go and no-go trials, respectively. The stimuli appear at the center of the display; unit spikes appear as dots. Activity 1 sec pre- and 1 sec post-stimulus is presented. Bottom row: frequency histogram representation of the same data. This superior colliculus unit shows a brief burst of increased activity shortly after the onset of both go and no-go stimuli. D. Stimulus histograms for go and no-go trials of a type 2b unit seen in the midbrain reticular formation; "tonic" increase of firing frequency is seen during both go and no-go trials. E. Stimulus rasters (top row) and frequency histograms (bottom row) of go (left) and no-go (right) trials of a type 2c unit seen at the borderline of central gray, deep superior colliculus and midbrain reticular formation. The unit shows a tonic increase of firing frequency during go trials; in the no-go trials, the similar initial increase is followed temporarily by reduced firing. F. Frequency histograms of a type 2d unit in the pretectal area. The unit shows a burst of increased firing around the reinforcement period, i.e., around the end of both go and no-go trials. G. Frequency histogram of an "anticipatory" unit in the pontine reticular formation. The unit shows a gradual increase of firing preceding the onset of both go and no-go stimuli, followed by a burst of activity evoked by the stimulus. The burst is more substantial for go stimuli. We classified this unit as type 2c (from Pragay et al., 1978).

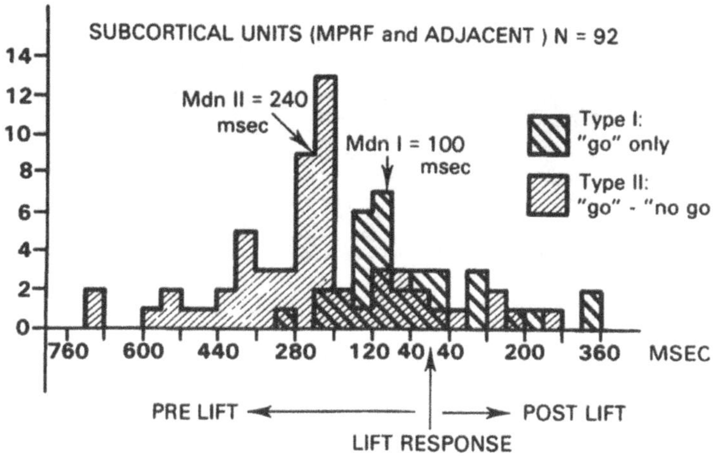

FIGURE 11.18. Latency to first significant change in firing for subcortical units during go trials. Type 1 (go only) and type 2 (go and no-go) units are plotted separately. Mdn = median latency. Significance levels were evaluated automatically by the computer program, by comparing each interval with the prestimulus frequency values. We called significant, changes which lasted 30 msec or longer. Interpretation of difference in mean type 1 and type 2 latencies is discussed in text (modified from Pragay et al., 1978).

related neurons, including those which responded in relation to an eye movement per se without regard to the position of the eyes, e.g., eye movement units in the superior colliculus, or "burst" eye movement units in the reticular formation.

The reticular/isodendritic areas contained a diffusely distributed population of go only units. This group included a small number of the eye movement-related "burst" neurons characteristic of the "oculomotor" reticular formation, i.e., the rostral paramedian pontine reticular formation (Henn & Cohen, 1976; Keller, 1972; Luschei & Fuchs, 1972) and a related projection area in the anterior mesencephalic area around the interstitial n. of Cajal (Büttner, Büttner-Ennever, & Henn, 1977). An example of a go response raster for one of these units is illustrated in Figure 11.19. The rest of the go only units in the reticular formation displayed robust responses which appeared to be related to sensorimotor events involved in limb and trunk movements, possibly a reflection of the widespread somatosensory afferentation of the reticular core.

Because of our long-standing interest in the attentive/discriminative functions of the reticular formation we were encouraged by the discovery of type 2 unit activity occurring early in the trial. We decided to test these units in relation to the same stimuli delivered in a nonreinforced, nonperformance

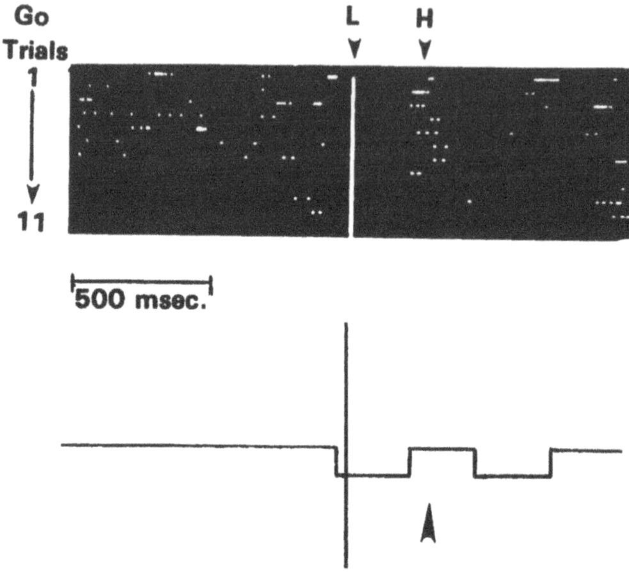

FIGURE 11.19. Paramedian pontine reticular formation (PPRF) eye movement-related unit. This shows the response raster for 11 go trials. Each horizontal line represents one trial. Each dot stands for the occurrence of a spike. The lower half of the figure represents the latency and direction of statistically significant changes in firing rate with respect to the one second pre-stimulus control period. This unit was localized in the rostral PPRF. It is an example of the "burst" saccade-related unit found in this region. Its firing rate decreased significantly during the interval between the lift (L) and the hit (H) components of the go response, returned to baseline during the hit, decreasing again after the conclusion of the response. These significant decreases in firing rate were associated with gaze fixation (from Ray, 1978).

context. This strategy has been used in a number of chronic single-unit investigations (Hyvärinen, Poranen, & Jokinen, 1974; Ryan & Miller, 1977; Sakai, 1974) to assess the effect of the "attention-attracting" properties of a stimulus. This involved training the animal to make a second-order discrimination between two conditions: the task condition in which responses to the stimuli were reinforced and a second condition in which responses to the same stimuli were not reinforced (noncontingent stimulus condition). The animal was said to have learned the discrimination when it responded to the stimuli in the task context and did not respond to them during the noncontingent condition. The reticular type 2 units we have tested thus far have shown a decrease in firing rate in relation to the noncontingent stimuli. An example of such a unit is shown in Figure 11.20 (Ray, 1978). There was

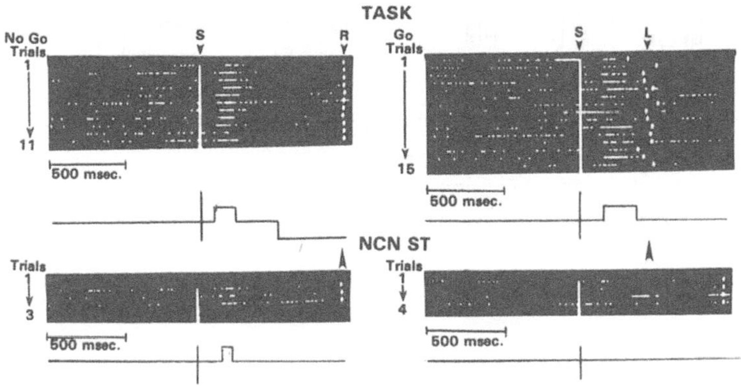

FIGURE 11.20. Type 2 unit from the anterior mesencephalic reticular formation. Stimulus rasters for contingent (task) and noncontingent stimuli (NCN ST) are shown. This unit showed a statistically significant increase in firing rate in conjunction with both the go (red) and no-go (green) stimuli. It also responded in relation to green noncontingent stimuli, but less strongly. No change in firing rate occurred in relation to the noncontingent red stimuli (from Ray, 1978).

also a small class of "anticipatory" units which appeared to begin their discharge many milliseconds before the onset of a trial. One such unit from the reticular formation is shown in panel G in Figure 11.16. Figure 11.21 shows the effect of the noncontingent stimuli on the spike discharges of such a unit. It is important to point out that this population did not include any burst eye movement units. Since we were concerned with studying the functions of the nonoculomotor reticular neurons we attempted to minimize stimulus-related eye movement activity as much as possible. Because the position of the animal's head was fixed, we found that putting the stimulus button at eye level effectively precluded the development of systematic stimulus-related eye movements, e.g., looking up or down at the stimulus when it came on. (There was, however, evidence that some of these units were related to the delivery of the reinforcer at the end of the trial, reflecting the tendency in some animals to orient to the juice drop.)

We believe our findings provide further support for the idea that a stimulus must be relevant in some way to the organism in order to induce and maintain optimal levels of reticular activity. This view has been emphasized and developed by both Jasper (1958) and Lindsley (1960). Lindsley (1958) in particular has proposed that this type of reticular activation may be necessary for optimum neurophysiological and behavioral efficiency. This is probably at the heart of what attention is all about.

The presence of anticipatory unit activity in the reticular system has been

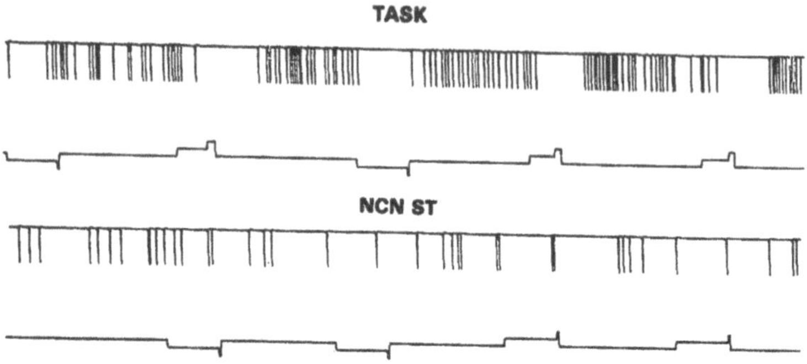

FIGURE 11.21. Firing pattern of "anticipatory" unit during contingent (task) and noncontingent stimulus (NCN ST) conditions. This shows two successive 14-sec segments from a polygraph recording. Unit spikes are displayed as standard 1 V pulses on channel 1. Associated task events are indicated on the second channel (no-go trials: downward deflections; go trials; upward deflections). The third and fourth channels show the behavior of this unit during the noncontingent stimulus condition (from Ray, 1978).

reported previously by Sparks and Travis (1968) in the monkey and by Olds, Mink, and Best (1972) in the rat. Because we used a constant two-second foreperiod in our experiments, the animals clearly could learn when to expect a stimulus. If it is assumed cellular activity which anticipates the cue stimulus reflects readiness to respond, it should be related in some way to reaction time. In our studies anticipatory units tended to be found most frequently in those animals which performed at a regular, rhythmic, unbroken pace. Typically, these animals also exhibited the least variation in reaction time; consequently, it has been difficult without a more sophisticated computerized method to document a trial-by-trial relation with reaction time. Although we can only speculate, perhaps the anticipatory firing type of activity is critical for the facilitatory post-synaptic effects discussed by Lindsley.

The role of the mesencephalon in the pathophysiology of petit mal epilepsy has obviously still to be established; however, these data on the cellular inhabitants of this region of the brain appear to add to the reasonableness of the inference of involvement.

It is to be hoped that Don Lindsley will consider this review of research from our laboratory an acceptable gift on the occasion of this *Festschrift*.

ACKNOWLEDGMENT

The research described here was supported in part by grants MH-12568, MH-14915 (Research Scientist Award to AFM) and NS-12201 from the U.S. Public Health Service. Dr. Ray was supported by a National Research

Service Award in Biobehavioral Research (MH-15189).

REFERENCES

Büttner, U., Büttner-Ennever, J., & Henn, V. (1977). Vertical eye movement related unit activity in the rostral mesencephalic reticular formation of the alert monkey. *Brain Research, 130*, 239-252.

Evarts, E. V. (1968). A technique for recording activity of subcortical neurons in moving animals. *Electroencephalography and Clinical Neurophysiology, 24*, 83-86.

Geller, M., & Geller, A. (1970). Brief amnestic effects of spike-wave discharges. *Neurology, 20*, 1089-1095.

Gloor, P. (1969). Neurophysiological bases of generalized seizures termed centrencephalic. In H. Gastaut, H. Jasper, J. Bancaud, & A. Waltregny (Eds.), *The physiopathogenesis of the epilepsies* (pp. 209-236). Springfield: Charles C. Thomas.

Gloor, P. (1978). Generalized epilepsy with bilaterally synchronous spike and wave discharge: New findings concerning its physiological mechanisms. *Electroencephalography and Clinical Neurophysiology (Suppl.), 34*, 245-249.

Henn, V., & Cohen, B. (1976). Coding of information about rapid eye movements in the pontine reticular formation of alert monkeys. *Brain Research, 108*, 307-325.

Hyvärinen, J., Poranen, A., & Jokinen, Y. (1974). Central sensory activities between sensory input and motor output. In F. Schmitt & F. Worden (Eds.), *The neurosciences: Third study program* (pp. 311-318). Cambridge, MA: MIT. Press.

Jasper, H. (1958). Recent advances in our understanding of ascending activities of the reticular system. In H. Jasper, L. Proctor, R. Knighton, W. Noshay, & R. Costello (Eds.), *Reticular formation of the brain* (pp. 319-331). Boston: Little, Brown.

Keller, E. (1972). Participation of medial pontine reticular formation in eye movement generation in monkey. *Journal of Neurophysiology, 35*, 445-461.

Lindsley, D. B. (1952). Psychological phenomena and the electroencephalogram. *Electroencephalography and Clinical Neurophysiology, 4*, 443-456.

Lindsley, D. B. (1958). The reticular system and perceptual discrimination. In H. Jasper, L. Proctor, R. Knighton, W. Noshay, & R. Costello (Eds.), *Reticular formation of the brain* (pp. 513-534). Boston: Little, Brown.

Lindsley, D. B. (1960). Attention, consciousness, sleep and wakefulness. In J. Field (Ed.), *Handbook of physiology, section III: Neurophysiology* (Chapter LXIV). Baltimore: Williams & Wilkins.

Lindsley, D. B., Bowden, J., & Magoun, H. W. (1949). Effect upon the EEG of acute injury to the brainstem activating system. *Electroencephalography and Clinical Neurophysiology, 1*, 475-486.

Lindsley, D. B., Schreiner, L. H., Knowles, W. B., & Magoun, H. W. (1950). Behavioral and EEG changes following chronic brainstem lesions in the cat. *Electroencephalography and Clinical Neurophysiology, 2*, 483-498.

Luschei, E., & Fuchs, A. (1972). Activity of brainstem neurons during eye movements of alert monkeys. *Journal of Neurophysiology, 35*, 445-461.

Mirsky, A. F., & Bloch, S. (1967). Effects of chlorpromazine, secobarbital and sleep deprivation on attention in monkeys. *Psychopharmacologia, 10*, 388-399.

Mirsky, A. F., Bloch, S., Tecce, J. J., Lessell, S., & Marcus, E. (1973). Visual evoked potentials during experimentally induced spike-wave activity in monkeys. *Electroencephalography and Clinical Neurophysiology, 35*, 25-37.

Mirsky, A. F., & Oshima, H. I. (1973). Effect of subcortical aluminum cream lesions on attentive behavior and the electroencephalogram in monkeys. *Experimental Neurology, 39*, 1-18.

Mirsky, A. F., Pragay, E. B., & Harris, S. (1977). Evoked potential correlates of stimulation-induced impairment of attention in *Macaca mulatta*. *Experimental Neurology, 57*, 242-256.

Mirsky, A. F., & Tecce, J. (1966). The relationship between EEG and impaired attention following administration of centrally acting drugs. In *Proceedings of the 5th International Congress of the Collegium Internationale Neuropsychopharmacologicum* (pp. 638-645).

Mirsky, A. F., & Van Buren, J. M. (1965). On the nature of the "absence" in centrencephalic epilepsy: A study of some behavioral, electroencephalographic and autonomic factors. *Electroencephalography and Clinical Neurophysiology, 18*, 334-348.

Olds, J., Mink, W., & Best, P. (1972). Single unit patterns during anticipatory behavior. *Electroencephalography and Clinical Neurophysiology, 35*, 144-158.

Orren, M. (1974). *Visuomotor behavior and visual evoked potentials during petit mal seizures*. Unpublished doctoral dissertation, Boston University, Boston, MA.

Orren, M. (1978). Evoked potential studies in petit mal epilepsy: Visual information processing in relation to spike and wave discharges. *Electroencephalography and Clinical Neurophysiology, 34*, 251-257.

Otero, J. (1975). Comparison between red nucleus and precentral neurons during learned movements in the monkey. *Brain Research, 101*, 37-46.

Penfield, W., & Jasper, H. (1954). *Epilepsy and the functional anatomy of the human brain*. Boston: Little, Brown.

Pragay, E. B., Mirsky, A. F., Fullerton, B., Oshima, H., & Arnold, S. (1975). Effect of electrical stimulation of the brain on visually controlled (attentive) behavior in *Macaca mulatta*. *Experimental Neurology, 60*, 83-95.

Pragay, E. B., Mirsky, A. F., Ray, C., Turner, D., & Mirsky, C. (1978). Neuronal activity in the brainstem reticular formation during performance of a "go no-go" visual attention task in the monkey. *Experimental Neurology, 60*, 83-95.

Ray, C. (1978). *Functional analysis of attention-related neurons in the monkey reticular formation*. Unpublished doctoral dissertation, Boston University, Boston, MA.

Ricks, N. L., & Mirsky, A. F. (1977). Unpublished observations.

Rosvold, H. E., Mirsky, A. F., Sarason, I., Bransome, E. D., Jr., & Beck L. H. (1956). A continuous performance test of brain damage. *Journal of Consulting Psychology, 20*, 343-350.

Russell, G. C. (1961). Hypothalamic, preoptic and septal regions of the monkey. In D. Sheer (Ed.), *Electrical stimulation of the brain* (pp. 232-250). Austin, TX: University of Texas Press.

Ryan, A., & Miller, J. (1977). Effects of behavioral performance on single unit firing patterns in inferior colliculus of the rhesus monkey. *Journal of Neurophysiology, 40*, 943-956.

Sakai, M. (1974). Prefrontal unit activity during visually guided level pressing reaction in the monkey. *Brain Research, 81*, 297-309.

Sparks, D., & Travis, R. (1968). Patterns of reticular unit activity observed during the performance of a discriminative task. *Physiology and Behavior, 3*, 961-967.

Weir, B. (1964). Spikes-wave from stimulation of the reticular core. *Archives of Neurology, 11*, 209-218.

12 · PUPILLOMETRIC SIGNS OF SELECTIVE ATTENTION IN MAN

J. Beatty

University of California at Los Angeles

The execution of information-processing functions in the human brain reliably elicits momentary increases in the output of brainstem activation systems, the magnitude of which is a function of the cognitive load imposed by the task (Beatty, 1977; Kahneman, 1973). This relation is true not only for complex, load-demanding processes such as mental arithmetic (Hess & Polt, 1964; Payne, Parry, & Harasymiw, 1968), serial memory (Beatty & Kahneman, 1966; Kahneman & Beatty, 1966; Kahneman, Beatty, & Pollack, 1967), and logical reasoning (Ahern, 1978), but for simpler perceptual processes as well (Kahneman & Beatty, 1967). We have demonstrated increases in signs of small event-related activation accompanying the visual encoding of letter pairs, and that the magnitude of this activation is directly related to the hierarchical level within the cognitive system in which the letter-pair is processed (Beatty & Wagoner, 1978). I now report that activation responses accompanying the processing of sensory stimuli may be controlled by attention.

The idea of activation as a physiological variable related to cognitive processing is a prominent feature of Lindsley's neuropsychology (Lindsley, 1960). As a neurophysiological concept, Lindsley and his co-workers identified activation with electrocortical desynchronization and autonomic arousal in their work on the physiological functions of the brainstem reticular activating system (Moruzzi, 1972). Much of this and more recent work demonstrates quite forcefully that to view activation as a single, one-dimensional description of central nervous system arousal is a serious oversimplification: the brainstem activation systems are complex, interdependent and subject to external regulation, particularly from telencephalic structures (Schlag, 1974). Nonetheless, a simplified concept of activation has proven to be of heuristic value in exploring the dynamics of complex information processing in the human brain.

Of the several available methods for measuring CNS activation in man, perhaps the most sensitive and reliable is the measurement of sympathetic and parasympathetic activity as reflected in pupillary diameter (Goldwater, 1972). Further, pupillometric measures have shown a striking correspondence to electrophysiological indices of activation in a long series of studies on corticoreticular interactions (Moruzzi, 1972). Pupillary dilations

DOI: 10.4324/9781003164647-16

indicative of CNS activation may be mediated by either forebrain inhibition of the Edinger-Westphal nuclei or increased sympathetic discharge (Lowenstein & Loewenfeld, 1952). Orderly relations between pupillary dilation and presumed cognitive load have been reported for a wide variety of cognitive tasks (Ahern, 1978; Beatty & Kahneman, 1966; Beatty & Wagoner, 1978; Hess & Polt, 1964; Kahneman & Beatty, 1966,1967; Kahneman, Beatty, & Pollack, 1967; Payne et al., 1968). Thus a convincing argument may be made that the degree of pupillary dilation observed during the execution of a particular cognitive function is directly proportional to the demands which that function places on the cognitive system.

METHODS

To test for the effects of attentional processes on pupillometric signs of task-induced activation to sensory stimuli, an experimental procedure similar to that previously reported by Hillyard (Hillyard, Hink, Schwent, & Picton, 1973) was employed. Eight young adults were required to monitor one channel of an auditory display for targets and report them by depressing a microswitch. A series of 50 msec 800 Hz tone bursts were presented to one ear and 1500 Hz tone bursts to the other by earphones. A channel was defined by frequency of the standard tone bursts, with the assignment of frequency to ear randomized over subjects. Infrequently presented targets were tone bursts of slightly higher frequency (860 and 1575 Hz for the low- and high-frequency channels respectively).

A computer-generated sequence of probabilistic decisions controlled the sequence of stimuli. Every 50 msec a decision was made to present a stimulus with a probability of .15. If this decision was positive two additional choices were made: first, the stimulus was assigned to the high-frequency channel with probability .50; otherwise the low-frequency channel was selected. Second, the standard stimulus was replaced by a signal on the selected channel with probability .066. This procedure generated an average of one stimulus every 333 msec and one target on the attended channel every 10 sec. The entire experiment was controlled by a general purpose digital computer.

This complex, probabilistic method of stimulus sequencing was necessary to permit unbiased recording of averaged evoked pupillary responses. Pupillary diameter was measured using a Whittaker 1053 TV pupillometer and was digitized at 50-msec intervals. Single evoked pupillary responses were computed to non-target stimuli for a period of 2 sec following stimulus presentation. Thus, on the average, six other stimulus events might be expected to occur as the response to any given stimulus was being recorded. To prevent confounding of the pupillometric data, the probability of stimulus presentation must not vary systematically in this period. By completely randomizing the schedule of stimulus presentation, the response to several stimuli could be acquired simultaneously, with each response at different stages of completion. To further reduce artifacts, all data acquired within 2.5 sec preceding either the presentation of a signal or the registration of a detection response were discarded automatically. This control was initiated to remove the large pupillary signs of detection and response initiation from the background monitoring data.

The experiment consisted of six blocks of 64 stored 2-sec records obtained according to the described procedure. Four listeners were instructed to detect targets on the high-frequency channel in the first three blocks and on the low-frequency channel in the last three blocks. These instructions were reversed for the remaining listeners.

All behavioral and pupillometric data were stored on disk for later artifact removal and analysis. The pupillometric data from each stimulus were individually inspected for artifacts without knowledge of stimulus type. All trials containing major artifacts were discarded. Trials with small artifacts occurring in noncritical periods were corrected by linear interpolation. For each subject four averaged evoked pupillary responses were then computed for each combination of auditory channel and instructional condition.

RESULTS

Figure 12.1 presents the group averaged evoked pupillary responses for background events as a function of stimulus channel and attentional instruction. A clear pupillary dilation is present at a latency of about 600 msec following presentation of background stimuli on the attended channel, which is completely absent on the unattended channel. An analysis of variance confirms this observation. The mean amplitude of the pupillary dilation in the 1-sec period following stimulus presentation did not differ significantly between stimulus channels ($F(1,7) = 0.60$, $p = .463$) nor between attentional instructions ($F(1,7) = 1.74$, $p = .229$). However, the interaction of stimulus channel and attentional instruction was highly significant ($F(1.7) = 11.59$, $p = .011$). Thus a small pupillary dilation appears following presentation of non-target stimuli when a channel is attended and not otherwise.

DISCUSSION

Several aspects of these data deserve comment. First, although the event-related pupillary response observed following stimuli on the attended channel is quite reliable and replicable, it is of extremely small magnitude (approximately 0.015 mm). In contrast, the dilation observed under comparable recording conditions for a six-digit short-term memory task is approximately .55 mm. Similarly, mentally multiplying a pair of two-digit

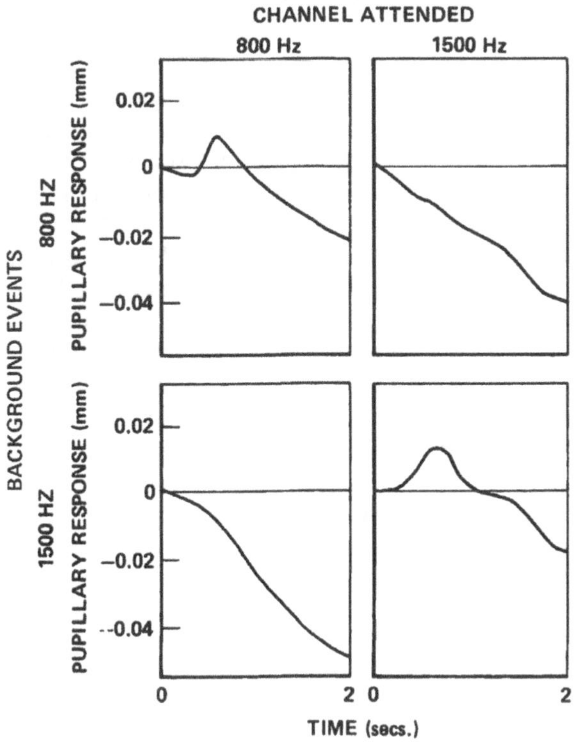

FIGURE 12.1. Averaged evoked pupillary responses for 8 subjects in a selective attention experiment. A post-stimulus pupillary dilation appears following non-signal stimuli in the channel to which attention is directed and is not present following stimuli in the unattended channel.

numbers such as 13 and 18 results in a dilation of approximately 0.50 mm. If the degree of pupillary dilation evoked by the performance of a cognitive task is proportional to the load imposed by that task on the individual's processing capacity (Beatty, 1977; Kahneman, 1973), then it would appear that the load imposed in processing information from the attended auditory channel is extremely small. In contrast, this same logic leads to the conclusion that the presentation of stimuli on the unattended channel imposes no measurable load upon the nervous system.

Second, these results are congruent with data previously reported by Hillyard and his colleagues (Hillyard, Hink, Schwent, & Picton, 1973; Hink, Van Voorhis, & Hillyard, 1977) using the same experimental paradigm to study the effects of selective attention on cortical event-related potentials.

These authors report that an early (60-70 msec) component of the cortical event-related potential is selectively enhanced for the attended channel and suppressed for the unattended channel. They suggest that the magnitude of this component is indicative of a stimulus set that admits all sensory input to the attended channel for further perceptual analysis. Extending this reasoning, the magnitude of the event-related pupillary response may index the degree to which further capacity-demanding processing is undertaken.

Third, it must be noted that all of the averaged pupillary responses are characterized by a descending baseline. This is probably a consequence of a sawtooth pattern in the raw data, with large and rapid pupillary dilations following the detection of a target, which diminish gradually as a function of time. Since all data surrounding target presentation or detection responses are removed from these averages, the averages presented in Figure 12.1 are drawn primarily from periods in which the tonic pupillary diameter is slowly decreasing. This property of the data is of little concern, as the short-latency event-related pupillary dilations and the slow baseline changes are of markedly different form.

Finally, these data demonstrate the utility of employing relevant physiological measures in the study of cognitive processing. From a behavioral point of view, they provide a useful converging operation for the analysis of complex mental functions. From a physiological perspective, the data help clarify the dynamics of forebrain-brainstem interactions and the role of activation in cognitive processing. Further, these data lend specific support to the idea that attended stimuli are processed at a capacity-demanding level, but that unattended stimuli are not.

REFERENCES

Ahern, S. K. (1978). *Activation and intelligence: Pupillometric correlates of individual differences in cognitive abilities.* Unpublished doctoral dissertation, University of California, Los Angeles.

Beatty, J. (1977). Activation and attention. In M. C. Wittrock, J. Beatty, J. E. Bogen, M. S. Gazzaniga, H. J. Jerison, S. D. Krashen, R. D. Nebes, & T. Teyler (Eds.), *The human brain* (pp. 63-85). Englewood Cliffs, NJ: Prentice-Hall.

Beatty, J., & Kahneman, D. (1966). Pupillary changes in two memory tasks. *Psychonomic Science, 5*, 371-372.

Beatty, J., & Wagoner, B. L. (1978). Pupillometric signs of brain activation vary with level of cognitive processing. *Science, 199*, 1216-1218.

Goldwater, B. C. (1972). Psychological significance of pupillary movements. *Psychological Bulletin, 77*, 340-355.

Hess, E. H., & Polt, J. H. (1964). Pupil size in relation to mental activity during simple problem solving. *Science, 143*, 1190-1192.

Hillyard, S. A., Hink, R. F., Schwent, V. L., & Picton, T. W. (1973). Electrical signs of selective attention in the human brain. *Science, 182*, 177-180.

Hink, R. F., Van Voorhis, S. T., & Hillyard, S. A. (1977). The division of attention and the human auditory evoked potential. *Neuropsychologia, 15*, 597-605.

Kahneman, D. (1973). *Attention and effort.* Englewood Cliffs, NJ: Prentice-Hall.

Kahneman, D., & Beatty, J. (1966). Pupil diameter and load on memory. *Science, 154*, 1583-1585.

Kahneman, D., & Beatty, J. (1967). Pupillary responses in a pitch-discrimination task. *Perception and Psychophysics, 2*, 101-105.

Kahneman, D., Beatty, J., & Pollack, I. (1967). Perceptual deficit during a mental task. *Science, 157*, 218-219.

Lindsley, D. B. (1960). Attention, consciousness, sleep and wakefulness. In J. Field (Ed.), *Handbook of physiology* (Vol. III, pp. 1553-1593). Washington, DC: American Physiological Society.

Lowenstein, O., & Loewenfeld, I. E. (1952). Disintegration of central autonomic regulation during fatigue and its reintegration by psychosensory controlling mechanisms. I. Disintegration. Pupillographic studies. *Journal of Nervous and Mental Disease, 115*, 1-21.

Moruzzi, G. (1972). The sleep-waking cycle. *Reviews of physiology: Biochemistry and experimental pharmacology.* New York: Springer-Verlag.

Payne, D. T., Parry, M. E., & Harasymiw, S. J. (1968). Percentage pupillary dilation as a measure of item difficulty. *Perception and Psychophysics, 4*, 139-143.

Schlag, J. (1974). Reticular influences on thalamo-cortical activity. In O. Creutzfeldt (Ed.), *Handbook of electroencephalography and clinical neurophysiology: Vol. 2. Electrical activity from the neuron to the EEG and EMG.* Amsterdam: Elsevier.

13 · REGIONAL CEREBRAL BLOOD-FLOW: SUPPLEMENT TO EVENT-RELATED POTENTIAL STUDIES OF SELECTIVE ATTENTION

R. Näätänen

University of Helsinki

In the 1960s, D. B. Lindsley was an important pioneer in the use of event-related potentials (ERPs) to study problems of human information processing and attention. His studies with Spong and Haider (Haider, Spong, & Lindsley, 1964; Spong, Haider, & Lindsley, 1965), both being foreign visitors in his laboratory, represent important early landmarks in this research endeavor. Spong et al. (1965) claimed that the N1-P2 amplitude of the ERP is enhanced by selectively attending to the eliciting stimuli. Although this result was questioned on methodological grounds (Näätänen, 1967, 1975), it has been supported by subsequent research on human selective attention. Most notably, Hillyard, Hink, Schwent, and Picton (1973) unequivocally demonstrated that the N1 amplitude to attended stimuli was enhanced (and hence N1-P2 since the measure from baseline to P2 peak was unchanged), thus confirming the early result of Lindsley's research group. Hillyard et al. (1973) used a dichotic-listening paradigm with short inter-stimulus intervals (ISIs), the subject's task being to detect occasional pitch changes in the input to the designated ear.

The exact nature of the cerebral event causing the N1 effect has been a matter of major disagreement. Hillyard et al. (1973) interpreted the effect as being caused by a "genuine" N1 enhancement, i.e., an increased response of the generator mechanism of the exogenous N1 component.[1] Their published data were indeed consistent with this interpretation, the attention effect appearing very much like a growth of the N1 component.

[1] An ERP is a sequence of negative and positive waves, or deflections. The latter do not generally represent any unitary brain event (generator process), but are rather composed of temporally overlapping components of the same or opposite polarity. An ERP *component* should be understood as a contribution of some single generator process (i.e., activation of some brain center) to the total waveform. Exogenous ERP components are mainly determined by physical and temporal stimulus factors; endogenous components are more associated with "psychological" factors and may occur even with no stimulus. The N1 component is usually regarded as an exogenous component. More recent research (e.g., Hari, Kaila, Katila, Tuomisto, & Varpula, 1982) has shown that the auditory N1 deflection is in fact

DOI: 10.4324/9781003164647- 17

However, Näätänen, Gaillard, and Mantysalo (1978), in a somewhat similar experiment but using longer (and constant) ISIs, found a different selective-attention effect. It consisted of a protracted negativity, beginning 150 msec from stimulus onset (at the descending slope of N1) and lasting for several hundred milliseconds. This protracted negativity seemed to emanate from a cerebral source different from that of the N1 component, and to be endogenous rather than exogenous in origin. This attention-related component was termed "processing negativity" by the authors, who further suggested that under short-ISI conditions, such as those of Hillyard et al. (1973), the latter may begin before the N1 peak, thus artificially making N1 appear larger (see Figure 13.1).

Several subsequent studies confirmed the processing negativity in various paradigms and demonstrated that it may indeed begin before the N1 peak (e.g., Hansen & Hillyard, 1980; Okita, 1979; Parasuraman, 1980). However, processing negativity cannot account for all of the observed effects. For example, in the visual modality, selective spatial attention seems to enhance exogenous components of the ERP such as N1 and P2 (Hillyard & Münte, 1984; Hillyard, Münte, & Neville, 1985; Van Voorhis & Hillyard, 1977). Moreover, it is not yet precluded that the auditory spatial attention effect, for large spatial separations, such as that in dichotic listening, could similarly be a "genuine" N1 effect, i.e., caused by enhancement of the generator process of some N1 component. For these reasons, Hillyard et al.'s (1973) N1 effect and Näätänen et al.'s (1978) processing negativity may be separate attention effects. Possibly both effects can even exist in parallel in the same ERP to attended stimuli (see Donald, 1983; Hansen & Hillyard, 1980; Hillyard, 1981).

Processing negativity has generally been interpreted in two ways: (1) it reflects further processing performed for the stimuli selected in the initial fast preliminary stimulus-set selection, or (2) it reflects this selection process itself. The available data strongly support the latter alternative. The crucial evidence is provided by the result that processing negativity is elicited even by stimuli to be ignored (Alho, Sams, Paavilainen, & Näätänen, 1986; for a review, see Näätänen, 1982). If the processing negativity reflected further processing, such stimulus generalization should not occur. The portion of the processing negativity common to both stimuli is longer and larger, the smaller the separation between the to-be-attended and to-be-ignored stimuli. Hence, processing negativity appears to be a real-time physiological sign of a matching process which requires longer times the more similar the stimuli. Presumably this matching process terminates at the moment after stimulus

---

composed of two components, of which one is modality-specific and the other appears to be nonspecific. Both of these components are probably exogenous in nature. Possibly there are even more negative components at this latency range (see McCallum & Curry, 1980.)

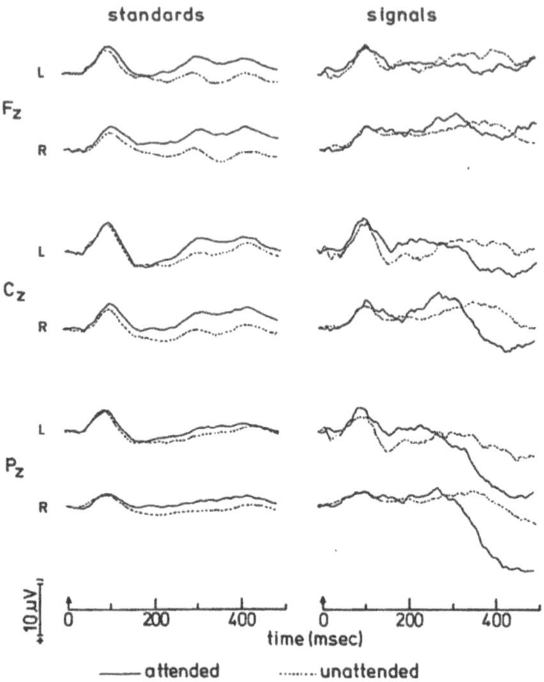

FIGURE 13.1. Frontal (F_z), vertex (C_z), and parietal (P_z) across-subject averaged ERPs to standard stimuli and deviant stimuli (signals) for left-ear (L) and right-ear (R) input when the input was attended and when the input was not attended (data from an unpublished study by Näätänen, Gaillard, and Varey, 1977); figure reproduced from Näätänen and Michie (1979) by permission of North-Holland Publishing Company.

onset when enough information for a difference between the current input and the stimulus to be attended has accumulated. Thereafter, no processing negativity is generated (for consistent data, see Hansen & Hillyard, 1983). When this matching process is not prematurely terminated, i.e., when the stimulus corresponds to that to be attended, the stimulus is selected for response or further processing, depending on the task.

If the processing negativity is generated by a cerebral matching process, there must be some kind of relatively accurate neurophysiological representation of the to-be-attended stimulus with which the matching process occurs. This representation was called the "attentional trace" by Näätänen (1982). He proposed that this voluntarily maintained trace provides fast recognition of the stimuli to be attended, and rapid rejection of all others. "Accepted" stimuli gain entry to further processing or response, according to the task to be performed.

How is the attentional trace developed and maintained? Näätänen (1982) proposed that in a "multi-channel" stimulus situation, the subject tries to maintain as precise an image of the stimulus to be attended as possible in order to recognize these stimuli. In the very beginning of the session, however, this image is weak, but becomes more vivid with increased presentations of the stimuli. It was further suggested that the development of this subjective image is paralleled by the emergence of the attentional trace in the sensory system, this trace being the neurophysiological basis of the subjective image concerned (see also Näätänen, in revision).

Unfortunately, ERPs are measures of transient cerebral responses only, providing no opportunity for directly measuring any underlying state factors (such as the attentional trace) which might somehow explain the differential ERP response to the attended and unattended stimuli. In the following, regional cerebral blood-flow (rCBF) studies on human selective attention are reviewed. Such data provide information on more tonic aspects of brain activity than that provided by the ERPs. It is hoped that these studies reveal some indication of a pre-set attentional bias in the state of the sensory systems. Such data would be of great value in explaining the ERP data on attentionally selective responses.

Regional Cerebral Blood-Flow (rCBF) Studies of Selective Attention. It is now possible to measure the regional distribution of the cerebral blood flow during task performance (Ingvar, 1979; Ingvar & Risberg, 1965, 1967; Lassen & Ingvar, 1963). It is maintained that increases in the regional cerebral metabolic rate of oxygen lead to proportional increases in rCBF (Ingvar & Lassen, 1975, 1977). The rCBF is suitable for studying cerebral states lasting, say, at least 30 sec.

The normal resting pattern of rCBF has been characterized as "hyperfrontal." Thus, premotor and frontal regions as well as the posterior sylvian region appear to be relatively active, as estimated on the basis of oxygen utilization, whereas the temporal, parietal, and the postcentral rolandic regions are relatively inactive. This resting hyperfrontal pattern is diminished during anesthesia and spontaneous sleep (Ingvar, 1979).

In a study by Ingvar (1979), weak cutaneous electrical stimulation of the thumb produced a moderate rCBF increase over the contralateral rolandic region (more pre- than post-centrally) and some increased frontal activation. However, when the stimulus intensity was augmented to a level of discomfort and/or slight pain, a notable general increase of the flow was observed which was most pronounced over the entire rolandic and frontal regions. This strong activation may be due to affective or motivational elements. This general flow increase is not observed during speech, reading (Ingvar &

Schwartz, 1974), or problem solving using Raven's matrices or a backward digit-span memory test (Risberg & Ingvar, 1973). In both problem solving tests, the flow increased in the premotor and frontal regions. The Raven test, which involves visual input, also activated parieto-occipital regions. Ingvar (1979) concluded that abstract thinking, memorizing, and problem solving especially involve the precentral and postcentral association cortex of the dominant hemisphere. He further suggested, more generally, that brain activities associated with consciousness primarily engage precentral and frontal cerebral structures. According to Pribram and Luria (1973), the activation of the anterior parts of the brain is associated with programming goal-directed intentional behavior.

Roland (1981) measured rCBF in 254 cortical regions of the same hemisphere in two situations: (1) while subjects were at rest, and (2) while they focused their attention on the tip of their index fingers where they expected a very weak touch stimulus. During focused attention (40 sec; no stimulus was actually given), the rCBF increased by 25% in the contralateral somatosensory finger area (and in the surrounding area). Smaller increases occurred in the superior prefrontal and midfrontal regions and very weak ones in the middle part of the posterior parietal region. Additionally there was a diffuse rCBF increase involving the frontal and parietal association areas. When attention was directed toward the upper lip instead of the finger, rCBF increased in the contralateral somatosensory mouth area and in the other regions mentioned above.

Roland (1981) concluded that preparatory tuning of the relevant somatotopical postcentral area occurs when attention is focused on a spot of the skin surface. He proposed that this tuning mainly resulted from the local sum of many EPSPs (excitatory post-synaptic potentials) integrated during the attention period of 40 sec and that this tuning reflected increased local excitability which might improve detection performance. He regarded the tuning of the sensory area as being most probably controlled by the superior and midfrontal cortical areas which, as he thought, probably did not participate in the analysis of somatosensory signals. On the other hand, a part of the midfrontal region which was also activated probably participates, according to him, in the specific somatosensory analysis (Roland & Larsen, 1976; Roland, Larsen, Skinhøj, & Lassen, 1977).

Hence it appears that during selective attention, the rCBF is most increased in the frontal and in the sensory-specific areas of the brain, both increases showing a well-defined spatial pattern, the frontal increase actually showing two separate patterns. Unfortunately, the poor temporal resolution of the rCBF method does not permit determining the order of emergence of the different rCBF increases. Such information might make it possible to determine the order of emergence of the underlying increases in neuronal activity. However, it appears plausible that the increase in frontal neuronal activity is in the lead, producing and commanding that in the sensory-specific areas. It is known that the frontal areas are essential to voluntarily directed attention (Luria & Homskaya, 1970; Pribram & Luria, 1973).

In a subsequent study, Roland (1982) delivered overlapping stimulus streams in the visual, auditory, and somatosensory modalities. The subject performed a discrimination task in one modality at a time, ignoring the stimuli of the two other modalities. The total stimulus situation was the same during all three discrimination tasks. The pattern of results was very complex and I mention here only those most relevant to the present topic. In the visual task, the rCBF in (a part of) the visual association cortex was increased in comparison to activity during auditory or somatosensory attention (although this activity was above rest levels). In the auditory task, the rCBF in the auditory association cortex was increased from the level prevailing during attention to one of the two other modalities. On the other hand, the somatosensory association cortex showed no selective-attention modulation of rCBF. (However, a part of that area could not be monitored with the methodology used.)

The primary projection area could be monitored directly only in the somatosensory modality. In the somatosensory hand area (the contralateral hand was stimulated), there was an increase (from the rest level) of the same magnitude regardless of the direction of attention. The same was true for the perisensory hand area. Since the increase of rCBF in the somatosensory hand area, even when another modality was attended, was larger than when the subjects performed the same somatosensory task in another study (Roland & Larsen, 1976) in which no other stimuli were presented, Roland (1982) thought one could infer that "there was no significant subcortical inhibition of this sensory input no matter to what modality the subjects turned their attention" (p. 1074) and that "*a priori* irrelevant signals were processed, at least preliminarily, by the cortex" (p. 1060). (See also Hyvärinen, 1982; Hyvärinen, Poranen, & Jokinen, 1980; Näätänen, 1967.)

Thus, there were two main sources of rCBF increase above the resting levels in Roland's (1982) study: (1) the increase due to sensory stimulation, and (2) that due to attention, with both nonspecific and specific components. Consequently the author concluded that

> there must be a mechanism that is anticipatory, task dependent, independent of stimulus intensity and stimulus rate, that differentially enhances and inhibits the metabolism in cortical areas. This mechanism is called differential tuning. Differential tuning is the physiological manifestation of modality-specific attention. Since short-term rCBF variations presumably are secondary to metabolic changes related to ion pumping, it is conceivable that differential tuning has an electrophysiological correlate, for example, in EPSP and IPSP changes in cortical areas of at least a few square centimeters in size. (p. 1076)

Roland (1982) assumed that the superior anterior prefrontal cortex participated in the control of this differential tuning. Referring also to the results of several previous studies, he stated that this region

> participated in every task that so far has been carried out according to a prior instruction. If this area, therefore, contributed functionally at all to all these tasks, it could only have been in the (internal) organization of brain activity according to the specifications in the instruction. (p. 1075)

On the other hand, a certain frontal region (the posterior superior lateral) modulated its rCBF according to the attended sensory modality in Roland's (1982) study. Consistently, this region showed a different pattern of rCBF increase in separate unisensory stimulus situations (a discrimination task) involving each of the three modalities (Roland & Larsen, 1976; Roland & Skinhøj, 1981; Roland, Skinhøj, & Lassen, 1981). Hence it appears that during selective attention, there are two frontal rCBF patterns: (1) the superior anterior prefrontal which is independent of the attended modality, and (2) the posterior superior lateral whose pattern varies with the attended modality.

Risberg and Prohovnik (1983) have obtained rCBF data suggesting that the right frontal lobe has a more important role in the control of selective attention than does the left frontal lobe. These authors were able to record rCBF simultaneously from both hemispheres whereas Roland (1981, 1982) was not. On the other hand, their spatial resolution was less than Roland's, which probably explains why in their somatosensory selective-attention task they obtained no contralateral rCBF focus in the somatosensory area corresponding to the hand attended. The right-hemisphere dominance in frontal attentional control was very clear when the right hand was attended, but was also observed when the left hand was attended. However, particularly in the latter situation, there was a notable diffuse rCBF increase in the whole frontal lobe superimposed on the more specific rCBF pattern. Also, parietal rCBF was increased by attention. Also Risberg, Maximilian, and Prohovnik (1977) have obtained rCBF results indicating nonspecific frontal activation during mental activity. This diffuse rCBF component of selective attention, also observed by Roland (1981, 1982) as mentioned above, is of great interest in light of activation-level theories (Duffy, 1962; Malmo, 1959; see however Lacey, 1967; Näätänen, 1973) and selective-attention theories which propose both nonspecific and specific components of attention (e.g., Posner & Boies, 1971).

Comparison of ERP and rCBF Studies of Selective Attention. The review of the rCBF literature revealed several well-defined, patterned blood-flow foci during selective attention in addition to more diffuse flow increases. One of these blood-flow foci occurred in the modality-specific sensory areas corresponding to the direction of attention. Roland (1981) showed such a focus during somatosensory attention as well as analogous effects in the visual and auditory modalities (Roland, 1982). Of these selective-attention effects, the somatosensory focus did not only reflect the modality of the attended stimulus but also more specific features of the latter. This focus seemed to cover the hand-representation area when attention was directed to the tip of the index finger and the mouth area when the upper lip was to be attended. The methodology used does not, however, make it possible to determine the exact boundaries of the effect in the somatosensory cortex.

The development of such a blood-flow focus reflects increased local metabolic rate. An appealing possibility is that the neuronal change causing the rCBF focus was even more specific than the rCBF focus recorded, i.e., when attending to a possible touch on the tip of the index finger, only the neurons corresponding to this finger tip were affected (producing, however, a somewhat more widely spread rCBF focus surrounding them). Such a highly specific pre-set selective-attention effect on cortical sensory neurons would of course correspond to the concept of the attentional trace, developed on the basis of ERP results, which was characterized as a voluntary pre-set neuronal representation of the physical features of the stimulus to be attended.

More generally, both methods show that selective attention influences sensory neuronal activity, i.e., they are "intra-sensory" rather than only "post-sensory" (see Johnston & Dark, 1982; Näätänen, 1985). Processing negativity has been shown to be modality-specific in scalp topography in the visual modality (Harter & Guido, 1980) and in the somatosensory modality (Michie, 1984; see also the data of Desmedt & Robertson, 1977). Some indication in this direction exists for the auditory modality also (Renault, Baribeau-Braun, Dalbokova, & El Massioui, in press). The ERP data suggesting the stimulus-specificity (beyond the modality-specificity of the effect) of the pre-set intra-sensory attentional bias are abundant (for a review, see Näätänen, 1982) whereas analogous rCBF data appear to be provided only by Roland (1981).

A review of the literature indicated that there are attention-related rCBF patterns, especially in the frontal regions of the brain. The underlying neuronal activity, or at least that causing the modality-independent frontal rCBF pattern, might be associated with directing and maintaining selective attention, i.e., presumably, controlling the neuronal change creating the stimulus-specific blood-flow focus in the sensory-specific areas during selective attention (see Roland, 1982). Interestingly, even ERP data point to the involvement of the frontal cortex in selective attention. Most studies have demonstrated that the auditory processing negativity is in fact composed of two components differing in their midline distribution: an early central component and an overlapping, somewhat later and slower, frontal component (Hansen & Hillyard, 1980; Näätänen, Gaillard, & Varey, 1981; see also Näätänen & Michie, 1979). However, two separate components in the processing negativity cannot be observed in all the auditory studies (e.g., Okita, 1979), and the evidence for a frontal component in the visual and somatosensory processing negativity is quite meager (for a review, see Näätänen, 1982).

When ISIs are relatively long, the frontal component of the auditory processing negativity can be large and have a long duration as illustrated in Figure 13.1. This is of course consistent with the idea of the frontal involvement in control and maintenance of selective attention. According to Risberg and Prohovnik's (1983) results, the right frontal lobe might be more important in this attentional control than the left frontal lobe. Consistently, in Okita, Konishi, and Inamori's (1983) study, the frontal processing negativity was larger when recorded over the right frontal cortex (F_4) than over the left frontal cortex (F_3).

We have seen that the ERP and rCBF studies of selective attention supplement one another in a very important way. Their mutual usefulness stems from the fact that whereas the ERP methodology, because of its excellent temporal resolution, is primarily aimed at investigating phasic responses and comparable "spontaneous" changes, the rCBF techniques monitor, and provide detailed distributional information of tonic changes and the background state of the cerebral activity which in part determine the phasic responses (see, e.g., Näätänen, 1975; Tecce, 1970). The rCBF method has a particularly good spatial resolution in comparison to that of the ERPs (recall, for instance, the highly detailed frontal patterns observed by Roland during the various kinds of selective-attention tasks).

ACKNOWLEDGMENTS

Supported by The Academy of Finland. The author is grateful to Miss Taina Kettunen for producing the manuscript.

REFERENCES

Alho, K., Sams, M., Paavilainen, P., & Näätänen, R. (1986). Small pitch separation and the selective-attention effect on the ERP. *Psychophysiology, 23*, 189-197.

Desmedt, J. E., & Robertson, D. (1977). Differential enhancement of early and late components of the cerebral somatosensory evoked potentials during forced-paced cognitive tasks in man. *Journal of Physiology, 271*, 761-782.

Donald, M. W. (1983). Neural selectivity in auditory attention: Sketch of a theory. In A. W. K. Gaillard & W. Ritter (Eds.), *Tutorials in ERP research: Endogenous components* (pp. 37-77). Amsterdam: North-Holland.

Duffy, E. (1962). *Activation and behavior*. New York: Wiley.

Haider, M., Spong, P., & Lindsley, D. B. (1964). Attention, vigilance and cortical evoked-potentials in humans. *Science, 145*, 180-182.

Hansen, J. C., & Hillyard, S. A. (1980). Endogenous brain potentials associated with selective auditory attention. *Electroencephalography and Clinical Neurophysiology, 49*, 277-290.

Hansen, J. C., & Hillyard, S. A. (1983). Selective attention to multidimensional auditory stimuli. *Journal of Experimental Psychology: Human Perception and Performance, 9*, 1-19.

Hari, R., Kaila, K., Katila, T., Tuomisto, T., & Varpula, T. (1982). Interstimulus interval dependence of the auditory vertex response and its magnetic counterpart: Implications for their neural generation. *Electroencephalography and Clinical Neurophysiology, 54*, 561-569.

Harter, M. R., & Guido, W. (1980). Attention to pattern orientation: Negative cortical potentials, reaction time, and the selection process. *Electroencephalography and Clinical Neurophysiology, 49*, 461-475.

Hillyard, S. A. (1981). Selective auditory attention and early event-related potentials: A rejoinder. *Canadian Journal of Psychology, 35*, 85-100.

Hillyard, S. A., Hink, R. F., Schwent, V. L., & Picton, T. W. (1973). Electrical signs of selective attention in the human brain. *Science, 182*, 177-180.

Hillyard, S. A., & Münte, T. F. (1984). Selective attention to color and location: An analysis with event-related brain potentials. *Perception and Psychophysics, 36*, 185-198.

Hillyard, S. A., Münte, T. F., & Neville, H. J. (1985). Visual-spatial attention, orienting and brain physiology. In M. I. Posner & O. S. Marin (Eds.), *Attention and performance XI* (pp.). Hillsdale, NJ: Lawrence Erlbaum associates.

Hyvärinen, J. (1982). *The parietal cortex of monkey and man*. Heidelberg: Springer-Verlag.

Hyvärinen, J., Poranen, A., & Jokinen, Y. (1980). Influence of attentive behavior on neuronal responses to vibration in primary somatosensory cortex of the monkey. *Journal of Neurophysiology, 43*, 870-882.

Ingvar, D. H. (1979). Brain activation patterns revealed by measurements of regional cerebral blood flow. In J. E. Desmedt (Ed.), *Cognitive components in cerebral event-related potentials and selective attention: Progress in clinical neurophysiology* (Vol. 6, pp. 200-215). Basel: Karger.

Ingvar, D. H., & Lassen, N. A. (Eds.). (1975). *Brain work: The coupling of function, metabolism and blood flow in the brain.* Copenhagen: Munksgaard.

Ingvar, D. H., & Lassen, N. A. (Eds.). (1977). *Cerebral function, metabolism and circulation.* (Acta Neurologica Scandinavica, Suppl.)

Ingvar, D. H., & Risberg, J. (1965). Influence of mental activity upon regional cerebral blood flow in man. In D. H. Ingvar & N. A. Lassen (Eds.), *Regional cerebral blood flow.* (Acta Neurologica Scandinavica, Suppl., 14, 183-186).

Ingvar, D. H., & Risberg, J. (1967). Increase of regional cerebral blood flow during mental effort in normals and in patients with focal brain disorders. *Experimental Brain Research, 3,* 195-211.

Ingvar, D. H., & Schwartz, M. S. (1974). Blood flow patterns induced in the dominant hemisphere by speech and reading. *Brain, 96,* 274-288.

Johnston, W. A., & Dark, V. J. (1982). In defense of intraperceptual theories of attention. *Journal of Experimental Psychology: Human Perception and Performance, 8,* 407-421.

Lacey, J. I. (1967). Somatic response patterning and stress: Some revisions of activation theory. In M. H. Appley & R. Trumbull (Eds.), *Psychological stress: Issues in research* (pp. 14-44). New York: Appleton.

Lassen, N. A., & Ingvar, D. H. (1963). Regional cerebral blood flow measurements in man. *Archives of Neurological Psychiatry, 9,* 615-622.

Luria, A. R., & Homskaya, E. D. (1970). Frontal lobes and the regulation of arousal processes. In D. I. Mostofsky (Ed.), *Attention: Contemporary theory and analysis* (pp. 303-330). New York: Appleton.

Malmo, R. B. (1959). Activation: A neuropsychological dimension. *Psychological Review, 66,* 367-386.

McCallum, W. C., & Curry, S. H. (1980). The form and distribution of auditory evoked potentials and CNVs when stimuli and responses are lateralized. In H. H. Kornhuber & L. Deecke (Eds.), *Motivation, motor and sensory processes of the brain: Electrical potentials, behaviour and clinical use. Progress in Brain Research* (Vol. 54, pp. 767-775). Amsterdam: Elsevier.

Michie, P. T. (1984). Selective attention effects on somatosensory event-related potentials. In R. Karrer, J. Cohen, & P. Tueting (Eds.), *Brain and information: Event-related potentials* (pp. 250-255). (Annals of the New York Academy of Sciences, 425.) New York: New York Academy of Sciences.

Näätänen, R. (1967). Selective attention an evoked potentials. *Annales Academiae Scientiarum Fennicae, B151,* 1, 1-226.

Näätänen, R. (1973). The inverted-U relationship between activation and performance - A critical review. In S. Kornblum (Ed.), *Attention and performance IV* (pp. 155-174). New York: Academic Press.

Näätänen, R. (1975). Selective attention and evoked potentials in humans - a critical review. *Biological Psychology, 2*, 237-307.

Näätänen, R. (1982). Processing negativity: An evoked-potential reflection of selective attention. *Psychological Bulletin, 92*, 605-640.

Näätänen, R. (1984). In search of a short-duration memory trace of a stimulus in human brain. In L. Pulkkinen & P. Lyytinen (Eds.), *Human Action and Personality - Essays in honour of Martti Takala: Jyvaskyla studies in education, psychology and social science* (pp. 22-36). Jyvaskyla: University of Jyvaskyla.

Näätänen, R. (1985). Selective attention and stimulus processing: Reflections in event-related potentials, magnetoencephalogram, and regional cerebral blood flow. In M. I. Posner & O. S. Marin (Eds.), *Attention and performance XI* (pp. 355-373). Hillsdale, NJ: Lawrence Erlbaum Associates.

Näätänen, R. (in revision). Theory of auditory selective attention based on event-related potentials in man. *The Behavioral and Brain Sciences.*

Näätänen, R., Gaillard, A. W. K., & Mantysalo, S. (1978). Early selective attention effect on evoked potential reinterpreted. *Acta Psychologica, 42*, 313-329.

Näätänen, R., Gaillard, A. W. K., & Varey, C. A. (1981). Early attention effect on evoked potential as a function of interstimulus interval. *Biological Psychology, 13*, 173-187.

Näätänen, R., & Michie, P. T. (1979). Early selective attention effects on the evoked potential. A critical review and reinterpretation. *Biological Psychology, 8*, 81-136.

Okita, T. (1979). Event-related potentials and selective attention to auditory stimuli varying in pitch and localization. *Biological Psychology, 9*, 271-284.

Okita, T., Konishi, K., & Inamori, R. (1983). Attention-related negative brain potentials for speech words and pure tones. *Biological Psychology, 16*, 29-47.

Parasuraman, R. (1980). Effects of information processing demands on slow negative shift latencies and N100 amplitude in selective and divided attention. *Biological Psychology, 11*, 217-233.

Posner, M. I., & Boies, S. W. (1971). Components of attention. *Psychological Review, 78*, 391-408.

Pribram, K. H., & Luria, A. R. (Eds.). (1973). *Psychophysiology of the frontal lobes*. New York: Academic Press.

Renault, B., Baribeau-Braun, J., Dalbokova, D., & El Massioui, F. (in press). Differential topographical analysis of auditory components in a selective attention task. *Electroencephalography and Clinical Neurophysiology* (Suppl.).

Risberg, J., & Ingvar, D. H. (1973). Patterns of activation in the grey matter of the dominant hemisphere during memorization and reasoning. *Brain, 96*, 737-756.

Risberg, J., Maximilian, V. A., & Prohovnik, I. (1977). Changes of cortical activity patterns during habituation to a reasoning test. *Neuropsychologia, 15*, 793-798.

Risberg, J., & Prohovnik, I. (1983). Cortical processing of visual and tactile stimuli studied by non-invasive rCBF measurements. *Human Neurobiology, 2*, 5-10.

Roland, P. E. (1981). Somatotopical tuning of postcentral gyrus during focal attention in man. A regional cerebral blood flow study. *Journal of Neurophysiology, 46*, 744-754.

Roland, P. E. (1982). Cortical regulation of selective attention in man. A regional cerebral blood flow study. *Journal of Neurophysiology, 48*, 1059-1077.

Roland, P. E., & Larsen, B. (1976). Focal increase of cerebral blood flow during stereognostic testing in man. *Archives of Neurology, 33*, 551-558.

Roland, P. E., Larsen, B., Skinhøj, E., & Lassen, N. A. (1977). Regional cerebral blood flow increase due to treatment of somatosensory and auditive information in man. *Acta Neurologica Scandinavica, 64* (Suppl.), 540-541.

Roland, P. E., & Skinhøj, E. (1981). Focal activation of the cerebral cortex during visual discrimination in man. *Brain Research, 222*, 166-171.

Roland, P. E., Skinhøj, E., & Lassen, N. A. (1981). Focal activation of human cerebral cortex during auditory discrimination. *Journal of Neurophysiology, 45*, 1139-1151.

Spong, P., Haider, M., & Lindsley, D. B. (1965). Selective attentiveness and cortical evoked responses to visual and auditory stimuli. *Science, 148*, 395-397.

Tecce, J. J. (1970). Attention and evoked potentials in man. In D. I. Mostofsky (Ed.), *Attention: Contemporary theory and analysis* (pp. 331-365). New York: Appleton.

Van Voorhis, S., & Hillyard, S. A. (1977). Visual evoked potentials and selective attention to points in space. *Perception and Psychophysics, 22*, 54-62.

14 · RELATIONSHIP BETWEEN UNILATERAL NEGLECT AND SENSORY EXTINCTION

A. S. Schwartz, P. Marchok, & C. Kreinick

Barrow Neurological Institute

Despite the wealth of written material on unilateral neglect and sensory extinction, few systematic studies of the relationship between them have been undertaken. Perhaps this is because some authors have used the terms interchangeably (Denny-Brown, Meyer, & Horenstein, 1952; Watson, Heilman, Cauthen, & King, 1973), while others propose that extinction is a mild form of neglect (Denny-Brown & Banker, 1954; Heilman & Watson, 1977; Kinsbourne, 1974; Mesulam, 1981), or that neglect is a severe case of extinction (Bender, 1952; Friedland & Weinstein, 1977). We earlier suggested that the two phenomena may not be identical although they may share certain mechanisms (Schwartz, Marchok, & Flynn, 1977; Schwartz, Marchok, Kreinick, & Flynn, 1979).

The latter view would appear to have some validity in that neglect may take many behavioral forms in contrast to extinction (Bender, 1952; Brain, 1941; Critchley, 1949, 1953; Denny-Brown & Banker, 1954; Denny-Brown et al., 1952; Friedland & Weinstein, 1977; Holmes, 1919), and in that the operations used to elicit these signs differ markedly (Bender, 1952; Critchley, 1949). Sensory extinction refers to the lack of response to a suprathreshold stimulus only in the presence of a second, simultaneous stimulus; the definition requires that when the first stimulus is presented alone to the so-called affected area, it is perceived adequately. Unilateral sensory neglect refers to the tendency to ignore suprathreshold stimuli on one side of the body or extrapersonal space whether or not other stimuli are simultaneously present. Importantly, the diagnosis of neglect requires the absence of primary sensorimotor deficits severe enough to account for the neurological signs, a criterion often difficult to establish.

In this report we explored several aspects of neglect and extinction which might serve to clarify the relationship between the two phenomena. First we asked whether the well-known multimodal nature of neglect also applies to extinction. For example, several authors (Battersby, Bender, Pollack, & Kahn, 1956; Critchley, 1953; Denny-Brown et al., 1952) have noted the frequent association between visual neglect and somatosensory deficits. In animals, neglect is characteristically multimodal (Kennard, 1939; Sprague, Chambers, & Stellar, 1961; Welch & Stuteville, 1958). As for the human,

 DOI: 10.4324/9781003164647- 18

Bender (1977) has stated that he never saw a patient with unilateral neglect who had a disturbance only in vision; these patients have disturbances also in the somatosensory and auditory modalities. The multimodal character of neglect fits well with current theories which attribute neglect to impairment of an attention mechanism, predominantly in the right hemisphere and whose main function is a general arousal and orientation to the contralateral field (Heilman, 1979; Mesulam, 1981). If extinction is to be considered as a mild form of neglect, we would then expect that a strong association exists between visual and/or auditory (i.e., non-tactile) extinction and somatosensory deficit, and that few patients would extinguish in only one or two sensory modalities. Second, neglect is overwhelmingly a consequence of right- rather than left-hemisphere damage (Friedland & Weinstein, 1977; Heilman, 1979; McFie, Piercy, & Zangwill, 1950; Mesulam, 1981; Oxbury, Campbell, & Oxbury, 1974). We therefore asked whether a similar asymmetric distribution applied to extinction.

METHODS AND PROCEDURES

Patients admitted to the Barrow Neurological Institute over a period of several months were screened for further study. Only right-handed patients with radiographically verified unilateral cerebral hemisphere lesions, who were able to cooperate in the testing procedure, are included here. One hundred and twenty-nine showed either neglect or extinction or both as defined below.

Unilateral sensory neglect was tested in the visual modality by means of the crossing-out-lines test (Albert, 1973), bisection of horizontal lines, and drawing a daisy and a clock. Somatosensory function was examined by 2-point discrimination, light touch sensitivity, graphesthesia, depthesthesia, stereognosis, pin prick, position sense, and vibration, although not all these tests were administered to each patient. Extinction upon bilateral simultaneous stimulation was tested in the visual, auditory, and tactile modalities by visual confrontation, by light auditory taps near the ears, by the Quality Extinction Test (QET), or by touching the patient's hands, respectively. The QET consists of a set of common, distinctively textured materials which are briefly presented to both hands simultaneously, but with a different texture to each hand, while the patient is blindfolded (Schwartz et al., 1977). An extinction trial is scored when the subject fails to report one of the materials. In no case is a subject examined for extinction if he cannot discriminate each stimulus when presented alone, to either side. The presence of neglect, extinction, or sensory deficits was defined as any performance on the various tests which did not fall within the range achieved by a group of non-brain-damaged control subjects (N = 83).

RESULTS

The association between neglect, extinction, and somatosensory disturbances is shown in Table 14.1. Twenty-four patients exhibited visual neglect; of these, about 83% also had somatosensory deficits. This agrees with previous observations (see above) that visual neglect and disturbances in other sensory modalities are often associated. We then examined the relationship between somatosensory deficit, neglect, and non-tactile (i.e., visual, auditory, or combined visual and auditory) extinction. A Chi-square analysis showed that somatosensory deficit was significantly more frequent in patients with visual neglect than in patients with extinction in the non-tactile modalities ($p<.001$).

TABLE 14.1. Distribution of visual neglect, non-tactile extinction, and somato-sensory deficit in 129 consecutive admissions to the Barrow Neurological Institute.

	Visual Neglect	Visual and/or Auditory Extinction Only
Somatosensory Deficit	20	5
No Somatosensory Deficit	4	23

The incidence of extinction in the three sensory modalities tested here is shown in Table 14.2. The majority of patients (58%) extinguished in only one modality, and many (37%) extinguished in only two modalities; only a small fraction (less than 5%) extinguished in all three modalities. These results suggest that extinction is not multimodal in character, but rather that it is usually exclusive to one or two sensory systems. It should be pointed out, however, that the tactile (QET) test is more sensitive than the conventional "finger-touch" test used to examine for tactile extinction (Schwartz et al., 1977). Utilization of comparably sensitive extinction tests in the visual and auditory modalities might elicit a greater incidence of extinction in these modalities as well, and show a higher incidence of multimodal extinction than indicated in Table 14.2. We have recently examined a large group of unilaterally brain-damaged patients for visual and auditory extinction with more sophisticated tachistoscopic and dichotic listening techniques, as well as for tactile extinction with the QET. Our results agree with the finding reported here: approximately 50% of these patients extinguished in only one

TABLE 14.2. Number of patients with tactile (T), visual (V) and/or auditory (A) extinction (N = 105; patients with visual neglect not included).

T + V + A	5
T + V	20
T + A	11
V + A	8
T	41
V	11
A	9

or two modalities while performing normally in the third. The evidence thus indicates that the mechanism of extinction is a selective one, and does not support the notion that it is a manifestation of mild neglect due to an impaired lateralized general arousal mechanism.

The number of patients showing visual neglect and those with extinction exclusively were separated according to the side of hemispheric damage (Table 14.3). A Chi-square analysis revealed that patients with visual neglect significantly differed from patients with extinction in regard to laterality of lesion ($p<.01$); neglect was rare among left brain-damaged (LBD) patients, while extinction was commonly seen in this group. A selective bias may have entered into the above sampling insofar as untestable patients (mainly due to aphasia) were predominantly left brain damaged. If such patients were somehow testable, the number of LBDs with neglect could have been much larger and reduced the hemispheric differences. However, Schwartz et al. (1979) showed that the incidence of tactile extinction among *testable* aphasics was not different from that in non-aphasics. We also observed in the present group that testable aphasics rarely exhibited neglect. It is therefore likely that the distribution of neglect and extinction among the non-testable aphasics would have maintained the relative proportions shown in Table 14.3.

TABLE 14.3. Number of left- and right-brain-damaged patients with visual neglect or extinction.

	LBD	RBD
Visual Neglect	3	21
Extinction	45	60

DISCUSSION

Our results do not support the view that sensory extinction and neglect represent mild and severe impairment, respectively, of cerebral functional mechanisms such as attention or arousal. A unilateral deficit in attention or arousal would theoretically involve all sensory modalities such as is usually seen in patients with neglect, and would not be expected to be selective or specific for certain classes of stimuli. By contrast, patients with extinction in one sensory modality are often quite capable of performing normally in another modality. Furthermore, the present data, in conjunction with results previously reported (Schwartz et al., 1979), indicate that damage to the left hemisphere is equally effective in producing extinction as right hemisphere damage, in contrast to their asymmetric propensity seen in neglect.

In addition to the findings presented here, other observations may be cited to support the view that neglect and extinction are more than quantitatively different expressions of dysfunction in a unitary mechanism. First, the operations used to elicit each phenomenon differ (see introduction); in extinction, a second sensory system is activated with all its possible physiological ramifications, whereas in neglect this procedure is irrelevant. Second, extinction but not neglect may be produced by lesions in the spinal cord (Bender, 1952; Bors, 1979; Eidelberg & Schwartz, 1971) or corpus callosum (Milner, Taylor, & Sperry, 1968; Sparks & Geschwind, 1968). It is difficult to see how any of the current theories of neglect (Heilman, 1979; Mesulam, 1981) can accommodate these observations.

We propose that the etiology of extinction is qualitatively different from that of neglect. More specifically, we propose that extinction involves competition for recruitment of a response mechanism by two information channels of unequal potency. Although each channel is capable of eliciting a response when activated alone, the "weaker" channel becomes ineffective in the presence of simultaneous activity in the "stronger" channel. Many implications of this proposal, such as the operation of occlusive processes and the characteristics of a "weak" versus a "strong" channel, have yet to be explored. It would seem important, however, that in order to understand the mechanism of arousal and attention and their disorders in the form of neglect and extinction, it is first necessary to differentiate between the two phenomena.

REFERENCES

Albert, M. L. (1973). A simple test of visual neglect. *Neurology (Minneap.)*, *23*, 658-664.

Battersby, W. S., Bender, M. B., Pollack, M., & Kahn, R. S. (1956). Unilateral "spatial agnosia" (inattention) in patients with cerebral lesions. *Brain, 79*, 68-93.

Bender, M. B. (1952). *Disorders in perception*. Springfield, IL: C. C. Thomas.

Bender, M. B. (1977). Extinction and other patterns of sensory interaction. In E. A. Weinstein & R. P. Friedland (Eds.), *Hemi-inattention and hemispheric specialization* (pp. 107-110). New York: Raven Press.

Bors, E. (1979). Extinction and synesthesia with spinal cord injuries. *Paraplegia, 17*, 21-31.

Brain, W. R. (1941). Visual disorientation with special reference to lesions of the right cerebral hemisphere. *Brain, 64*, 244-272.

Critchley, M. (1949). The phenomenon of tactile inattention with special reference to parietal lesions. *Brain, 72*, 538-561.

Critchley, M. (1953). *The parietal lobes*. New York: Hafner.

Denny-Brown, D., & Banker, B. Q. (1954). Amorphosynthesis from left parietal lesions. *Archives of Neurology, 71*, 302-313.

Denny-Brown, D., Meyer, J. S., & Horenstein, S. (1952). The significance of perceptual rivalry. *Brain, 75*, 433-471.

Eidelberg, E., & Schwartz, A. S. (1971). Experimental analysis of the extinction phenomenon in monkeys. *Brain, 94*, 91-108.

Friedland, R. P., & Weinstein, E. A. (1977). Hemi-inattention and hemispheric specialization: Introduction and historical review. In E. A. Weinstein & R. P. Friedland (Eds.), *Hemi-inattention and hemispheric specialization* (pp. 1-31). New York: Raven Press.

Heilman, K. M. (1979). Neglect and related disorders. In K. M. Heilman & E. Valenstein (Eds.), *Clinical neuropsychology* (pp. 268-307). New York: Oxford University Press.

Heilman, K. M., & Watson, R. T. (1977). The neglect syndrome--a unilateral defect of the orienting response. In S. Harnad (Ed.), *Lateralization in the nervous system* (pp. 285-302). New York: Academic Press.

Holmes, G. (1919). Disturbances of visual space perception. *British Medical Journal, 2*, 230-233.

Kennard, M. A. (1939). Alteration in response to visual stimuli following lesions in the frontal lobes in monkeys. *AMA Archives of Neurology and Psychiatry, 41*, 1153-1165.

Kinsbourne, M. (1974). Lateral interactions in the brain. In M. Kinsbourne & W. L. Smith (Eds.), *Hemispheric disconnection and cerebral function* (pp. 239-259). Springfield, IL: C. C. Thomas.

McFie, J., Piercy, M. F., & Zangwill, O. L. (1950). Visual-spatial agnosia associated with lesions of the right cerebral hemisphere. *Brain, 73*, 167-190.

Mesulam, M-M. (1981). A cortical network for directed attention and unilateral neglect. *Annals of Neurology, 10*, 309-325.

Milner, B., Taylor, L., & Sperry, R. W. (1968). Lateralized suppression of dichotically presented digits after commissural section in man. *Science, 161*, 184-186.

Oxbury, J. M., Campbell, D. C., & Oxbury, S. M. (1974). Unilateral spatial neglect and impairment of spatial analysis and visual perception. *Brain, 97*, 551-564.

Schwartz, A. S., Marchok, P. L., & Flynn, R. E. (1977). A sensitive test for tactile extinction: Results in patients with parietal and frontal lobe disease. *Journal of Neurology, Neurosurgery, and Psychiatry, 40*, 228-233.

Schwartz, A. S., Marchok, P., Kreinick, C., & Flynn, R. E. (1979). The asymmetric lateralization of tactile extinction in patients with unilateral cerebral dysfunction. *Brain, 102*, 669-684.

Sparks, R., & Geschwind, N. (1968). Dichotic listening in man after section of neocortical commissures. *Cortex, 4*, 3-16.

Sprague, J. M., Chambers, W. W., & Stellar, E. (1961). Attentive, affective and adaptive behavior in the cat. *Science, 133*, 165-173.

Watson, R. T., Heilman, K. M., Cauthen, J. C., & King, F. A. (1973). Neglect after cingulectomy. *Neurology (Minneap.), 23*, 1003-1007.

Welch, K., & Stuteville, P. (1958). Experimental production of unilateral neglect in monkeys. *Brain, 81*, 341-347.

15 · THE NONSPECIFIC INTRALAMINAR SYSTEM OF THE THALAMUS AND VISUAL ATTENTION

M. Schlag-Rey and J. Schlag

University of California at Los Angeles

A fundamental feature of visual attention is the ability to direct one's gaze to a point in space and to fixate that point, no matter what conditions of visual stimulation might prevail.

Undoubtedly the forebrain must regulate the powerful visuomotor reflexes organized at cerebellar and lower brainstem levels, but how is this higher control exerted? In contrast to the enlightening results obtained by microelectrode explorations in the midbrain and pons, the search in the forebrain for neurons firing specifically before self-initiated eye movements (EM) has yielded puzzling results. In several cortical or thalamic structures, correlations were found between neuronal activity and EM, but either the changes did not lead the movements or they failed to occur in total darkness. Such EM-related units could not likely send command signals to the ocular motoneurons. In particular -- and in contradiction with expectations based on stimulation studies -- extensive microelectrode investigations have shown that frontal eye field (area 8) neurons do not fire before self-initiated saccades (Bizzi, 1968; Mohler, Goldberg, & Wurtz, 1973). Inferior parietal neurons (area 7) were also claimed to assume command functions of gaze (Mountcastle, 1976), but the parietal neurons so far described do not fire before any spontaneous saccade in darkness; even in light, whether their firing is related to stimuli or saccades has not been unequivocally determined (see Robinson, Goldberg, & Stanton, 1978).

Curiously enough, neurons firing before and during EM, in darkness as well as in light, were found in the thalamic internal medullary lamina (IML) of the cat (Schlag, Lehtinen, & Schlag-Rey, 1974). This region is part of a diffuse, nonspecific thalamo-cortical system; however, it has privileged connections with the inferior parietal lobule or its suprasylvian homologue in the cat (see, for instance, Mesulam, Van Hoesen, Pandya, & Geschwind, 1977; Robertson & Rinvik, 1973) and with the frontal eye fields (Rinvik, 1968). In addition, the IML receives inputs from the deep layers of the superior colliculus (Niimi, Miki, & Kawamura, 1970), from the ventral lateral geniculate nucleus (Kawamura, Fukushima, Hattori, & Tashiro, 1978), from parts of the pretectum reached by retinal fibers (Benevento, Rezak, & Santos-Anderson, 1977), and from the pontine reticular formation involved in horizontal EM (Büttner-Ennever & Henn, 1976). The IML has been shown

DOI: 10.4324/9781003164647- 19

also to project caudally, toward the brainstem (Scheibel & Scheibel, 1967). Thus, anatomical connections place IML neurons at the crossroads of many structures involved in the guidance of gaze. The firing characteristics of IML units were correlated with parameters of EM (Schlag et al., 1974) and with parameters of visual stimuli (Schlag-Rey & Schlag, 1977). The present report, based on approximately 300 units, focuses on properties related to visual attention.

Methods of recording were described in earlier reports. The cats, implanted with EOG d.c. electrodes, were trained to look alternately to unmarked sites on a tangent screen. When the anticipatory fixation was on the correct site, it was reinforced by a light signal invariably followed by milk delivery. This paradigm, originally developed in D. B. Lindsley's laboratory (Schlag-Rey & Lindsley, 1970), served to elicit spontaneous saccades or anticipatory EM. Targeting saccades were elicited by another stimulus, a dim pattern of light never paired with reinforcement. It was presented in schedules designed to maximize the natural curiosity of the cats. Its position on the screen was continuously monitored as well as the position of the eyes.

Typical results obtained with one IML unit in the conditioning situation are shown in Figure 15.1. Each dot represents a spike, each row, a correct, reinforced trial. In Figure 15.1A, the trials are synchronized on the onset of the reinforcing light. In Figure 15.1B, the same trials are synchronized on the anticipatory saccades. Four properties of IML cells are illustrated in this figure.

1. *IML cells were activated before spontaneous EM in a specific direction.* They were unresponsive or inhibited before EM in the opposite direction. Presaccadic changes of activity, in darkness as in light, were so conspicuous that no averages were necessary. The directional selectivity present in more than 80% of the cells was generally coarse, often extending to a 180° span. However, for many cells, the number of spikes during a saccadic burst appeared quantitatively related to the amount of deviation of the eyes in the preferred direction. As Henn and Cohen (1976) have argued with regard to other EM neurons, if saccade cells can specify the amount of extra-ocular muscle pull in their preferred direction required to reach any position, collectively these cells can program a saccade with more accuracy than can any single cell.

Figure 15.1 makes it clear that presaccadic activation, observed with rightward saccades in this case, was not dependent on the cat expecting a reinforcement since leftward saccades, equally reinforced, were not preceded by neuronal activation. We have compared presaccadic firing under various rules of reinforcement: single alternation, reversal learning and no reinforcement at all. In all cases, the changes of firing were always predicated on the direction of the gaze (if the cell was selective), not on the pay-off of the EMs. The spatial competence of each cell differed; thus, by

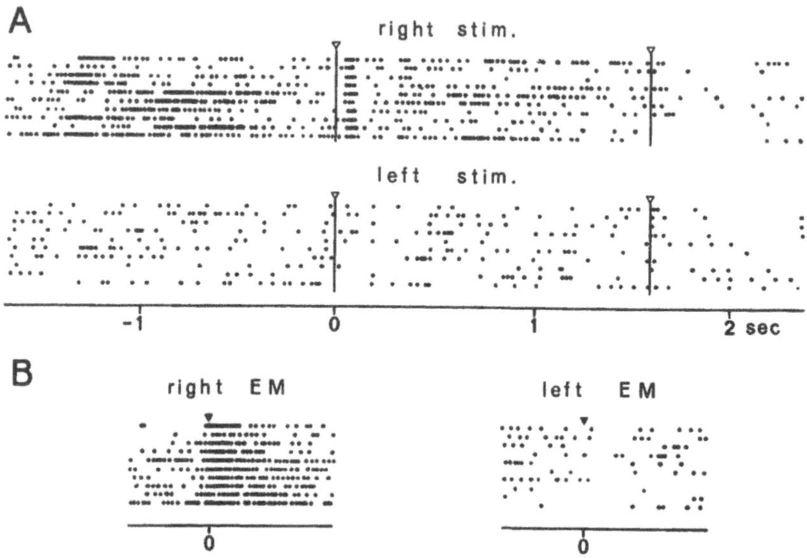

FIGURE 15.1. Neuron in the left thalamic internal medullary lamina discharging while the cat was making anticipatory saccades, aimed at constant locations on the screen, where the appearance of a spot of light announced milk delivery. Each dot represents a spike, each row, a correct trial. In A, the neuronal activities are synchronized on the onset of the light signal; the cell increased its firing with rightward saccades and right stimuli (A, upper raster) but not with leftward saccades nor left stimuli (A, lower raster). In B, for the same trials, the neuronal activities are synchronized on the beginning of the eye movements (EM).

interacting, these cells may contribute to the process culminating in the decision of looking, here or there. The hypothesis assigning a role in the initiation of EM is further substantiated by the length of presaccadic activation (up to 400 msec) observed when the cat did not immediately orient to a visual stimulus.

2. *The same cells were activated by EM and by visual stimuli.* This occurred whether EM followed or preceded the onset of a light. Most EM cells were capable of responding to visual stimuli in the absence of EM, as well as altering their firing before EM in the absence of stimuli. Only a small minority of IML cells complied with the traditional partition in sensory versus motor cells. Response properties time-locked to the visual stimuli were best investigated with the test stimulus (not paired with reinforcement) because nonreinforced stimuli produced more variable reaction times. Not infrequently, targeting EM occurred several seconds after the onset of the stimulus. Under these circumstances, stimulus-related and EM-related activities of a given cell could be separated without ambiguity.

Receptive fields were plotted by computing the position of effective stimuli with respect to point of fixation of the gaze. Most receptive fields were large, often covering most of a half field. They were found generally close to the center of the visual field with a restricted area of responsiveness. As stimuli fell closer to this restricted area, the latency of transient spike bursts decreased, the number of spikes in the bursts increased and the bursts were progressively transformed in sustained responses. These gradients of responsiveness suggest that IML cells may be coding stimulus position with much more precision than could be inferred from overall receptive field size.

3. *A spatial correspondence always existed between visual and EM-related patterns of firing.* Receptive fields were always on the side toward which EM were preceded by increased firing. This rule applied to all cells, including the few cases where increased firing was associated with stimuli and EM ipsilateral to the hemifield recorded. Consequently, the change of activity time-locked to the stimulus may already contain a program for movement, although the movement itself may be postponed. This assumption is supported by the relative insensitivity of IML cells to visual parameters such as shape or brightness. If the assumption is correct, the double activation shown by IML cells when a stimulus elicits a targeting EM implies that the cells are involved in programming the EM as soon as the stimulus appears and as late as the targeting saccade occurs. If the same cells are also responsible for setting up the temporal priorities according to which visual targets are selected for fixation, these cells have all the properties expected from a system controlling target acquisition reflexes.

4. *Some IML cells recognized left from right and up from down.* The identification of neurons discriminating between absolute rather than retinal locations of stimuli would provide a physiological basis for the ability to locate objects in space. The existence of such neurons is also postulated by models of the oculomotor system (e.g., Robinson, 1975). The first indication that some of these neurons could be found in IML was obtained during conditioned anticipatory fixations. When these fixations were perfectly on target, the conditioned light signals, whether right or left, excited the same retinal receptors. In this case, the difference in evoked responses, shown in Figure 15.1A for right and left stimuli, must have resulted from their position on the screen. But anticipatory fixations often undershot reinforcing lights, thus bringing about discrepancies in retinal activation sufficient to account for the observed activity. To ascertain the sensitivity of a cell to the absolute position of a stimulus, it is necessary to vary the absolute coordinates of the stimulus while their retinotopic coordinates remain constant. For this purpose, our dim test stimulus proved most useful because we could present it at various locations on the screen while the cat was fixating different locations. Thus we found IML cells discharging in a sustained manner when the cat fixated or tracked a stimulus in one hemifield only. Moreover the rate

of firing of the cell could be related to the degree of eccentricity of the stimulus (Schlag & Schlag-Rey, 1977). For some of these cells, fixation was necessary, i.e., the stimulus had to fall on the central field of vision for the activity to occur. In this case, of course, we could not determine if their activity depended on the absolute coordinates of the stimulus or on the position of the gaze. Other cells did not require fixation in order to show responsiveness to absolute position. An example is given in Figure 15.2 where, for the same trials, the rate of firing of one IML unit is plotted as a function of gaze direction (A), stimulus retinotopic coordinates (B), and stimulus absolute position on the screen (C).

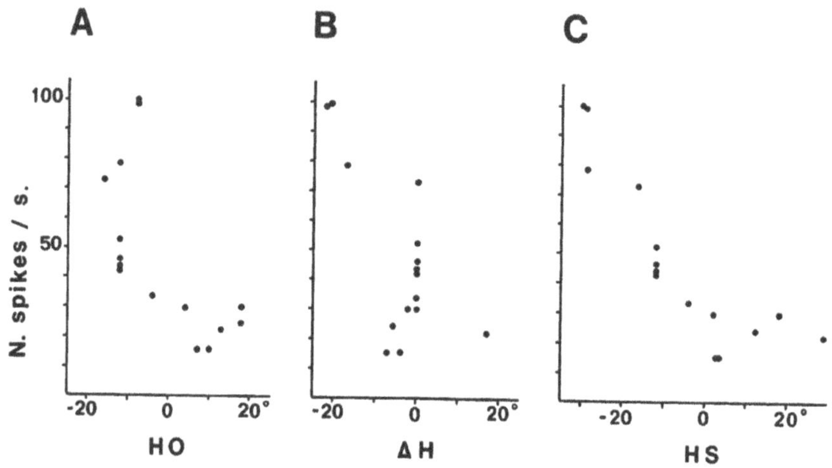

FIGURE 15.2. Neuron in the right thalamic internal medullary lamina discharging while the cat observed, but not always fixated, a small stationary pattern of light. The rate of firing (N. spikes/s.), computed for epochs of 600 msec following visually triggered saccades, is shown for the same trials. A: as a function of horizontal deviations of the gaze (HO); B: horizontal retinotopic coordinates of the stimulus, i.e., its distance from the point of fixation of the gaze (ΔH); C: horizontal absolute position of the stimulus on the screen (HS).

How could cells extract information on the absolute position of a stimulus from retinotopically coded visual signals? Obviously, the same cells must receive both information of retinal origin and information on the position of the eyes.

Our analysis of absolute position-related activities in the IML is far from complete; but their occurrence, in a structure involved in the initiation of EM, makes this structure all the more worthwhile studying in relation to visual attention.

ACKNOWLEDGMENT

This work was supported by Grant NS-04955 from the U. S. Public Health Services.

REFERENCES

Benevento, L. A., Rezak, M., & Santos-Anderson, R. (1977). An autoradiographic study of the projections of the pretectum in the rhesus monkey (Macaca mulatta): Evidence for sensorimotor links to the thalamus and oculomotor nuclei. *Brain Research, 127*, 197-218.

Bizzi, E. (1968). Discharge of frontal eye field neurons during saccadic and following eye movements in unanesthetized monkeys. *Experimental Brain Research*, 69-80.

Büttner-Ennever, J. A., & Henn, V. (1976). An auto-radiographic study of the pathways from the pontine reticular formation involved in horizontal eye movements. *Brain Research, 108*, 155-164.

Henn, V., & Cohen, B. (1976). Coding of information about rapid eye movements in the pontine reticular formation of alert monkeys. *Brain Research, 108*, 307-325.

Kawamura, S., Fukushima, N., Hattori, S., & Tashiro, T. (1978). A ventral lateral geniculate nucleus projection to the dorsal thalamus and the midbrain in the cat. *Experimental Brain Research, 31*, 95-106.

Mesulam, M-M., Van Hoesen, G. W., Pandya, D. N., & Geschwind, N. (1977). Limbic and sensory connections of the inferior parietal lobule (area PG) in the rhesus monkey: A study with a new method for horseradish peroxidase histochemistry. *Brain Research, 136*, 393-414.

Mohler, C. W., Goldberg, M. E., & Wurtz, R. H. (1973). Visual receptive fields of frontal eye field neurons. *Brain Research, 61*, 385-389.

Mountcastle, V. B. (1976). The world around us: neural command functions for selective attention. *Neurosciences Research Progress Bulletin, 14*, 1-47.

Niimi, K., Miki, M., & Kawamura, S. (1970). Ascending projections of the superior colliculus in the cat. *Folia Anatomica Japonica, 47*, 269-287.

Rinvik, E. (1968). The corticothalamic projection from the gyrus poreus and the medial wall of the rostral hemisphere in the cat: An experimental study with silver impregnation methods. *Experimental Brain Research, 5*, 129-152.

Robertson, R. T., & Rinvik, E. (1973). The corticothalamic projections from parietal regions of the cerebral cortex: Experimental degeneration studies in the cat. *Brain Research, 51*, 61-79.

Robinson, D. A. (1975). Oculomotor control signals. In G. Lennerstrand & P. Bach-y-Rita (Eds.), *Basic mechanisms of ocular mobility and their clinical implications* (pp. 337-378). New York: Pergamon Press.

Robinson, D. L., Goldberg, M. E., & Stanton, G. B. (1978). Parietal association cortex in the primate: Sensory mechanisms and behavioral modulations. *Journal of Neurophysiology, 41*, 910-930.

Scheibel, M. E., & Scheibel, A. B. (1967). Structural organization of nonspecific thalamic nuclei and their projection toward cortex. *Brain Research, 6*, 60-94.

Schlag, J., Lehtinen, I., & Schlag-Rey, M. (1974). Neuronal activity before and during eye movements in thalamic internal medullary lamina of the cat. *Journal of Neurophysiology, 37*, 982-995.

Schlag, J., & Schlag-Rey, M. (1977). Visuomotor properties of cells in cat thalamic internal medullary lamina. In R. Baker & A. Berthoz (Eds.), *Control of gaze by brainstem neurons, developments in Neuroscience* (Vol. 1, pp. 453-462). Amsterdam: Elsevier North Holland.

Schlag-Rey, M., & Lindsley, D. B. (1970). Effects of prefrontal lesions on trained anticipatory attending in cats. *Physiological Behavior, 5*, 1033-1041.

Schlag-Rey, M., & Schlag, J. (1977). Visual and presaccadic neuronal activity in thalamic internal medullary lamina of cat: A study of targeting. *Journal of Neurophysiology, 40*, 156-173.

SECTION V. LATENCIES AND MOTOR PERFORMANCE

A distinguishing feature of Professor Lindsley's research with both animals and humans has been the simultaneous measurement of both electrophysiological and behavioral responses. Such research is very demanding and difficult to do, requiring much greater effort than investigating only electrophysiological responses or only behavioral responses. As an example, several of his published studies have measured both electrophysiological latencies and their corresponding motor latencies. Such comprehensive, detailed reporting of both electrophysiology and behavior inevitably gives a more complete picture of what is going on. This combined experimental approach involving the measurement of multiple response systems constitutes one of Professor Lindsley's important contributions.

The first paper of this section by Donchin et al., nicely illustrates in humans this combined electrophysiological-behavioral approach. In it, Donchin and his colleagues employ an oddball stimulus paradigm involving the common names of individuals in the American culture. Twenty percent of the names were male while the other 80% were female. Results showed several things: stimulus probability manipulations affected subjects' response strategies; error trials were generally associated with faster RTs but later P300s than correct trials; stimulus probability affected RT latency but not P300 latency. Overall, the data support the suggestion that the magnitude of the P300 can serve as a measure of the degree of revision in the systems' biases. By studying P300 latency and amplitude, it becomes possible to show how the human information processing system deals with error trials.

The next three chapters, by Lansing et al., Kietzman et al., and Galbraith and Gliddon, are all concerned primarily with motor performance and response latencies in a variety of different ways. Thus, Lansing, Meyerink, and Thomas focus on the topic of human respiratory reaction times, about which little has previously been published. Specific questions considered are the nature of the respiratory reaction times (RTs) to both external and internal stimuli and the conditions that may affect them. The results of this study show that the length and variability of respiratory RTs are similar to those typically found for simple limb movements. Also discussed is the determination of the shortest RTs possible, i.e., the physiological limits of the circuits mediating simple learned responses.

Kietzman and co-workers' RT report investigates stimulus energy integration, also known as temporal integration or temporal summation. The question considered is whether temporal integration, which has already been

 DOI: 10.4324/9781003164647- 20

demonstrated for both electrophysiological and psychological responses in several modalities, is also demonstrated for response latency measures such as visual RT. It concludes that complete temporal integration is displayed for simple RTs, but the period of integration is quite brief (between 10 and 22 msec), which is briefer than complete integration for detecting the presence of a visual stimulus (50-100 msec). Other studies show longer periods of complete integration for detection and discrimination than for RT. Temporal integration is widely generalizable as is demonstrated by the fact that it is obtained in several modalities for both humans and animals using different response measures under a wide variety of stimulus conditions.

The Galbraith and Gliddon report involves a dual sensorimotor task, namely the tapping of the index finger with and without the simultaneous speech. In addition, EEG data were collected with eyes opened and eyes closed using both left and right hemisphere electrode placements. The subjects were 36 institutionalized mentally retarded individuals. The purpose of these procedures was to investigate patterns of time-shared motor performance and EEG asymmetry. The primary results indicate that speaking with tapping equally disrupted both left- and right-hand tapping performances. It is concluded that cerebral dominance for speech in a sample of mentally retarded subjects is poorly established, and that in these subjects baseline hemisphere EEG patterns correlate significantly with interference during time-shared motor performance.

16 · AFTER A RASH ACTION: LATENCY AND AMPLITUDE OF THE P300 FOLLOWING FAST GUESSES

E. Donchin, G. Gratton, D. Dupree and M. Coles

University of Illinois at Urbana-Champaign

A persistent theme in Lindsley's writings has been the focus on the temporal characteristics of psychophysiological signals. His second published paper (Herren & Lindsley, 1931) was concerned with the latencies of tendon reflexes. This interest in latencies continually appears in subsequent studies (e.g., Chalupa, Rohrbaugh, Gould, & Lindsley, 1974; Donchin & Lindsley, 1965; Lindsley, 1944; Lindsley, 1954; Lindsley, 1982; Lindsley & Emmons, 1958; Lindsley, Fehmi, & Adkins, 1967; Lindsley & Rubenstein, 1937; Lindsley, Seales, & Wilson, 1973; Schwartz & Lindsley, 1964). This chronometric approach, that has played a critical role in many areas of psychology (Posner, 1978), makes the seemingly paradoxical assumption that even if we do not know what happened it is useful to know when it happened. The relative timing of events, the rhythm with which they occur and the factors that increase, or decrease, the speed with which the events are triggered, can be useful in the analysis of a system even if we are not quite ready to provide a full account of the processes that underlie the events. It is clear, for example, that when Shakespeare, in the 16th century, has Leontes in the Winter's Tale (Act I sc. 2) say: "I have a tremor cordis on me: my heart dances; But not for joy; not joy...," he is expressing a recognition of a common relationship between an accelerated heart rate and "joy." It is also evident that Shakespeare is aware that even though the dancing heart is commonly an indicator of "joy" this relationship can break down. The accuracy and utility of this "psychophysiological" implication of Shakespeare's does not depend on his knowing, as he obviously did not, the physiology of the heart and of the control of the heart's rhythms. The validity of such psychophysiological observations depends on the careful observation of the temporal characteristics of the physiological events ("my heart dances"), on the proper definition of the psychological concepts ("not joy"), and on the theory within whose framework these relationships play a role.

In his work on excitability cycles, in studies of backward masking and in the examination of the various rhythms of the EEG, with particular emphasis on the alpha rhythm, Lindsley has shown how it is possible to use the timing of psychophysiological signals in the study of the mind even if the nature of the processes observed by the psychophysiologist is not fully, or even partially,

 DOI: 10.4324/9781003164647- 21

understood. The point is simple. A well-defined psychophysiological response is an event. Once we have identified the occurrence of an event, we are in possession of information about its time of occurrence, even if we do not know its causes or its nature. These temporal data can serve as dependent variables in our studies. If our theories generate differential predictions about the variation of these temporal variables as a function of properly selected independent variables, these data can play a critical role in understanding the mind.

This point of view is illustrated by the widespread use of the latency of components of event-related brain potentials (ERPs). Thus, for example, Hillyard and his associates demonstrated that a differentiation between the ERPs elicited by attended and ignored stimuli appears as early as the first 100 msec following the eliciting stimuli (Hillyard, 1984). The observation of Hillyard's that makes a difference for a theory of attention is the observation about the timing of the ERP component, and this theoretical point is largely unaffected by debates about the precise nature of the ERP components observed (Näätänen, 1982). Another ERP component whose latency has proven a useful tool in the analysis of cognitive function is P300. In this chapter we illustrate the richness of information that P300 latency yields regarding processes that are essentially opaque to the more traditional tools of cognitive psychology. We will, for the purposes of this illustration, provide a partial description of a study that will be reported in full elsewhere (Gratton, Dupree, Coles, & Donchin, 1986). The analysis of P300 timing will be made, in this chapter, from a perspective of a theory of P300 that makes specific predictions regarding the consequences of changes in an internal process, changes that are manifested at the scalp by changes in the amplitudes of P300 (Donchin, 1981). As the predictions are confirmed by the data, the study lends support to our interpretation of the P300.

The study we discuss is one in a series employing the "oddball" paradigm in which the stimuli are names of individuals commonly used in the American culture. In all cases the series are constructed so that 20% of the names were names of males (e.g., David, Henry, Thomas). The other names used in the series are commonly associated with females (e.g., Nancy, Helen, Susan). On some occasions, the subject is required to count the number of names that fell in one or another category (*count* condition). On other occasions the subject indicates the occurrence of one of the categories by pressing one of two buttons (*reaction time*, or RT, condition).

The initial study in this series was reported by Kutas, McCarthy, and Donchin (1977). Their subjects were presented with three different oddball series. A "variable names" series was constructed from names of males and females as described in the previous paragraph. A "fixed names" series included just the names "David" and "Nancy." The third series was a sequence of words, 20% of which were synonyms of "prod." In the latter case, the subject's task was to press one button in response to such synonyms and to press another button in response to all other words. The rare events in each series elicited a large P300. This was true regardless of the specific task assigned to the subject.

The results showed that P300 latency varied across the three conditions. This was particularly evident when the subjects were instructed to be accurate. The shortest latency was observed when the subject discriminated between the two names, David and Nancy. A longer latency was seen when the names varied from trial to trial. The longest latency was associated with the need to decide whether each of a rather disparate list of words is a synonym of "prod." These, and a considerable amount of additional data, lead us to suggest that the latency of the P300 depends on the time required for the evaluation of the stimulus and is independent of response selection. Subsequent work (McCarthy & Donchin, 1981) provided strong support to the assertion that the latency of P300 is largely independent of the duration of processes that are involved in the execution of the response. The interesting conclusion from the data reported by Kutas et al. (1977) has been that the latency of P300 is proportional to the time it takes to categorize the stimuli. If this is the case, the P300 latency may be used as a tool in mental chronometry to measure timing uncontaminated by "motor" processes (Coles, Gratton, Bashore, Eriksen, & Donchin, 1985; Donchin, 1981; Magliero, Bashore, Coles, & Donchin, 1984; McCarthy & Donchin, 1981). For studies in which P300 latency is indeed utilized in this fashion see Duncan-Johnson and Donchin (1982), Goodin, Squires, and Starr (1978), and Pfefferbaum, Ford, Roth, and Kopell (1980).

Kutas et al. (1977) examined the relationship between the latency of P300 and the RT associated with each of the trials in an oddball study using names, sorted according to gender (see also McCarthy, Kutas, & Donchin, 1979). Their analysis capitalized on a filtering technique that allowed the measurement of the latency of P300 on individual trials (Woody, 1967). The principal finding was that the correlation between P300 latency and RT depends on the strategy adopted by the subjects. When the subjects were instructed to be accurate the correlation between P300 latency and RT was larger (.61) than it was when they were instructed to be fast (.48). These data support the suggestion (Donchin, 1979, 1981) that the P300 and the motor response may each be the product of a series of processing activities and that these streams of processing can, in principle, be quite independent of each other.

Since the invocation of P300 is dependent on the evaluation of information conveyed by the stimulus, the latency of P300 must be at least as long as the duration of these evaluative processes. The overt responses, on the other hand, may well be released "prematurely" on the basis of limited information. The correlation between RT and the latency of the P300 will therefore depend on the degree to which the overt responses that define the RT are made contingent on the evaluation of the stimulus. The more inclined the subject is to respond prematurely, the poorer the correlation between the latency of P300 and RT.

One striking aspect of the data acquired by Kutas et al. (1977) was observed in the trials in which subjects made errors. These were mostly trials on which the subject responded to a rare event as if it were a frequent event. That is, even though a male name appeared on the screen, the subject pressed the button associated with female names. There were but a few such trials in the study reported by Kutas et al. (1977). However, in virtually all these trials, the pattern was the same: the RTs were (relatively) short and the P300 latency was (relatively) long. It was as if on these trials the subjects first acted and then thought! A partial report on these data can be seen in McCarthy (1984). In 10 of the 11 subjects the pattern obtained was identical. The incorrect responses were associated with very short RTs and relatively long P300 latencies. One of several possible interpretations of these data is that, when the information processing system detects an error, the invocation of the P300 is delayed. According to this suggestion, the delay is required to allow further processing of the trial's data. While this interpretation is consistent with the data, it is not the only possible account for the increased latency of the P300 on error trials. Several other possible mechanisms need be considered. Another interpretation of the data is based on the fact that on all the error trials the subject responded rather fast to the stimulus. In other words, these are clearly trials on which a variety of factors are injected into the stream of processing. How do we know that it is the recognition of the error, rather than the fact that a very fast response was emitted on the trial, that accounts for the delay? A different, but related possibility is that it is not that P300 is delayed on error trials, but rather that errors may be more likely on trials on which P300 latency is long. Finally, it can be suggested (see McCarthy, 1984) that the positive peak that is observed on the error trials is *not* a "P300," at least not that elicited by the names, but is rather a different component, perhaps even a different "P300," which may be elicited by the recognition of the error.

The last interpretation raises the issues of how we define "P300." There is no doubt that one of the major difficulties presented by ERP data is associated with the definition and the proper identification of components of the ERP. As we noted above, if our strategy depends on the measurement of the timing of events we must be sure that the timing we measure on different occasions is indeed the timing of the same event. The approach calls for considerable care in the definition of components if the features of the waveform we measure can be affected, as they no doubt are, by different components (Donchin & Heffley, 1978; Fabiani, Gratton, Karis, & Donchin, in press). Consider, for example, each of the positive going peaks observed by Kutas et al. (1977). Each of these positive peaks has been labeled "P300" even though the peaks differ in latency by as much as 100 msec. What leads us to believe that these three peaks are indeed instances of a component whose latency is shifted by the duration of the processing preceding its invocation? How do we know that the peaks with the longer latencies are not entirely new components that are elicited by the presentation of a word, or by the search for a synonym? The issue is generally resolved on the basis of the similarity of waveshapes, of the scalp distribution of the potentials and of the manner in which they respond to experimental manipulations (Donchin, Ritter, & McCallum, 1978; Sutton & Ruchkin, 1984).

To resolve some of the doubts that remained regarding the ERPs elicited on error trials we replicated, and extended, the study reported by Kutas et al. (1977). While in structure the study reported here followed closely the "variable name" phase of the Kutas et al. (1977) study, we expanded the design in a number of ways. To assure that the number of error trials was sufficient to allow the needed comparisons, the sessions were greatly extended and we recorded 800 trials in each of the experimental conditions. As before, each subject performed the task under an accuracy and under a speed regime. Furthermore, in order to determine the extent to which the observations depend on the imbalance between the probabilities with which the two categories of names appear in the series, the Speed and the Accuracy conditions were run twice. In one case Male names were rare (p=.20) while in the other case Male and Female names appeared with equal probability. The equal probability condition (which was not present in the Kutas et al. study) allowed us also to assess the parameters (RT, accuracy, P300 latency, etc.) of fast guesses, where there was no particular advantage for either response.

METHOD

Seven right-handed male students at the University of Illinois were paid $3.50 an hour for their participation in the study. The subject was positioned in front of a PLATO terminal with the fingers of each hand resting on a zero displacement dynamometer. Male and female names were presented on the screen for 200 msec every 2000 msec, and the subject was required to squeeze one of the dynamometers following the presentation of a name.

Subjects were shown the names in blocks of 100 trials. Blocks were composed of either 80 female and 20 male names or 50 of each. In different trial blocks, the subjects were instructed to respond either as quickly as possible (Speed instructions), or as quickly as possible but without making errors (Accuracy instructions). Eight hundred trials were run for each of the probability x instruction conditions, with half the trials run during one session and the remaining half run during a second session.

The EEG was recorded using Burden electrodes at F_z, C_z, P_z, placed according to the International 10-20 system (Jasper, 1958) and referred to linked mastoids. The signals were amplified and filtered on-line (8 sec time constant, and 35 Hz upper half amplitude cutoff point). EOG was recorded for purposes of subtracting out ocular artifact from the EEG. The subtraction of the ocular artifact was accomplished by means of a procedure described by Gratton, Coles, and Donchin (1983). The EEG and EOG signals were digitized at 100 Hz for a period of 1400 msec, starting 100 msec before each stimulus presentation.

Average ERP waveforms were computed for each instruction, probability, stimulus category, subject and electrode. P300 latency and amplitude were assessed on each single trial according to a procedure described by Gratton et al. (1986).

RESULTS

The results will be divided into several sections. First, we will present data supporting the claim that the stimulus probability manipulation did indeed affect the subjects' response strategy. Second, we will describe the relationships between P300 latency and RTs, and between these two variables and response accuracy. These data replicate the results obtained by Kutas et al. (1977). Then, we will describe a procedure devised to interpret these relationships. Finally, we will analyze some of the consequences of the processes involved in P300 generation. Note that the present chapter is but a partial report of the study. It is intended to illustrate the chronometric use of psychophysiological signals rather than to serve as a comprehensive report of the study. Therefore, we shall ignore, in this discussion, the data obtained in the accuracy instruction condition. Furthermore, we shall ignore many of the detailed analyses of the data that are required to fully support our interpretations. For a full description of the study see Gratton et al. (1986). Note also that some of the analyses were based on five subjects only, because of the small number of error trials in the frequent female condition.

Effects of Manipulations on Response Strategy. The frequency and latency of correct and incorrect overt responses for each probability x stimulus x response condition are shown in Figure 16.1. The subjects' overt response was affected by the probability of the stimulus. Indeed, the response to the rare male stimuli was less accurate than that to the frequent female stimuli (under speed instructions), as revealed by the instruction x stimulus x response interaction, $F(1,4)=9.22$, $p<.05$. The latency of the correct response for male stimuli was slower than that for female stimuli, while the latency of the incorrect response was faster, as revealed by the stimulus x response interaction, $F(1,4)=11.69$, $p<.05$. This was particularly evident for the 20/80

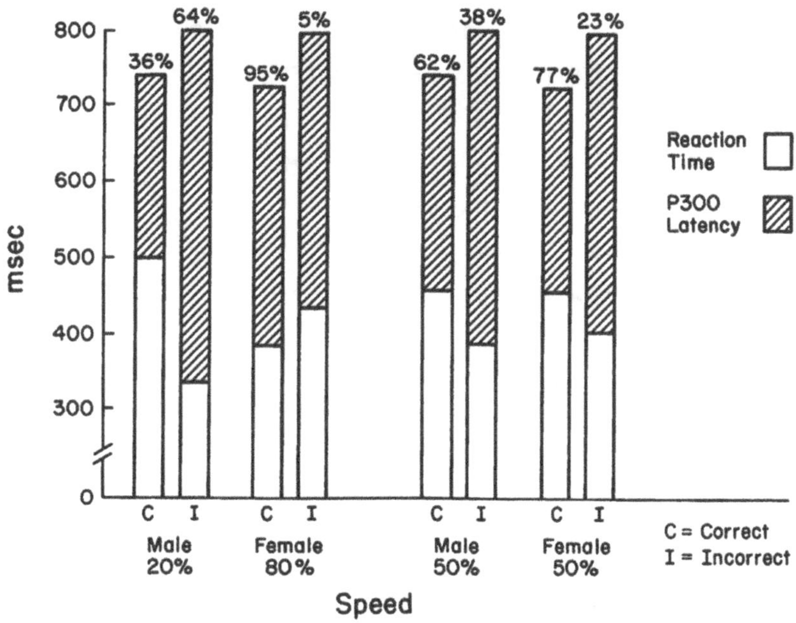

FIGURE 16.1. Reaction time and P300 latency for each probability (20/80 and 50/50), stimulus (male and female) and response (correct, C, and incorrect, I) condition. The frequency of correct and incorrect responses for each condition is also indicated.

condition, $F(1,4)=91.68$, $p<.001$. In particular, in the 20/80 condition, for male stimuli (rare) the incorrect response was faster than the correct response by 134 msec, while for female stimuli (frequent) the correct response was 50 msec faster than the incorrect response.

These findings support the conclusion that the subjects indeed conformed their response strategy to the probability manipulation. In particular, when the female stimulus was presented more often (the 20/80 condition) the subjects tended to execute fast female responses whatever stimulus was presented. In fact, the error rate for male stimuli under these conditions is 64%, while the error rate for female stimuli is only 5%.

P300, RT, and Accuracy. The grand average waveforms at P_z, for each probability, stimulus, and response condition, are shown in Figure 16.2. Inspection of this figure reveals several interesting points. A large positivity is the most dominant feature of the waveforms. We interpret this positivity as P300. The latencies of the P300 peaks (shown in Figure 16.2) for trials where a correct response was given were 60 msec shorter than for trials where an incorrect response was given, $F(1,4)=24.87$, $p<.01$. The category of the stimulus did not affect P300 latency, $F(1,4)=0.66$ (N.S.), nor did the

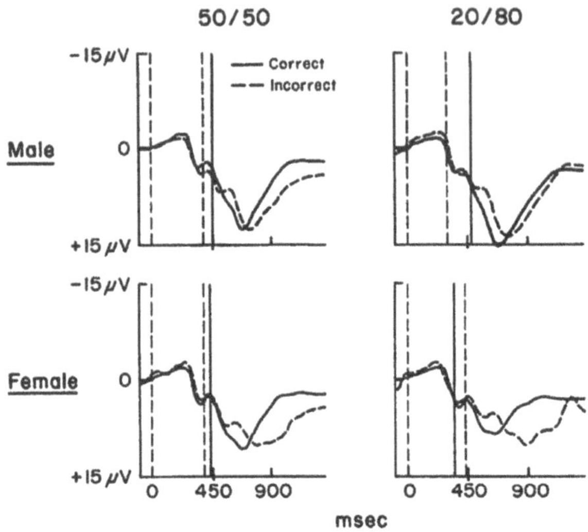

FIGURE 16.2. Grand average waveforms at P_z for each probability, stimulus and response condition. The solid lines refer to the grand average waveforms for the correct responses, and the dashed lines to the waveforms for the incorrect responses. The average reaction times for correct (solid) and incorrect (dashed) responses are indicated by vertical lines.

probability manipulation, $F(1,4)=1.31$ (N.S.), or the stimulus x probability interaction, $F(1,4)=0.00$ (N.S.). Thus, the time the subject takes to emit a P300 does not depend on whether the stimulus is male or female, or, in fact, on whether the probability of the stimulus is manipulated. These results contrast with those obtained for the RT. They indicate that the timing of those processes on which the emission of the P300 depends is not affected by variations in the criteria for overt response emission, which were introduced by the experimental manipulations. However, the amplitude of the P300 (see Figure 16.2) was affected by some of these variables. In particular, the male stimuli, when rare, elicited a larger P300 than the female stimuli (in this case, frequent). This produced a significant stimulus x probability interaction, $F(1,4)=15.34$, $p<.05$. On the other hand, P300 for incorrect responses was only slightly (and not significantly) larger than for correct responses, $F(1,4)=2.99$ (N.S.).

Summarizing these findings, we note that we have replicated and extended the Kutas et al. (1977) study. Error trials are generally associated with faster RTs, but later P300s than correct trials. The stimulus probability affects RT but not P300 latency.

Error Recognition. The results presented above are consistent with the interpretation that P300 latency is influenced by variables affecting the time required to evaluate the stimulus, but relatively independent of the variations of the response criteria adopted by the subject. However, the observation that the overt response is fast and P300 late on error trials requires some explanation. In fact, two explanations are possible. First, it may be that, in analyzing the error trials, we select those trials in which fast guessing and/or delay in the evaluation of the stimulus occur. Thus, errors may occur because stimulus evaluation (i.e., P300) is late in comparison with response activation processes. Second, it may be that processing of the error may delay P300. In this case, P300 latency would not only reflect the processing of the external stimulus (male or female name), but also the processing of internal events leading to the response.

To choose between these two interpretations, we focused our attention on a condition in which errors are particularly frequent (the Speed condition). Our procedure was based on an analysis of the speed/accuracy functions for this condition. In addition, we were interested in distinguishing among trials with different P300 latency. Speed/accuracy functions for different P300 latency bins are presented in Figure 16.3, for 20/80 and 50/50 conditions separately. For the 20/80 conditions, the functions were computed separately for male and female names. For the 50/50 condition, data from the two name categories were pooled together, since they had the same probability. These functions reveal several interesting points. First, the longer the RT the higher the accuracy. Second, accuracy is higher for trials on which P300 latency is relatively short. These findings suggest that accuracy is largely dependent on the relative timing of P300 latency and RT.

We also note the very low accuracy when the rare male names are presented. This is especially true when RTs are fast and P300 latency is long. In this case, we might speculate that the subject's basic strategy is to emit the "female" response to the "male" stimulus. In fact, virtually no errors can be observed in response to frequent female names. It may be possible to suggest that this pattern of results is due to the fact that errors occur when stimulus evaluation time is for some reason slow so that the male stimulus is not presented fast enough to inhibit the female response. If this explanation is valid then the delay in P300 on error trials is not due to the processing of the consequences of the error.

However, a third observation is not compatible with the interpretation that P300 is solely dependent on the time required to decide whether the stimulus was male or female. In fact, if P300 is sensitive only to the stimulus categorization process, and a delay in P300 indicates only a delay in this categorization process, then responses given before this process is sufficiently established should have a chance level of being correct. In the 50/50 condition this chance level is 0.5. Thus, this interpretation should predict that, in the 50/50 Speed condition, the error rate would never exceed 0.5 even in cases of long P300 latency and short RTs. Actually, Figure 16.3 reveals

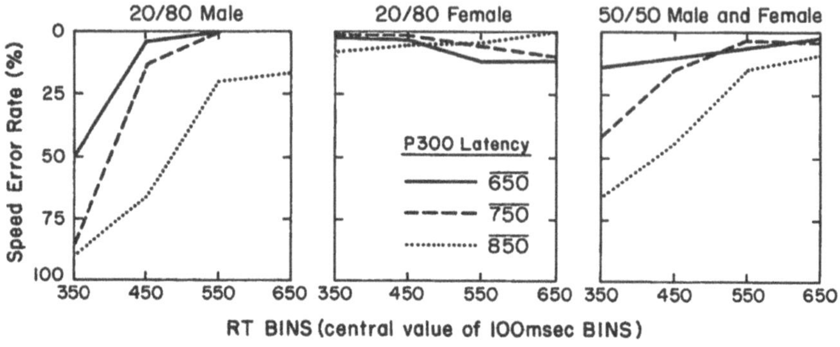

FIGURE 16.3. Speed-accuracy functions for rare male stimuli, frequent female stimuli and all stimuli in the 50/50 condition. Separate speed-accuracy functions were computed for each of three P300 latency bins (600 to 699 msec, 700 to 799 msec, and 800 to 899 msec).

that, in this condition, the accuracy for trials with fast RTs and slow P300 latency is lower than the chance level. This indicates that, by looking at trials with fast RTs and long P300 latency, we are "selecting" error trials. We interpret this finding as demonstrating that the association between incorrect response and long P300 latency is not due solely to the fact that errors occur because of a delay in stimulus evaluation. We must also propose that the processing of the incorrect responses causes a delay in P300.

The P300 and Future Action. We have demonstrated that P300 is delayed on incorrect trials. This delay indicates that, before emitting the P300, the subject must have not only categorized the name, but also compared the stimulus category with the current response. Presumably, the delay in P300 reflects some process that occurs when the system processes the commission of an error (a recognition that need not reach the subject's awareness). Given the relationship between P300 and schema updating (Donchin, 1981; Karis, Fabiani, & Donchin, 1984), we hypothesized that in the present experiment this process was related to adjustments in the subject's strategy subsequent to the recognition of an error. If this hypothesis is correct then the characteristics of the P300 elicited on the error trials should predict variations in the response criteria in the following trials.

To test this hypothesis we used the following procedure. First, we identified the male trials in which an incorrect response was given. Then, we sorted these trials on the basis of the response made on the following trial on which a male name was presented. We assumed that when an incorrect response to a male trial was given, the subject was biased to emit the "female" response regardless of the stimulus. If the subject responded again incorrectly to the following male trial, then we assumed that the subject's bias remained the same. On the other hand, if the subject responded correctly to the following male trial, then we assumed that the subject had revised his strategy. We labeled the latter sequences "switch" sequences, and the former, "no-switch" sequences. Note that we assumed that the switch in response bias occurred as a consequence of the recognition of an error after the first trial of a sequence. In particular, we predicted that the P300 to the first stimulus of the sequence (incorrect male name) was larger for switch than no-switch sequences. For both the 20/80 and 50/50 conditions P300 elicited on the first trial of a switch sequence was larger than P300 elicited on the first trial of a no-switch sequence, $F(1,5)=6.66$, $p<.05$, and $F(1,5)=14.25$, $p<.05$, respectively. To test further the hypothesis that the switch in response bias does indeed occur immediately following the recognition of an incorrect response to a male trial, we examined the female trials which intervened between the first and last male trials of the sequence described above. The prediction was that the response to the intervening female trial should be slower for switch than no-switch sequences. The RT for these trials are shown in Figure 16.4, as a function of sequence (switch vs. no-switch), lag from the first trial of the sequence, and condition (20/80 vs. 50/50). Accuracy is higher and RTs are slower for female trials in a switch sequence than in a no-switch sequence, $F(1,6)=13.36$, $p<.05$, $F(1,5)=7.50$, $p<.05$, respectively. This indicates that the subjects did indeed modify their response strategy at the beginning of a switch sequence.

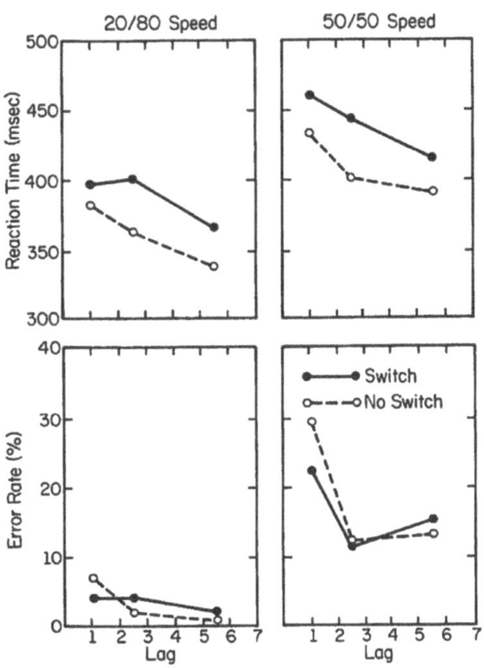

FIGURE 16.4. Reaction time and error rate for female trials in SWITCH and NO SWITCH sequences, as a function of the lag from the incorrect male trial initiating the sequences, for each probability condition.

DISCUSSION

The data presented above indicate that when a subject chooses the wrong alternative in a two-choice discrimination task, and that error is more than likely due to a bias to respond to the "wrong" stimulus, this recognition tends to introduce a delay of about 60 msec in the invocation of the P300 by that same stimulus. This delay in P300 by the occurrence of an error appears to be related to an evaluation by the system of the context within which it operates. The data appear consistent with the suggestion that the magnitude of the P300 can serve as a measure of the degree of revision in the system's biases. This assertion is inferred from the fact that the larger the P300 elicited following the error the less likely the error on the next error-prone trial. Moreover, the larger the P300 that is elicited on error trials the slower will be the subject to respond on the immediately succeeding trial. A shift in response bias, or a tendency to place the response under controlled, rather than automatic, processing mode are plausible interpretations of these data.

These results, and the interpretation proposed above, indicate how the P300 and the study of its latency and amplitude can reveal aspects of the manner in which the human information processing system deals with error trials. These aspects are opaque to the more traditional tools largely because they permit a view of information processing activities that do not have an overt manifestation in performance on the trial in which they are elicited. The view that emerges is one in which at least two information processing streams proceed in parallel. Both depend for their initiation on the initial detection and encoding of the stimulus. However, the processing that leads to the overt response may be completed, yielding the actual response, independently of the evaluative processes whose role is the maintenance of the operating environment.

The metaphor that captures our intent is that of an organization whose operating and administrative arms operate in a highly interactive, but nevertheless independent manner. Actions by the organization's staff are taken in the light of the local interpretation of ongoing events and under the constraints established by the administration's policy. Each event outside the organization, and each action by the organization, trigger in the administration an evaluative process that may long outlast the staff's actions as the administration must optimize its operating policies given the consequences of its own actions and the events in its surround. The time course of the administrative processing may be quite independent of the time course of the processing required by the operating staff before it takes action. Indeed, if staff action was patently erroneous the administration may require additional time before it closes the book on the action, files the reports and makes the necessary adjustments in policy.

The P300 component can, we believe, be viewed as a manifestation of "administrative" rather than "operational" information processing. Donchin et al. (1978) labeled these classes of information processing "strategic" and "tactical," respectively. Evidence is accumulating that the process manifested by the P300 is "future oriented" (see, for example, Donchin, 1981; Fabiani, Karis, Coles, & Donchin, 1983; Karis et al., 1984; Klein, Coles, & Donchin, 1984). The data we reviewed in this chapter are consistent with this view. It seems clear that the magnitude of the P300 elicited on an error trial is related to the performance of the subject on a subsequent trial. Such an effect implies, almost by definition, that the process manifested by the P300 has consequences for future performance. It is, of course, possible that the relationship we observed is fortuitous and both P300 amplitude on trial N, and the subject's performance on trial N+1, are correlated with yet a third factor accounting for both variations. To address this issue we must continue seeking the elucidation of the functional significance of the P300.

In many ways the study described in this report is a direct descendant of the work that Lindsley and his colleagues undertook as electroencephalographic techniques made their way from Europe to the United States in the 1930s. The EEG has, of course, become a standard clinical tool and much of the research utilizing the EEG is clinical in nature and in orientation. There is, however, a flourishing research enterprise in which the EEG, and the ERPs embedded in it, are used as tools in the study of cognitive function. Lindsley's work, spanning more than half a century was, and continues to be, an outstanding illustration of the way a scientist bringing the skills and sensibilities of a psychologist can turn the record of a bodily function so it provides a window on the mind. A key element in this enterprise is the chronometric approach that has been so important in Lindsley's research program.

REFERENCES

Chalupa, L. M., Rohrbaugh, J., Gould, J. E., & Lindsley, D. B. (1974). Evoked potentials and reaction time correlates in monkeys during a simultaneous visual discrimination task. *Society for Neurosciences Abstracts*, 164.

Coles, M. G. H., Gratton, G., Bashore, T. R., Eriksen, C. W., & Donchin, E. (1985). A psychophysiological investigation of the continuous flow model of human information processing. *Journal of Experimental Psychology: Human Perception and Performance, 11*, 529-553.

Donchin, E. (1979). Event-related brain potentials: A tool in the study of human information processing. In H. Begleiter (Ed.), *Evoked brain potentials and behavior* (pp. 13-75). New York: Plenum.

Donchin, E. (1981). Surprise! . . . Surprise? *Psychophysiology, 18*, 493-513.

Donchin, E., & Heffley, E. F. (1978). Multivariate analysis of event-related potential data: A tutorial review. In D. Otto (Ed.), *Multidisciplinary perspectives in event-related brain potential research* (pp. 555-572). Washington: US Government Printing Office.

Donchin, E., & Lindsley, D. B. (1965). Visual evoked response correlates of perceptual masking and enhancement. *Electroencephalography and Clinical Neurophysiology, 19*, 325-335.

Donchin, E., Ritter, W., & McCallum, W. C. (1978). Cognitive psychophysiology: The endogenous components of the ERP. In E. Callaway, P. Tueting, & S. H. Koslow (Eds.), *Event-related brain potentials in man* (pp. 349-411). New York: Academic Press.

Duncan-Johnson, C. C., & Donchin, E. (1982). The P300 component of the event-related brain potential as an index of information processing. *Biological Psychology, 14*, 1-52.

Fabiani, M., Gratton, G., Karis, D., & Donchin, E. (1987). The definition, identification, and reliability of measurement of the P300 component of the event-related brain potential. In P. K. Ackles, J. R. Jennings, & M. G. H. Coles (Eds.), *Advances in psychophysiology* (Vol. 2). Greenwich, CT: JAI. (in press).

Fabiani, M., Karis, D., Coles, M. G. H., & Donchin, E. (1983). P300 and recall in an incidental memory paradigm. *Psychophysiology, 20*, 439.

Goodin, D. S., Squires, K. C., & Starr, A. (1978). Long latency event-related components of the auditory evoked potential in dementia. *Brain, 101*, 635-648.

Gratton, G., Coles, M. G. H., & Donchin, E. (1983). A new method for off-line removal of ocular artifact. *Electroencephalography and Clinical Neurophysiology, 55*, 468-484.

Gratton, G., Dupree, D., Coles, M. G. H., & Donchin, E. (1986). *The consequence of an error: P300 and future action.* Manuscript in preparation.

Herren, R. T., & Lindsley, D. B. (1931). Central and peripheral latencies in some tendon reflexes of the rat. *American Journal of Physiology, 99*, 167-171.

Hillyard, S. A. (1984). Event-related potentials and selective attention. In E. Donchin (Ed.), *Cognitive psychophysiology: Event-related potentials and the study of cognition* (Vol. 1, pp. 51-72). The Carmel Conferences. Hillsdale, NJ: Lawrence Erlbaum Associates.

Jasper, H. H. (1958). The ten-twenty electrode system of the International Federation. *Electroencephalography and Clinical Neurophysiology, 10*, 371-375.

Karis, D., Fabiani, M., & Donchin, E. (1984). P300 and memory: Individual differences in the von Restorff effect. *Cognitive Psychology, 16*, 177-216.

Klein, M., Coles, M. G. H., & Donchin, E. (1984). People with absolute pitch process tones without producing a P300. *Science, 223*, 1306-1309.

Kutas, M., McCarthy, G., & Donchin, E. (1977). Augmenting mental chronometry: The P300 as a measure of stimulus evaluation time. *Science, 197*, 792-795.

Lindsley, D. B. (1944). Electroencephalography. In J. McV. Hunt (Ed.), *Personality and behavior disorders* (Vol. 2, pp. 1037-1103). New York: Ronald Press.

Lindsley, D. B. (1954). Electrical response to photic stimulation in visual pathways of the cat. *Electroencephalography and Clinical Neurophysiology, 6*, 690-691.

Lindsley, D. B. (1982). Neural mechanisms of arousal, attention and information processing. In J. Orbach (Ed.), *Neuropsychology after Lashley* (pp. 315-407). Hillsdale, NJ: Lawrence Erlbaum Associates.

Lindsley, D. B., & Emmons, W. H. (1958). Perceptual blanking evoked potentials and perception time. *Electroencephalography and Clinical Neurophysiology, 10*, 359.

Lindsley, D. B., Fehmi, L. G., & Adkins, J. W. (1967). Visually evoked potentials during perceptual masking in man and monkey. *Electroencephalography and Clinical Neurophysiology, 23*, 79.

Lindsley, D. B., & Rubenstein, B. B. (1937). Relationship between brain potentials and some other physiological variables. *Proceedings of the Society of Experimental Biology and Medicine, 35*, 558-563.

Lindsley, D. B., Seales, D. M., & Wilson, G. F. (1973). Changes in the late components of visual evoked potentials with visual information processing. *Society for Neuroscience Abstracts*, 422.

Magliero, A., Bashore, T. R., Coles, M. G. H., & Donchin, E. (1984). On the dependence of P300 latency on stimulus evaluation processes. *Psychophysiology, 21*, 171-186.

McCarthy, G. (1984). Stimulus evaluation time and P300 latency. In E. Donchin (Ed.), *Cognitive psychophysiology: Event-related potentials and the study of cognition* (Vol. 1, pp. 254-285). The Carmel Conferences. Hillsdale, NJ: Lawrence Erlbaum Associates.

McCarthy, G., & Donchin, E. (1981). A metric for thought: A comparison of P300 latency and reaction time. *Science, 211*, 77-80.

McCarthy, G., Kutas, M., & Donchin, E. (1979). Detecting errors with P300 latency. *Psychophysiology, 16*, 175.

Näätänen, R. (1982). Processing negativity: An evoked-potential reflection of selective attention. *Psychological Bulletin, 92*, 605-640.

Pfefferbaum, A., Ford, J. M., Roth, W. T., & Kopell, B. S. (1980). Age differences in P3-reaction time associations. *Electroencephalography and Clinical Neurophysiology, 49*, 257-265.

Posner, M. I. (1978). *Chronometric explorations of the mind*. Hillsdale, NJ: Lawrence Erlbaum Associates.

Schwartz, A. S., & Lindsley, D. B. (1964). Critical flicker frequency and photic following in the cat. *Boletin Inst. Estudios Medicos y Biologicas, Mexico, 22*, 249-262.

Sutton, S., & Ruchkin, D. S. (1984). The late positive complex: Advances and new problems. In R. Karrer, J. Cohen, & P. Tueting (Eds.), *Brain and information: Event-related potentials* (pp. 1-23). New York: New York Academy of Sciences.

Woody, C. D. (1967). Characterization of an adaptive filter for the analysis of variable latency neuroelectric signals. *Medical and Biological Engineering, 5*, 539-553.

17· HUMAN RESPIRATORY REACTION TIMES

R. Lansing, L. Meyerink, and J. Thomas

University of Arizona

The precision with which respiratory muscles can be voluntarily controlled is evidenced in singing, speaking, playing wind instruments and coordinating breathing with skilled body movements (Bramble & Carrier, 1983; Proctor, 1980; Wyke, 1974). Respiratory force can be regulated within 1 to 2 cm H_2O, about 1% of maximum voluntary effort, and changes in force can be developed (half time, 45-60 msec) or repeated (upper frequency limit, 5-7 per sec) as rapidly as they are for limb muscle contractions (Agostoni, 1970). The major respiratory muscles (intercostal, diaphragm, abdominal) are capable of brief pulsatile contractions (50 to 200 msec) and their activity is accurately timed with other muscle contractions in complex movements such as speaking, instrument playing, or swimming (Berger, 1968; Draper, Ladefoged, & Whitteridge, 1960; Smith, 1973).

Very little is known, however, about respiratory reaction times (RT) to external or internal stimuli, and the conditions which affect them. This information is as important for understanding the voluntary control of breathing as it has been for understanding the neural control of other skeletal muscle groups. Respiratory RTs may affect the quality of a performance which requires that inspiratory or expiratory efforts be triggered by specific internal or external stimuli. It is possible, for example, that for some complex movement patterns the sequential activation of participating muscles depends on afferent triggering of each at its proper time rather than on the preprogramming of a central controller (Cole & Abbs, 1983). Respiratory RTs may also limit the speed with which corrective actions can be taken when voluntary breathing movements are perturbed. Finally, the partitioning of afferent, central, and efferent transmission times in simple RT experiments can give some insight into the neural control of skilled breathing acts.

We have studied simple respiratory RTs in young, practiced subjects under attentional conditions which favor short, stable RTs. We have measured inspiratory and expiratory RTs to external stimuli and studied how they are affected by a change in lung volume with its attendant alterations in muscle length, mechanical advantage, and respiratory pressures. Then, using expiratory tasks and recording from abdominal muscles we have attempted to estimate the portion of total RT which can be ascribed to sensory, central and peripheral motor delays. Finally, expiratory RTs to internal (proprioceptive)

 DOI: 10.4324/9781003164647- 22

stimuli were determined with an attempt to establish minimal RTs. This latter work is published elsewhere (Lansing & Meyerink, 1981) and only briefly summarized here.

Respiratory Reaction Times to External Stimuli. We studied inspiratory and expiratory reaction times to visual stimuli and tried to determine the importance of respiratory position (lung volume) at the time of the stimulus. Five subjects were trained to make brief inspiratory and expiratory movements as quickly as possible after a flash from a Grass photostimulator lamp (intensity 4; visual angle, 40°). The subject breathed through a mouthpiece from which airflow, volume (by integration), and CO_2 levels were continually recorded using appropriate transducers and Grass 7P1 amplifiers (flat from DC to 60 Hz). Reaction times were measured by starting a Hewlett Packard timer at the time of the flash and stopping it by the onset of airflow as the subject responded. There was a small constant delay between the beginning of airflow and stopping of the timer which was taken into account to obtain the actual response latencies. A background noise masked the slight photoflash click.

The subjects fixated on a black cross in the center of the stimulus lamp and, after hearing a click, held their breath with the glottis open at their normal end-expiratory level (EL) or end-inspiratory level (IL), depending on which of these pre-stimulus positions was being studied in that series of trials. Within 2 to 4 sec (randomized foreperiod) after achieving the desired level, the stimulus was presented. Over a period of three days, 360 RTs were measured for each subject, 90 under each of the four experimental conditions: inspiratory responses from EL and IL positions, and expiratory responses from EL and IL positions. The RTs for each condition were obtained in three runs of 30 trials each; these runs occurred on separate days. The order of conditions was rotated across subjects and days.

The results are shown in Table 17.1. The average RT, for all conditions and all subjects, was 166 msec. Inspiratory RTs (158 msec) were shorter than expiratory RTs (173 msec). This difference, which was statistically significant ($p<.05$), may reflect differences in mechanical coupling between the muscles involved and the structures producing airflow. It is hazardous to compare absolute RTs from different studies, but clearly these simple respiratory RTs to light stimulation fall in the general range of those recorded for limb movements under similar conditions of attention and practice (Bartlett, 1963; Woodworth & Schlosberg, 1970). Voice RTs, where respiratory activity precedes phonation, have been reported to be from 190 to 370 msec (Baken, Cavallo, & Weissman, 1979; Starkweather, Hirschman, & Tannenbaum, 1976). Respiratory RTs are of necessity shorter than these (Draper et al., 1960) but probably by a variable amount depending on the nature of the phonatory act. Baken et al. (1979) found RTs for pre-phonatory inspirations

TABLE 17.1. Respiratory reaction times (RT) to light stimuli.

	Inspiratory RTs (msec)			Expiratory RTs (msec)		
Respiratory Position	Mean	S.D.	Range	Mean	S.D.	Range
End Expiratory	159.7	24.5	152.0-171.8	174.6	22.9	156.8-193.1
End Inspiratory	155.9	19.2	143.4-163.1	171.5	24.5	159.5-184.2

Note: Each mean and S.D. represents the combined data for all five subjects, 90 RTs for each subject.

of 245 msec, with phonation beginning 93 msec later. For some voice reactions, subglottal pressure may be built up well before the stimulus and the reaction is simply that of a glottal or articulatory release ("ga," "pa," etc.) with no respiratory RTs involved.

The finding that neither inspiratory nor expiratory RTs were affected by differences in lung volume agrees with the results of Baken et al. (1979) who found that pre-phonatory inspiratory RTs were unaffected by the respiratory volume or direction of breathing movements at the time of the stimulus. The importance of our findings for RT measures becomes evident when the physiological-mechanical differences prevailing at the EL and IL are compared. At greater lung volumes (IL) the inspiratory muscles (parasternals, external intercostals, diaphragm) are shorter than they are at lower lung volumes (EL) while the reverse is true of the expiratory muscles (abdominal). This places these muscles on different portions of their length-tension curves and alters their mechanical effectiveness even when neural activation is constant. Further, the elastic forces of the lung, rib-cage, and abdomen, and the resistance to airflow are different for the EL and the IL positions (Campbell, Agostoni, & Newsom Davis, 1970). It appears that changes in these mechanical conditions and muscle effectiveness do not alter respiratory RTs. Recent work (Banzett, Lansing, & Reid, 1985; Banzett & Mead, 1984) suggests that the differences in respiratory muscle effectiveness at EL and IL may have been compensated for by reflex increases or decreases in neural drive.

Abdominal Muscle RTs and Estimation of Central and Peripheral Delays. In studying the reaction time process for limb movements it has been helpful to estimate the delays associated with the sensory, central nervous, neuromuscular, and mechanical stages of the total response (Botwinick & Thompson, 1966; Luschei, Saslow, & Glickstein, 1967; Miller & Glickstein, 1967). We have attempted such a fractionation for expiratory RTs to light stimuli.

Four subjects were trained, and motivated by frequent urging, to make the fastest possible expiratory responses to light flashes. EMGs were recorded from abdominal muscles, a principle agonist for the response. The same stimuli and procedures were used as in the first experiment except these subjects were semi-reclining to eliminate postural activity of the abdominal muscles, and the stimulus was presented only in the end-expiratory position. Ninety reaction times were obtained for each subject. An estimate of the time required for sensory input to the brain was obtained by averaging visual evoked potentials with an active scalp electrode 3 cm up from the inion on the midline and a reference electrode on the right ear. An electrode over the eye controlled for eyeblinks. Abdominal muscle potentials were recorded from electrodes over the external oblique (5 cm apart, midway between the iliac crest and the lower costal margin, straddling the anterior axillary line) and the rectus abdominis (5 cm apart above and below the umbilicus). Airflow at the mouth was recorded as before. The EEG and EMG potentials were amplified with Grass 7P5 amplifiers (half amplitude points .1 and .3K Hz for EEG; 10 and .3K Hz for EMG), stored on magnetic tape, and averaged off-line with a Fabritek 1052 signal averager. EMG potentials were full-wave rectified before averaging. Latencies were measured from 30 trial averages; there were three of these averages for each subject.

The airflow responses were very rapid and brief; peak flow rates of 1.2 liter/sec were reached within 98 msec, on the average, and the responses lasted 330 msec. As shown in Figure 17.1, the earliest occipital response (N1 of the visual evoked potential) occurred 30 msec after the flash in the three subjects for whom we could obtain reliable measures. The earliest abdominal contractions, external oblique for three subjects and rectus abdominis for one, began 107 msec after the flash (the shortest abdominal latencies recorded from individual subjects ranged from 81 to 129 msec). Airflow at the mouth began 165 msec after the flash (earliest recorded responses ranged from 135 to 200 msec). We have tried to estimate the time required for the main physiological stages of the simple RT response using our averaged values (Figure 17.1). Assuming that the visual cortex is a critical link in the neural circuit mediating the response (Miller & Glickstein, 1967), a sensory transmission time to the cortex of 30 msec accounts for about a fifth of the total reaction time. Even if parallel processing occurred and the visual cortex

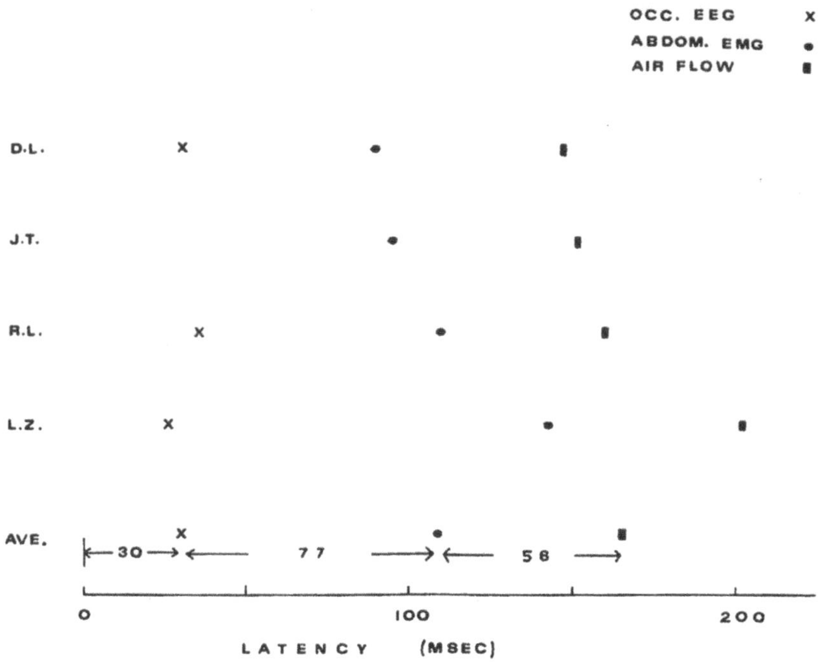

FIGURE 17.1. Latency of occipital EEG (evoked potential), abdominal EMG (external oblique or rectus abdominis), and air flow responses to photic stimuli. Latencies measured from computer averages of 90 responses.

was bypassed, about the same fraction could be ascribed to sensory input to the other parts of the brain since at least 25 msec is probably required for the retinal response and retino-geniculate conduction (Monnier, 1952; Wall, Remond, & Dobson, 1953). Subtracting this sensory time of 30 msec from the latency of the abdominal EMG response shows that 77 msec must be assigned to cerebral processing and efferent discharge. The fastest efferent conduction time from cortex to abdominal muscles is probably about 15 msec, extrapolating from values obtained for digital flexors (Pagni, Ettori, Infusa, & Marossero, 1964), with peripheral motor neuron transmission accounting for 6 to 8 msec of that time (Kugelberg & Hagbarth, 1958). This leaves 62 msec for cerebral processes. If slower corticospinal or peripheral nerve fibers are used, or if some motor unit recruitment time must be allowed before EMG activity is measurable from surface leads, the time remaining for intracerebral transmission would be even less. The minimal delays obtainable under our experimental conditions are shown in the reactions of subject DL. In his fastest 30-trial run (not shown in Figure 17.1) the external oblique responded 81 msec after the flash and the afferent conduction time was 31 msec.

Subtracting this 31 msec, and an estimated 15 msec for impulses to travel from cortex to muscle, leaves 35 msec for brain processing.

It must be remembered that other muscles (internal intercostals, internal oblique, transversus abdominis) not recorded from participate in the response and may have become active earlier than the external oblique. In certain RT tasks, for example, the internal oblique discharges 6 msec earlier. Available information on internal intercostals suggests that they do not precede abdominal muscle contraction in skilled expiratory efforts (Berger, 1968).

The time elapsing between the beginning of abdominal muscle contraction, measured by the EMG, and the onset of airflow was 56 msec. The earliest mechanical changes, an inward movement of the abdominal wall and a rise in intra-abdominal pressure, probably occurred about 40 msec after the first EMG discharge (Lansing & Meyerink, 1981). These values for the delay between contraction of respiratory muscles and their mechanical effects lie in the same general range as those measured for limb reaction time responses (Luschei et al., 1967; Woodworth & Schlosberg, 1970). The actual times recorded for a given movement would, of course, depend on which mechanical effect is registered (Bartlett, 1963), the mechanical advantage of the muscles, the rate of recruitment of motor units, and the inertial forces to be overcome. The data in Figure 17.1 suggest that large variations in total RT from subject to subject result from differences in central processing times rather than differences in muscular-mechanical linkage.

Respiratory Reaction Times to Proprioceptive Stimuli. During skilled breathing movements unexpected variations in the mechanical load against which the respiratory muscles are working must be adjusted quickly for the planned movement to be completed smoothly. Proprioceptive cues from muscles, tendons, joints, lungs or airways would be expected to trigger these compensatory responses. The only previous work done on this problem was that of Newsom Davis and Sears (1970) who found that when a subject's voluntary breathing movements were momentarily opposed by application of a pressure load in the external airway, a brief "excitatory response" was recorded from the intercostal muscles within 50 to 60 msec. The actual mechanical correction achieved by the response is not known but the amplitude variations with changing loads and lung volume suggested a load compensating action. This intercostal reflex may be the respiratory equivalent of the late stretch reflex (M2 component) of upper limb muscles which is also elicited 50 to 60 msec after a voluntary contraction has been blocked (Hammond, 1954; Lee & Tatton, 1975; Marsden, Merton, & Morton, 1972). These respiratory and limb muscle reflexes only appear if the subject is instructed to resist the disturbing load, and may operate through supraspinal or transcortical circuits (Desmedt, 1978). But they have not been considered learned responses of the type usually recorded in RT experiments on the assumption that 60 msec is much too short for a voluntary RT. The evidence of Crago, Houk, and Hasan (1976); Evarts and Granit (1976); Hufschmidt, Kilimov, and Linke (1977) and our findings, briefly described below, show that this assumption is not correct and that the distinction between learned and reflex reactions to the unexpected loading of a movement must be made on some other basis than response latency.

We studied compensatory responses and proprioceptive RTs of abdominal muscles following the perturbation of a voluntary contraction (Lansing & Meyerink, 1981). The six subjects in that experiment held a constant lung volume (end-expiratory level) against a steady pressure (6 cm H_2O) in the external airway. A few seconds after a warning signal, an additional pressure was suddenly applied to the airway; it drove the chest and abdomen in the inspiratory direction at the rate of .450 to 1.1 l/sec and the subject had to oppose this with an expiratory effort requiring abdominal muscle contractions. On half the trials the preparatory intervals were 2, 2.5, or 3 sec (used randomly) and for the other half 1, 2, 3, 4, and 5 sec. Catch trials occurred on 25% of the trials. The latency of abdominal muscle (external oblique, internal oblique, and abdominis rectus), abdomen movement, and mouth pressure changes were determined from inkwriter, CRO, and computer-averaged records. Figure 17.2 shows examples of responses recorded for different tasks: (1) maintaining the pre-load position as accurately as possible in spite of the load ("maintain position" task); (2) reacting as quickly as possible to the load without regard to accuracy ("load RT" task); (3) reacting as quickly as possible to a click with no load applied ("auditory RT" task); and (4) making no effort at all to respond to the load ("no response" task). During training and after every experimental trial, the response the subject had just made was displayed on a storage oscilloscope against a template showing the best performance possible.

When subjects were instructed not to respond to the loads, no responses were recorded. When trained to maintain their pre-load position in spite of the load, a two-phase compensatory response was found. The initial response (phase 1) was 100-300 msec in duration; it was followed by a continuous discharge (phase 2) which continued to the offset of the pressure load (Figure 17.2, A1, B1, C1, & D1). For the simple RT tasks the abdominal response to the pressure load or click stimulus was a single EMG burst.

The combined response latencies for all subjects and conditions are presented in Table 17.2. The shortest possible latencies are given in Table 17.3; data are shown for the earliest responding muscle (internal oblique), obtained on the second day of practice, under the condition of optimal readiness (the ready signal occurring 1 to 3 sec before the load onset). The latencies of the load responses were measured from rise in mouth pressure, the earliest possible source of proprioceptive stimulation. The latest proprioceptive stimuli would arise from muscle abdominal stretch; this

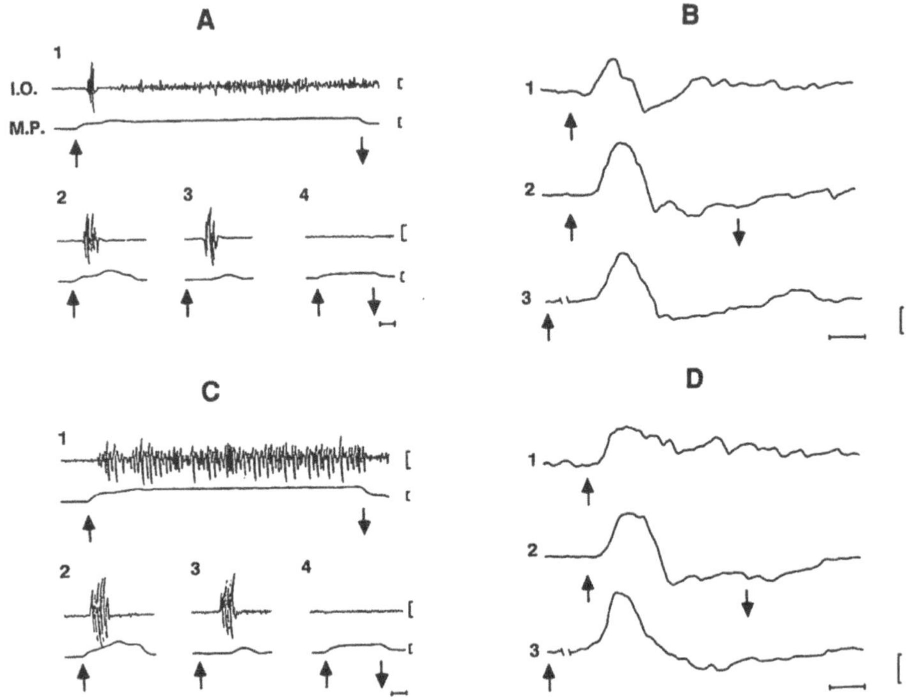

FIGURE 17.2. Sample reactions to airway loading for subject R. J. On the left, inkwriter recordings of internal oblique EMG (I. O.) and mouth pressure (M. P.) for airway stimulus loads of 6 cm H_2O (A) and 12 cm H_2O (C). The task conditions were (1) maintain position; (2) load RT; (3) auditory RT; and (4) no response. On the right (B and D) computer averages of 16 internal oblique responses for the same series of trials from which A (1 to 3) and B (1 to 3) were taken. Arrows show onset and offset of stimuli. Calibration bars: time is 100 msec; EMG is 100 μV; mouth pressure is 12 cm H_2O (Lansing & Meyerink, 1981).

occurred 10 to 15 msec after the mouth pressure change according to our accelerometer measures of abdominal wall movement.

For the maintain position task, phase 1 latencies (internal oblique) averaged 66-90 msec for individual subjects, but for single trials with optimal conditions of practice and preparatory intervals these ranged from 42 to 110 msec with a third of the reactions occurring within 50-60 msec.

The phase 1 EMG latencies for load RTs were about 5 msec shorter than for the maintain position task. The latency distributions for the two types of responses were similar, and both were affected to the same degree by practice, and by changing the length and variability of the preparatory interval. Because of these similarities we conclude that both the phase 1 burst of the maintain position task and the load RT response were mediated by the central systems operative in classical simple reaction time experiments.

TABLE 17.2. Mean latency (msec) of abdominal EMG and mouth pressure responses to airway loading during the different task conditions.

	Response Measure			
Task	Internal Oblique	External Oblique	Rectus Abdominis	Mouth Pressure
Maintain Position	80 (66-90)	86 (78-103)	102 (92-114)	178 (147-201)
Load RT	74 (64-87)	79 (70-98)	95 (83-107)	160 (139-187)
Auditory RT	156 (127-196)	162 (138-196)	175 (143-212)	219 (195-238)

Note: The range of means is given in parentheses (from Lansing & Meyerink, 1981).

TABLE 17.3. Latency of internal oblique responses to loading for individual subjects with optimal conditions of practice and preparatory set.

Subject	Maintain Position Task			Load RT Task		
	Mean	S.D.	Range	Mean	S.D.	Range
T.Z.	66.3	24.0	50-95	60.4	13.3	52-100
D.L.	68.6	12.9	42-100	57.1	11.3	40-75
L.M.	71.8	8.3	55-90	74.9	8.4	65-100
L.N.	73.1	9.3	50-88	67.0	9.5	50-130
R.J.	75.0	16.4	50-110	69.5	10.2	50-85
R.S.	75.6	18.8	45-110	60.0	12.5	40-90

Note: These data are for the second day of practice and with preparatory intervals of 2 to 3 sec. Each mean is based on 20 trials. (From Lansing & Meyerink, 1981.)

The phase 1 response in the maintain position task might be considered the abdominal counterpart of the 50- to 60-msec late stretch reflex of limb muscles (Lee & Tatton, 1975) or the "excitatory reflex response" of the intercostals (Newsom Davis & Sears, 1970) since they were elicited under similar task conditions, and over a fourth of the phase 1 latencies, measured on individual trials, were less than 65 msec. This interpretation cannot account for the continuous distribution of latencies ranging up to 130 msec or the susceptibility of the phase 1 burst to the effects of practice and attention. The short RT latencies we obtained cannot be attributed to anticipatory reactions, since these occurred only rarely on catch trials, and then with a very long latency.

How much central processing time could be assigned to cerebral circuits if the abdominal muscle RT to airway loading was as short as 50 msec? Estimating a minimum sensory conduction time of 10 msec from mouth stimulation (Marsden et al., 1972), or 20 msec from abdominal stimulation (Milner-Brown, Stein, & Lee, 1975), and minimum brain to muscle time of 15 msec (Pagni et al., 1964), the intracerebral processing time would be 15 to 25 msec. If slower conducting paths to and from the brain were used, this brain transmission time would be even less.

Evidence that these uncommonly short RTs are not peculiar to the respiratory system, or to situations in which responses are made to counteract a perturbing force, was obtained from further study of three of the subjects. The experimental conditions were identical to those used in the above experiment, except now subjects were trained to make fist clenches or forearm flexions in response to airway loading, and EMGs were recorded from the appropriate arm muscles. EMG RTs averaged 61, 61, and 70 msec, and the latencies for individual subjects ranged from 50-95, 42-90, and 45-97 msec. In a subsequent experiment (Meyerink & Lansing, 1981), we showed that EMG RTs as brief as 45 to 60 msec could be obtained in some subjects regardless of whether abdominal muscle contractions were made in response to arm movements, or biceps contractions were made to airway pressure changes.

Simple RTs of non-respiratory muscles to kinesthetic stimuli typically average 90 to 160 msec under favorable conditions (Chernikoff & Taylor, 1952; Poulton, 1981), but many studies report RTs in the 45- to 70-msec range on some trials. These RT studies include biceps responses to forearm extension (Crago et al., 1976), biceps responses to forearm torsion (Evarts & Granit, 1976), shoulder muscle contraction after dropping the relaxed arm (Wyrick & Duncan, 1974), mouth muscle responses to lip stimulation (Cole & Abbs, 1983) and biceps response to passive flexion of the ipsilateral or contralateral elbow (Hufschmidt et al., 1977).

The shortest possible RTs are of value in estimating the maximum speed with which learned respiratory responses can be made when optimal conditions prevail at the time of the stimulus. Important temporal properties of the neural circuits mediating the reaction are revealed when they are operating near their physiological limit. These short RTs can also be used to evaluate the possibility that learned, afferent-triggered responses participate in certain phases of complex movement control, for which absolute values of timing and delay are known (Cole & Abbs, 1983; Houk, 1978). It is important to recognize, though, that minimum RTs vary with the method of measurement. For example, they may represent the shortest single RT recorded, the first quartile mean for a series of reactions, the shortest mean RT obtained for several experimental runs or for several subjects, or the latency of a computer-averaged waveform derived by summing or averaging a number of rectified EMG responses. In this latter case, the latency is measured as the point at which the averaged waveform first departs from the baseline (see Figure 17.2) and this departure point will reflect the earliest, high amplitude responses. In a given experimental run then, the waveform latency can be determined by a single response, the responses of a few trials, or the responses of a large number of trials (see Evarts & Vaughn, 1978; Meyerink & Lansing, 1981).

SUMMARY

We studied respiratory RTs and some of the conditions affecting them, using classical simple RT methods. The length and variability of respiratory RTs were similar to those typically found for limb movements. Inspiratory RTs to light flashes were slightly shorter than expiratory RTs, but neither were affected by changes in lung volume and the differences in mechanical resistance and muscle effectiveness associated with these changes. Minimal afferent, cerebral, efferent, and peripheral mechanical delays were estimated using our EMG and mechanical RT measures and the published values for certain portions of the conducting pathways. The shortest cerebral processing time for expiratory abdominal muscle responses was 35 msec (group average 62 msec) when responding to light flashes, and from 15 to 25 msec when responding to mouth pressure changes or arm movement stimuli.

When expiratory efforts were opposed by sudden application of pressure at the mouth, the abdominal muscles made rapid compensatory responses. We conclude that these represent the action of RT processes since the latencies were similar to those obtained with RT methods, and they were affected in the same way by practice and foreperiod length. We found EMG RTs under these conditions to be much shorter than usually thought possible for responses to proprioceptive stimuli, and demonstrated that RTs of 45 to 70 msec are not restricted to respiratory reactions or to load compensating responses. Similar brief latencies have been obtained by other workers using limb muscle and face muscle responses. Our results also show that voluntary reactions to muscle stretch or movement perturbation cannot be distinguished from long-loop or late stretch reflexes on the basis of their latency, since both may occur in the 45- to 70-msec range. The value of determining the shortest RTs possible, and therefore the physiological limits of the circuits mediating simple learned responses, was pointed out.

REFERENCES

Agostoni, E. (1970). Dynamics. In E. J. M. Campbell, E. Agostoni, & J. Newsom Davis (Eds.), *The respiratory muscles: Mechanics and neural control* (pp. 80-114). Philadelphia: W. B. Saunders.

Baken, R. J., Cavallo, S. A., & Weissman, K. L. (1979). Chest wall movements prior to phonation. *Journal of Speech and Hearing Research, 22,* 862-972.

Banzett, R. B., Lansing, R. W., & Reid, M. B. (1985). Reflex compensation of voluntary inspiration when immersion changes diaphragm length, *Journal of Applied Physiology, 59,* 611-618.

Banzett, R. B., & Mead, J. (1984). Reflex compensation for changes in operational length of respiratory muscles. In C. Roussos & P. T. Macklem (Eds.), *The thorax: Vital pump (Lung Biology in Health and Disease Series)* (pp. 595-605). New York: Dekker.

Bartlett, N. R. (1963). A comparison of manual reaction times as measured by three sensitive indices. *Psychological Record, 13,* 51-56.

Berger, K. (1968). Electromyographic recording during wind instrument performance. In A. Bouhuys (Ed.), *Sound production in man* (pp. 297-302). (Annals of the New York Academy of Sciences, *155*). New York: New York Academy of Sciences.

Botwinick, J., & Thompson, L. W. (1966). Premotor and motor components of reaction time. *Journal of Experimental Psychology, 71,* 9-15.

Bramble, D. M., & Carrier, D. R. (1983). Running and breathing in mammals. *Science, 219,* 251-256.

Campbell, E. J. M., Agostoni, E., & Newsom Davis, J. (1970). *The respiratory muscles: Mechanics and neural central.* Philadelphia: W. B. Saunders.

Chernikoff, R., & Taylor, F. V. (1952). Reaction time to kinesthetic stimulation resulting from sudden arm displacement. *Journal of Experimental Psychology, 43,* 1-8.

Cole, K. J., & Abbs, J. H. (1983). Intentional responses to kinesthetic stimuli in orofacial muscles: Implications for the coordination of speech movements. *Journal of Neuroscience, 3,* 2660-2669.

Crago, P. E., Houk, J. C., & Hasan, Z. (1976). Regulatory actions of human stretch reflex. *Journal of Neurophysiology, 39,* 925-935.

Desmedt, J. (Ed.). (1978). *Cerebral mechanisms of motor control in man: Long loop mechanisms.* Basel: Karger.

Draper, M. H., Ladefoged, P., & Whitteridge, D. M. (1960). Expiratory pressures and airflow during speech. *British Medical Journal, 1*, 1837-1843.

Evarts, E. V., & Granit, R. (1976). Relations of reflexes and intended movements. In S. Homma (Ed.), *Progress in brain research: Understanding the stretch reflex* (pp. 1-14). Amsterdam: Elsevier.

Evarts, E. V., & Vaughn, W. J. (1978). Intended arm movements in response to externally produced arm displacements in man. In J. E. Desmedt (Ed.), *Cerebral motor control in man: Long loop mechanisms* (pp. 178-192). Basel: Karger.

Hammond, P. H. (1954). Involuntary activity in biceps following the sudden application of velocity to the abducted forearm. *Journal of Physiology (London), 127*, 23P.

Houk, J. C. (1978). Participation of reflex mechanisms and reaction-time processes in the compensatory adjustments to mechanical disturbances. In J. E. Desmedt (Ed.), *Cerebral motor control in man: Long loop mechanisms* (pp. 193-215). Basel: Karger.

Hufschmidt, H. J., Kilimov, N., & Linke, D. (1977). A very short reaction time in man. *Electroencephalography and Clinical Neurophysiology, 43*, 622.

Kugelberg, E., & Hagbarth, K. E. (1958). Spinal mechanism of the abdominal and erector spinal skin reflexes. *Brain, 81*, 290-304.

Lansing, R. W., & Meyerink, L. (1981). Load compensating responses of human abdominal muscles. *Journal of Physiology (London), 320*, 253-268.

Lee, R. G., & Tatton, W. G. (1975). Motor responses to sudden limb displacements in primates with specific CNS lesions and in human patients with motor system disorders. *Canadian Journal of Neurological Sciences, 2*, 285-293.

Luschei, E., Saslow, C., & Glickstein, M. (1967). Muscle potentials and reaction time. *Experimental Neurology, 18*, 429-442.

Marsden, C. D., Merton, P. A., & Morton, H. B. (1972). Servo action in human voluntary movement. *Nature, 238*, 140-143.

Meyerink, L. H., & Lansing, R. W. (1981). Short latency reaction times to loading of limb and respiratory systems: Within and across systems comparisons. *Neuroscience Abstracts, 7*, 476.

Miller, J., & Glickstein, M. (1967). Neural circuits involved in visuo-motor reaction time in monkeys. *Journal of Neurophysiology, 30*, 399-414.

Milner-Brown, H. S., Stein, R. B., & Lee, R. G. (1975). Synchronization of human motor units: possible roles of exercise and supraspinal reflexes. *Electroencephalography and Clinical Neurophysiology, 38*, 245-254.

Monnier, M. (1952). Retinal, cortical, and motor responses to photic stimulation in man: Retino-cortical time and optomotor integration time. *Journal of Neurophysiology, 15*, 469-486.

Newsom Davis, J., & Sears, T. A. (1970). The proprioceptive reflex control of the intercostal muscles during their voluntary activation. *Journal of Physiology, 209*, 711-738.

Pagni, L. A., Ettori, G., Infusa, L., & Marossero, F. (1964). EMG responses to capsular stimulation. *Experientia, 20*, 691.

Poulton, E. C. (1981). Human manual control. In V. B. Brooks (Ed.), *The nervous system: Motor Control, Part 2* (Vol. II). In S. R. Geiger (Ed.), *Handbook of physiology* (pp. 1337-1389). Bethesda: American Physiological Society.

Proctor, D. F. (1980). *Breathing, speech, and song*. New York: Springer-Verlag.

Smith, K. U. (1973). Physiological and sensory feedback of the motor systems: Neural metabolic integration for energy regulation in behavior. In J. D. Maser (Ed.), *Efferent organization and the integration of behavior* (pp. 19-65). New York: Academic Press.

Starkweather, C. W., Hirschman, P., & Tannenbaum, R. S. (1976). Latency of vocalization onset: Stutterers and non-stutterers. *Journal of Speech and Hearing Research, 19*, 481-492.

Wall, P. D., Remond, A. G., & Dobson, R. L. (1953). Studies on the mechanism of the action of visual afferents on motor cortex excitability. *Electroencephalography and Clinical Neurophysiology, 5*, 385-393.

Woodworth, R. S., & Schlosberg, H. (1970). *Experimental psychology*. New York: Holt, Rinehart & Winston.

Wyke, B. (1974). *Ventilatory and phonatory control systems*. London: Oxford University Press.

Wyrick, W., & Duncan, A. (1974). Electromyographical study of reflex premotion and simple reaction time of relaxed muscle to joint displacement. *Journal of Motor Behavior, 6*, 1-10.

18 · VISUAL TEMPORAL INTEGRATION FOR SIMPLE REACTION TIME

M. L. Kietzman, E. Shapiro, & G. E. Bruder

Queens College of the City University of New York
and
New York State Psychiatric Institute

Temporal integration (also called temporal summation, intensity-time reciprocity, or Bloch's law) refers to stimulus conditions for which the energy of the stimulus determines the response, regardless of how that energy is distributed over time. Traditionally, the experimental stimulus paradigm used to measure temporal integration is one in which the luminance needed to yield a constant level of performance, such as a 50% detection level, is determined for various stimulus durations (see, e.g., Barlow, 1958). This is an *equal-performance paradigm*. The longest stimulus duration displaying such complete integration of luminance over time is called *critical duration*. For longer stimulus durations, beyond critical duration, there is a breakdown in the intensity-time reciprocity relationship such that more luminance energy is needed to maintain a constant level of performance. It is as if some of the energy of the longer duration stimuli is not integrated, or is "lost."

The initial report (Bloch, 1885) of visual temporal integration was for the absolute threshold measurement of human subjects. Since then the empirical generality of this relationship has been demonstrated repeatedly in numerous psychophysical and electrophysiological studies for a number of different species and for a variety of different response measures.[1] However, Teichner and Krebs (1972) have concluded, based upon earlier studies (Grossberg, 1968; Raab & Fehrer, 1962; Raab, Fehrer, & Hershenson, 1961), that for simple RT, "It is clear that Bloch's law does not hold" (p. 352). Teichner and Krebs' conclusion that Bloch's law does not apply for simple RT has important implications. If true, then the processes underlying simple RT can be presumed to operate differently than processes underlying other response measures that have demonstrated temporal integration. If, on the other hand, intensity-time reciprocity occurs for RT, although perhaps differing in some respects from other response measures, then the generality of Bloch's law would be maintained.

[1] It is rarely noted that Hartline's early study (1934) of temporal integration in limulus reported that *both* frequency and latency of the spike discharge displayed complete temporal integration, with response latency having a shorter critical duration than response frequency.

 DOI: 10.4324/9781003164647- 23

Results from additional RT studies of temporal integration (Bernstein, Futch, & Schurman, 1973; Bruder & Kietzman, 1973; Kietzman & Gillam, 1972; Kietzman, Shapiro, & Bienstock, 1975; Mansfield, 1973; Pease, 1972; Ueno, 1976, 1977, 1978) do not support Teichner and Krebs' (1972) conclusion but give evidence that Bloch's law may be demonstrated for simple RTs.

An infrequently used stimulus paradigm that can also be employed to investigate temporal integration is to measure changes in performance to equal-energy stimuli of increasingly long stimulus durations. The rationale for using equal-energy stimuli is that for complete integration, the equal-energy stimuli should give the same response regardless of how that energy is distributed in time (Boynton, 1961). Thus, in the *equal-energy paradigm*, the level of performance for the shortest stimulus durations remains constant since the stimulus energy is constant; this is the region of complete temporal integration. However, at longer stimulus durations performance changes (e.g., RTs become slower) because these durations are longer than the critical duration and complete temporal integration for these stimuli is no longer obtained. There are two ways of varying the duration of the equal-energy stimuli: the first is to use two- or multiple-pulse stimuli of constant luminance but with varying dark intervals between the pulses. A second way of varying duration while keeping luminance energy constant is to use stimuli of increasingly longer durations but with reduced luminances. Both procedures have been used previously to investigate temporal integration for absolute threshold (e.g., Bouman & van den Brink, 1952; van den Brink & Bouman, 1954) or for simple reaction time (Bruder & Kietzman, 1973; Grossberg, 1968, 1970; Kietzman & Gillam, 1972; Ueno, 1976, 1977). The present experiment uses the first procedure.

For several years our laboratory has measured temporal integration for simple RTs using a number of different experimental paradigms. This report presents median RT data from several of these experiments, all of which employed an equal-energy, double-pulse stimulus paradigm. These data, collected under various diverse conditions, enable us to further test the generality of Bloch's law.

METHOD

Stimulus Manipulations. In the present study, RTs were measured to stimuli consisting of single and double pulses of light at energy levels slightly above threshold. The double-pulse stimuli consisted of pairs of 2-msec pulses separated by a variable dark interval. The use of dark intervals of different durations distributed the energy of the double-pulse stimulus over various stimulus durations. The independent variable was the total duration of the stimulus (for the double-pulse stimulus this was the duration of the entire stimulus package measured from the onset of the first pulse to the offset of the second pulse). The different double-pulse stimuli had total durations of from 5 to 35 msec. RTs were also measured to a 4-msec pulse of light equal in luminance and energy to all of the double-pulse stimuli. It was assumed that the 4-msec single-pulse stimulus, because of its brevity, was fully integrated, and therefore could serve as a baseline condition to evaluate the amount of integration of the double-pulse stimuli. Complete temporal integration is demonstrated, i.e., Bloch's law applies, when equal RTs are obtained to single- and double-pulse stimuli that have different total durations, but the same total luminous energy.

For most subjects, the luminance level chosen provided 90% frequency of seeing a single, 2-msec light pulse. Thus, the 4-msec pulse and the various double-pulse stimuli were seen essentially 100% of the time in all experiments, i.e., this was not a detection experiment.

Subjects. The data were collected from 26 subjects. Several of these subjects were tested in more than one experiment. All subjects were paid for their services and were uninformed as to the details of the experiment. Subjects either had normal vision or corrected normal vision.

Apparatus. Foveally fixated stimuli were presented either monocularly, using a Maxwellian view system, or binocularly, in a "free-viewing" situation. In the experiments using a Maxwellian view, stimuli were orange (573 nm), circular, and subtended a visual angle of 50 min. In the experiments using free viewing, the image of the crater of a gas-discharge tube was projected onto a diffusing glass to produce a white-appearing circular target, which subtended a visual angle of 22 min. Both optical systems used a glow-modular light source (Sylvania R1131C) which was operated at a constant current of 23 ma and was irradiated by an argon ultra-violet lamp (GE AR4). The light pulses had rise times of approximately 20 microseconds and decay times of 2 microseconds. Luminance was controlled by Kodak Wratten neutral density filters and Tiffen metallic filters. The duration of the light pulses and other events within each trial were controlled by a transistorized multivibrator timer with an indeterminacy of 1 part in 10,000 and a digital random time generator (Logical Instruments). The intensity and duration of the light pulses were constantly monitored on an oscilloscope display of the output of a photomulitplier tube (RCA 1P21). Reaction times to the nearest tenth of a millisecond, measured from the onset of the stimulus, were recorded on an electronic clock (Hewlett Packard, 5300A).

Procedure. Although the single- and double-pulse light stimuli were common to all the experiments used to collect these data, several other factors differed across the experiments. Some experiments used ancillary stimulation (such as electrocutaneous shocks or noise bursts) presented at brief intervals of time prior to or following the light pulses; other experiments employed a background noise; some experiments compared narrow and wide variable foreperiods; a few experiments compared RT and discrimination measures of temporal integration; other experiments used different probabilities of stimulus occurrence; some experiments had different levels of light adaptation; and there were experiments using various instructions and monetary payoffs.[2]

In all experiments, the subject was positioned either at a bite board (for the Maxwellian view) or a chin and forehead rest (for free viewing) and dark adapted for 10 minutes at the beginning of a session. A warning click was presented prior to the onset of the light stimulus. In the different experiments, the interval between the warning click and light stimulus was either a 2-sec fixed foreperiod, or a randomly variable foreperiod (in most cases the variable foreperiods ranged from 1.5 to 2.5 sec). Both single- and double-pulse light stimuli were presented in a random order within a block of trials. The subject was instructed to lift his finger from a telegraph key as fast as possible upon seeing the light.

RESULTS

Median RT data collected from all subjects who participated in the various experiments are shown in Figure 18.1. Each data point represents the median RTs of a single subject to the two common stimulus conditions: RTs to the 4.0-msec single pulse of light (abscissa) and RTs to a double pulse of light (ordinate). The total duration (TD) of the double-pulse stimuli are shown as the parameter of the figure. In most cases, the number of trials for the two stimulus conditions of each data point ranged from 60 to 100 trials with a maximum of 832 trials and a minimum of 36 trials. All together, Figure 18.1 depicts 36,624 RTs.

The 45° diagonal line in each quadrant of Figure 18.1 represents the prediction, based on Bloch's law, of equal RTs to the equal-energy, single- and double-pulse stimuli. Data points that fall above this equal-performance diagonal indicate slower RTs to the double-pulse stimuli than to the single-pulse stimulus, while points below the line indicate faster RTs to the double-pulse stimuli.

These data show a definite region of complete temporal integration for simple RT as can be seen in Figure 18.1a where data points of 20 subjects for double-pulse stimuli of 5-10 msec cluster closely around the equal-

[2] Although these additional parameters under which the data were collected frequently altered RT levels, they did not appear to systematically influence estimates of critical duration. Therefore, no attempt was made to indicate these parameters in Figure 18.1.

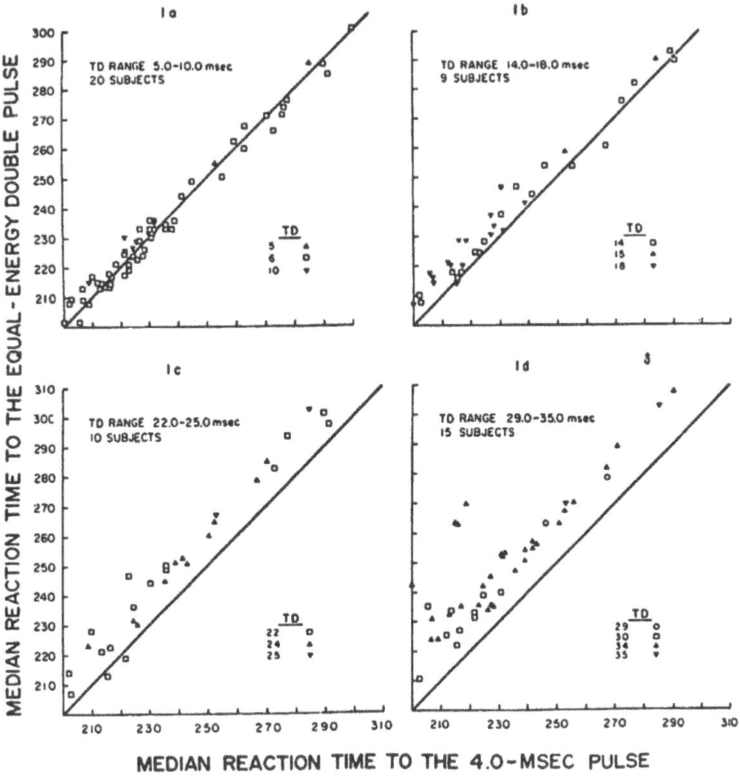

FIGURE 18.1. Scatter diagram of the median RTs (in msec) for the 4-msec stimulus and the equal-energy double-pulse stimuli to total durations (TDs) ranging from 5 to 35 msec.

performance diagonal (there are an equal number above and below the diagonal). In comparison, for total durations of 14-18 msec (Figure 18.1b), most of the data points fall above the equal-performance diagonal. Further increases in total duration of the double-pulse stimulus, as in Figure 18.1c and 1d, result in even greater deviations above the diagonal. Beyond 22 msec all subjects for all conditions show longer RTs to the double-pulse stimuli than to the equal-energy 4-msec stimulus. Therefore, critical duration for these median RT data must be between 10 and 22 msec.

DISCUSSION

These results lead to several conclusions. First, simple RT does display temporal integration but this integration is over short periods of time, i.e., simple RT has a brief critical duration. These data are in agreement with prior RT studies that have reported complete temporal integration for single-pulse stimuli (Bernstein et al., 1973; Bruder & Kietzman, 1973; Hildreth, 1973; Kietzman & Gillam, 1972; Lewis, 1964; Mansfield, 1973; Ueno, 1977, 1978), and for double-pulse stimuli (Kietzman & Gillam, 1972; Pease, 1972; Ueno, 1976). The previous studies also reported brief critical durations, a result that is consistent with our 10- to 22-msec estimate of critical duration and much briefer than the 50- to 100-msec value frequently cited as the critical duration for visual threshold measures (see also, Bruder & Kietzman, 1973). Thus, simple RT can be included among the response measures demonstrating Bloch's law. The diversity of the experimental conditions under which these data were obtained as well as the large number of RTs, and the numerous subjects tested, emphasizes the generality of this conclusion.

A second conclusion of this study is that temporal integration can be tested with different stimulus paradigms and conditions. Specifically, these results show that the equal-energy paradigm is a viable procedure for measuring temporal integration. Also, they indicate that the two-pulse stimulus condition can display temporal integration, lending additional weight to prior studies which have suggested a broadened definition of temporal integration to include multiple pulses of light (Davy, 1952; Herrick, 1974). The general conclusion is that temporal integration is a vigorous phenomenon that is not dependent upon specific experimental paradigms, stimulus conditions, or responses measures.

The fact remains that a certain number of experiments (Grossberg, 1968, 1970, 1974; Raab & Fehrer, 1962; Raab, Fehrer, & Hershenson, 1961) have not reported temporal integration for simple RT, and that Teichner and Krebs (1972) have concluded that Bloch's law does not apply to RT. There are several possible reasons for the apparent discrepancy between the results reported here and by others, and the negative results of the earlier studies. The discrepancies across studies are best understood by noting that temporal integration for RT is appropriately considered a "small effect," one which requires considerable attention to certain experimental details. First, the short critical durations of temporal integration for RT as compared to other responses make it essential that a sufficient number of very brief stimuli be tested so that integration may be displayed. A second factor is that the stimuli to be tested must be low-energy stimuli (near threshold). If the stimulus intensity is very far above threshold then the changes in RT are too small relative to the measure of RT variability to adequately show temporal integration. A third factor is the need for precise energy control of the stimulus; for example, the equal-energy stimuli must be as equal as possible. One advantage of the double-pulse procedure, with its changes in duration rather than intensity of the stimulus, is that the precision of manipulating time (the stimulus duration) is much better than the precision of manipulating intensity.

A potentially complicating fourth factor in this area of research is the possibility that non-sensory factors can be introduced into temporal integration experiments and confound the results. There is ample evidence (Greenbaum, 1963; Grice, 1968; Murray, 1970) that RT measures are sensitive to response criterion or biasing factors, and in reaction time-temporal integration experiments it is essential to eliminate, control, or measure such confounding variables. For example, in an integration-RT study, Hildreth (1973) used an experimental design that allowed for an asymmetry in the programming of the independent variable, and as a result non-sensory factors were introduced that modified the integration results. One way to control for such biasing is to use a "random by trial" programming procedure, in which any stimulus condition can occur on any given trial. This prevents subjects from adopting a different criterion or set towards the different stimulus conditions.

The stability and strength of the phenomena of visual temporal integration across the different experimental paradigms and stimulus conditions for different response measures as seen in this and other psychophysical and physiological investigations of human and subhuman species strongly suggest that temporal integration is a fundamental neural process. Future physiological and behavioral models of sensory and information processing should be able to incorporate the empirical evidence describing the characteristics of temporal integration. The data presented here suggest that the latency of response, in this case simple reaction time, can and should be included in such model building.

ACKNOWLEDGMENT

This research was supported by USPHS Grants MH-11688 and MH-18191 awarded to M. Kietzman. We want to acknowledge the assistance of R. Laupheimer and R. Simon for the design, construction and maintenance of the equipment, and I. Berenhaus, P. Collins, R. Levine, and M. Zalusky for their help with data collection and analysis. We also wish to thank M. Wallach, Director of the Kingsboro Psychiatric Center, for providing space for laboratory facilities.

REFERENCES

Barlow, H. B. (1958). Temporal and spatial summation in human vision at different background intensities. *Journal of Physiology, 141*, 337-350.

Bernstein, I. H., Futch, D. G., & Schurman, D. L. (1973). Some exposure duration effects in simple reaction time. *Journal of Experimental Psychology, 97*, 317-322.

Bloch, A. M. (1885). Experiences sur la vision. *Societe De Biologic Comptes Rendus, 11*, 493-496.

Bouman, M. A., & van den Brink, G. (1952). On the integrate capacity in time and space of the human peripheral retina. *Journal of the Optical Society of America, 42*, 617-660.

Boynton, R. (1961). Some temporal factors in vision. In W. A. Rosenblith (Ed.), *Sensory communication* (pp. 739-756). New York: Wiley.

Bruder, G. E., & Kietzman, M. L. (1973). Visual temporal integration for threshold, signal detectability, and reaction time measures. *Perception and Psychophysics, 13*, 293-300.

Davy, E. (1952). The intensity time relation for multiple flashes of light in the peripheral retina. *Journal of the Optical Society of America, 42*,937-941.

Greenbaum, H. B. (1963). *Simple reaction time: A case study in signal detection*. Unpublished doctoral dissertation, Columbia University.

Grice, G. R. (1968). Stimulus intensity and response evocation. *Psychological Review, 75*, 359-373.

Grossberg, M. (1968). The latency of response in relation to Bloch's law at threshold. *Perception and Psychophysics, 4*, 229-232.

Grossberg, M. (1970). Frequencies and latencies in detecting two-flash stimuli. *Perception and Psychophysics, 7*, 377-380.

Grossberg, M. (1974). Failure of Bloch's law for simple reaction time. *Bulletin of the Psychonomic Society, 4*, 147-149.

Hartline, H. K. (1934). Intensity and duration in the excitation of single photo receptor units. *Journal of Cellular and Comparative Physiology, 5*, 229-247.

Herrick, R. M. (1974). Foveal light detection threshold with two temporally spaced flashes. *Perception and Psychophysics, 15*, 316-367.

Hildreth, J. D. (1973). Bloch's law and a temporal integration model for simple reaction time to light. *Perception and Psychophysics, 14*, 421-432.

Kietzman, M. L., & Gillam, B. J. (1972). Visual temporal integration and simple reaction time. *Perception and Psychophysics, 11*, 333-340.

Kietzman, M. L., Shapiro, E., & Bienstock, B. (1975, April). *Visual temporal integration and simple reaction time for equal-energy stimuli of different luminances*. Paper presented at the meeting of the Eastern Psychological Association, New York.

Lewis, M. F. (1964). *Magnitude estimation and reaction time as a function of flash luminance and duration in the fovea*. Unpublished doctoral dissertation, Columbia University.

Mansfield, R. J. W. (1973). Latency functions in human vision. *Vision Research, 13*, 2219-2234.

Murray, H. G. (1970). Stimulus intensity and reaction time: Evaluation of a decision-theory model. *Journal of Experimental Psychology, 84*, 383-391.

Pease, V. P. (1972). Effect of luminance and duration of interstimulus interval upon human reaction time. *Journal of the Optical Society of America, 62*, 1505-1507.

Raab, D. H., & Fehrer, E. (1962). The effects of stimulus duration and luminance upon human reaction time. *Journal of Experimental Psychology, 64*, 326-327.

Raab, D. H., Fehrer, E., & Hershenson, M. (1961). Visual reaction time and the Broca-Sulzer phenomenon. *Journal of Experimental Psychology, 61*, 193-199.

Teichner, W., & Krebs, M. (1972). Laws of simple visual reaction time. *Psychological Review, 79*, 344-358.

Ueno, T. (1976). Temporal summation and reaction time to double-light pulses at suprathreshold levels. *Perception and Psychophysics, 19*, 399-404.

Ueno, T. (1977). Reaction time as a measure of temporal summation at suprathreshold levels. *Vision Research, 17*, 227-232.

Ueno, T. (1978). Temporal summation in human vision: Simple reaction time measurements. *Perception and Psychophysics, 23*, 43-50.

van den Brink, G., & Bouman, M. A. (1954). Variation of integrative actions in the retinal system: An adaptational phenomena. *Journal of the Optical Society of America, 44*, 616-620.

19 · VERBAL-MANUAL TIME-SHARING AND EEG ASYMMETRY IN MENTALLY RETARDED INDIVIDUALS

G. C. Galbraith and J. B. Gliddon

University of California at Los Angeles
and
Lanterman Developmental Center

Generally when two or more sensorimotor tasks are performed concurrently, i.e., the tasks are time shared, an overall performance decrement results. Moreover, dual motor activities that involve interactions within the same cortical hemisphere suffer greater disruption than dual activities involving different hemispheres. Thus, in normal, right-handed subjects, speaking while simultaneously performing a manual motor task depresses right-hand scores more than left-hand scores (Hicks, 1975; Kinsbourne & Cook, 1971). This effect is especially evident in children. Thus, Kinsbourne and McMurray (1975) studied the effects of speaking on tapping with the left or right index finger in preschool children. Although speaking diminished tapping rates for both fingers, the effect was greater on the right side. Hiscock (1982) reported that as many as 85.5% of right-handed children show the expected pattern of performance asymmetry.

Time-shared performance asymmetry in right-handed subjects is attributed to left hemisphere lateralization of both speech (Rasmussen & Milner, 1975; Zangwill, 1967) and motor control of the contralateral right hand (Brinkman & Kuypers, 1972). In left-handed adults, who evidence less cortical lateralization (Rasmussen & Milner, 1975), speaking may equally disrupt performance of either hand (Hicks, 1975). Time-shared performance asymmetries thus provide a useful technique for assessing hemispheric dominance and patterns of intra-hemispheric interaction.

Functional hemispheric asymmetry may also be inferred from EEG and evoked potential recordings. Thus, EEG studies have consistently reported task-related alpha suppression for recordings over the left hemisphere during linguistic processing, and suppression for recordings over the right hemisphere during visuospatial processing (Butler & Glass, 1974; Galin & Ellis, 1975; Galin & Ornstein, 1972; Robbins & McAdam, 1974). Although EEG asymmetry is most pronounced during performance of specific lateralizing tasks, resting (baseline) EEG asymmetry may also reflect hemispheric dominance (Aird & Gastaut, 1959; Creutzfeldt, Grünewald, Simonova, & Schmitz, 1969; Margerison, St. John-Loe, & Binnie, 1967).

DOI: 10.4324/9781003164647- 24

Evoked potential studies have also reported amplitude differences over hemispheres engaged in lateralized tasks (Brown, Marsh, & Smith, 1973, 1979; Buchsbaum & Fedio, 1970; Wood, Goff, & Day, 1971).

The present study investigated patterns of time-shared motor performance and EEG asymmetry in mentally retarded individuals. Such individuals frequently have defects in motor coordination and speech. Moreover, the lateralization of cerebral function for mentally retarded subjects is often less clear, as evidenced by their showing an increased left-hand preference as the severity of retardation increases (Hicks & Barton, 1975; Hicks & Kinsbourne, 1978). Finally, electrophysiological studies of mentally retarded individuals also show diminished or altered patterns of hemispheric asymmetry when compared with nonretarded controls (Bigham, Dustman, & Beck, 1970; Gliddon, Busk, & Galbraith, 1975; Rhodes, Dustman, & Beck, 1969).

This investigation of mentally retarded individuals used the behavioral paradigm of Kinsbourne and McMurray (1975). Eyes closed and eyes open EEG activity recorded from over left and right hemispheres were submitted to a detailed Fourier spectral frequency analysis. It was thus possible to compare the time-shared behavioral performance with patterns of asymmetry in the baseline EEG. In this manner, we were able to investigate several dimensions of lateralization of brain function in mentally retarded persons.

METHOD

Subjects. Thirty-six institutionalized mentally retarded individuals served as subjects. All subjects had previously come to the laboratory for an EEG recording and were known to be cooperative. Ages ranged from 13.1 to 54.2 years (mean = 24.8, S.D. = 10.4); IQs ranged from 12 to 73 (mean = 42.4, S.D. = 14.0).

EEG Recording and Analysis. Baseline EEG data were recorded from all subjects in a resting condition with eyes closed and eyes open. Scalp electrodes were placed over occipital and parietal areas of the left (O_1 and P_3) and right (O_2 and P_4) hemispheres. Interelectrode resistance was always kept below 10K ohms. Data were recorded on a standard ink oscillograph as well as on magnetic tape for later computer analysis. The ink tracings were evaluated for the presence of artifacts, and the corresponding artifact-free segments of the magnetic tape were digitized (100 samples/sec) and analyzed by means of fast Fourier techniques to determine spectral intensities (Walter, 1963). It was thus possible to quantify the intensity of EEG signals over both hemispheres, in a frequency range of 0.86 to 23.2 c/sec, using a band width of 0.86 c/sec.

By computing left/right spectral intensity ratios, it was possible to determine the degree of hemispheric asymmetry. Thus, O_1/O_2 and P_3/P_4 ratios were formed for eyes open and eyes closed EEGs. Ratios were computed in the following frequency bands: Theta (4.3-6.9 c/sec), $alpha_1$ (7.7-9.5 c/sec), $alpha_2$ (10.3-12.0 c/sec), $beta_1$ (12.9-15.5 c/sec) and $beta_2$ (16.3-18.9 c/sec).

Tapping Apparatus and Procedure. Using the index finger, subjects tapped a 1 x 14 cm spacing bar of a computer keyboard. The computer was programmed to determine, with an accuracy of 5 msec, the interval of time between successive taps. The computer also provided summary statistics as well as interval histogram plots of the distribution of intertap intervals.

Subjects were first evaluated in terms of their ability to cooperate and to tap consistently in the various conditions of the experiment. Subjects were asked to name such items as different colors, favorite television shows, friends at the hospital, animals, games, etc. Those categories which yielded the largest repertoire of verbal responses for a given individual were subsequently used in the experiment during the speech trials. When it was apparent that the subject understood and was able to perform the task, then the following four conditions were given: (a) tapping with left hand; (b) tapping with right hand; (c) tapping with left hand while speaking; and (d) tapping with right hand while speaking. Each subject received a different randomized sequence of these conditions. At least 2 min of data were obtained in each condition, not including rest pauses, if needed. Upon completion of the experiment, each subject was asked to carry out a series of tasks in order to assess handedness. These included picking up objects, drawing on the blackboard, printing his/her name (if possible), etc. A long play record album was presented as a reward for participation.

Tapping Analysis. Means and variances were computed for the time interval between taps. In addition, a statistic used by Tingley and Allen (1975) to assess speech timing control in children, the half mean-square successive difference (von Neumann, Kent, Bellison, & Hart, 1941), was computed as a means of quantifying a second parameter of tapping variability. This statistic, q^2, is defined as follows:

$$q^2 = \frac{1}{2(N-1)} \sum_{i=2}^{N} (t_i - t_{i-1})$$

where N represents the number of taps, and t_i is the time (in msec) of the ith tap. The q^2 measure was used to detrend the data and eliminate slow changes in tapping rate. It was also used to quantify the temporal stability of tapping, since q^2 is small when tapping is both rapid and consistent, and large during periods of erratic tapping.

In order to assess the possible disruptive effects of speech upon tapping, (speech+tapping)/tapping ratios were formed for left- and right-hand performances, for both the time interval and the q^2 statistics. Speech interference upon tapping would result in ratios larger than 1.0 since such interference would slow the overall tapping rate (increase the mean interval between taps) as well as produce greater variability in tapping patterns (i.e., a larger q^2).

Comparison of Tapping and EEG. Correlation coefficients were computed between EEG and tapping ratios. Since there was an age range among subjects that might affect both EEG and motor performance, the final statistical analysis involved a computation of partial correlation coefficients with age held constant.

RESULTS

Among the 36 subjects, 9 (25%) were predominantly left-handed. Analysis of variance of tapping scores showed that there were no significant differences between left- and right-handed subjects; hence, the data for all subjects are pooled in the analyses that follow.

Figure 19.1 graphically illustrates the tapping performance of two subjects. Figure 19.1A is that of a nonretarded control subject who shows the typical asymmetric pattern reported for normal right-handed children. Thus, the interval histogram distributions for left-hand tapping (top panel) are virtually identical during tapping alone and tapping while speaking. During right-hand tapping (bottom panel), however, the distributions reveal quite different means and variabilities. Thus, the generally faster and more consistent tapping alone pattern becomes relatively slower and more variable during time-shared tapping and speaking. Figure 19.1B illustrates the performance of a mentally retarded subject that is typical of most retarded subjects in the present study. It is apparent that speaking interfered equally with the tapping of *either* hand.

Table 19.1 presents overall means and standard deviations for the four tapping conditions. Both the intertap interval and the q^2 measures show sizable increases when speech and tapping are time shared, but the results for left- and right-hand tapping alone conditions are virtually identical. Thus, significant repeated measure F-ratios for the interval ($F = 26.713$, $df = 3{,}105$, $p < .001$) and q^2 ($F = 9.679$, $df = 3{,}105$, $p < .001$) measures were due entirely to the larger values obtained during the time-shared performance.

We were particularly interested in measures of variability as a means of assessing the interference of speech upon tapping. Table 19.2 shows the results of a series of t tests for equality of variance for dependent samples.

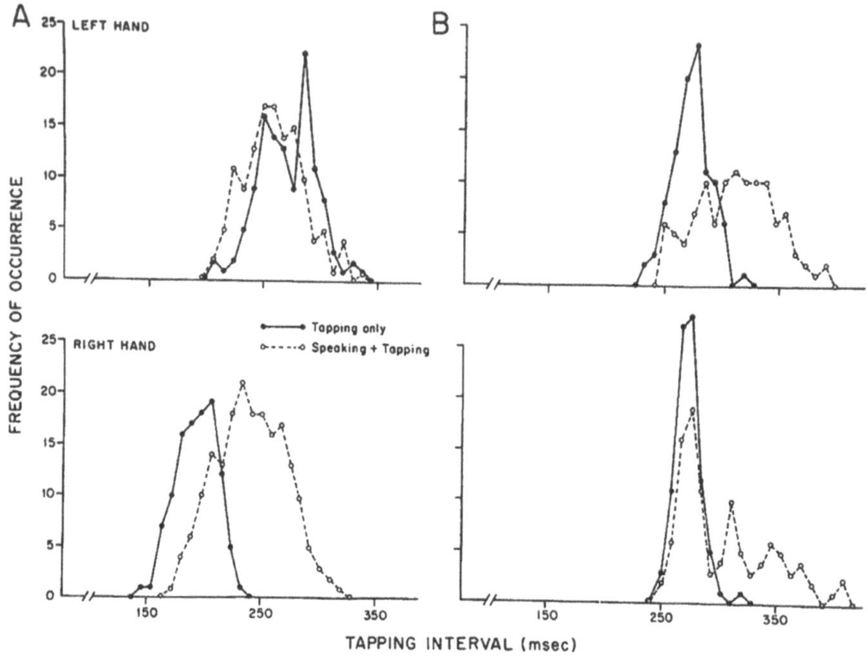

FIGURE 19.1. Frequency distribution of tapping intervals (measured to 5 msec accuracy). Left panel: 10-year-old, right-handed, control subject. Right panel: 37-year-old retarded man, right-handed, IQ 33.

TABLE 19.1. Summary statistics.[a]

Condition	Intertap Interval (in msec)	Half Mean Square Successive Difference (q^2)
Left Tap	217 (39)	125 (80)
Right Tap	218 (38)	118 (84)
Left Tap plus Speech	258 (60)	198 (148)
Right Tap plus Speech	260 (51)	203 (173)

[a] Group means with standard deviations in parentheses.

TABLE 19.2. *T* test for equality of variances (dependent samples).[a]

	Intertap Interval (in msec)			Half Mean Square Successive Difference (q^2)		
	Right tap	Left tap + Speech	Right tap + Speech	Right tap	Left tap + Speech	Right tap + Speech
Left tap	0.02	3.34*	2.12	0.50	4.78**	5.41**
Right tap		2.98*	2.34		4.32**	5.50**
Left tap + Speech			-1.60			1.32

[a] t test is based on $df = 35, p = .05/6 = .0083$ and $.01/6 = .0012$.
* $p < .0083$
** $p < .0012$

Since there were six pairwise comparisons, the Bonferroni correction (Kirk, 1982) was applied to all significance levels (Type I error rate divided by 6). For the interval measure only left-hand tapping during speech showed significantly greater variability than tapping alone. However, the q^2 measure shows that both left- and right-hand tapping during speech were more variable than tapping alone, with the largest t values occurring for right-hand tapping.

Table 19.3 presents the results of a correlational analysis of EEG spectral intensity ratios (left hemisphere/right hemisphere) and tapping ratios (tapping + speech)/(tapping alone). Eyes open and eyes closed EEG spectral ratios in 30 frequency bins were cross-correlated with the interval and q^2 tapping ratios, computed for each hand. Table 19.3 presents a frequency tally of the number of significant correlation coefficients (out of 30 possible in each cell). The results are clearly skewed in favor of the q^2 measure, which shows a grand total of 79 significant correlations versus only 7 for the intertap interval measure. Moreover, all significant correlations are positive in sign, indicating that larger tapping ratios (greater time-shared interference) are associated with larger left/right EEG asymmetries (relatively greater baseline activation in the right hemisphere).

Table 19.4 presents actual correlations between EEG asymmetry ratios in five frequency bands and q^2 tapping ratios. Eleven significant correlations out of 20 possible were obtained, all positive in sign. The greatest number of

TABLE 19.3. Number of significant correlation coefficients between EEG spectral intensity ratios (left hemisphere/right hemisphere) and verbal-manual tapping ratios ((Tapping + Speech)/Tapping only).[a]

		Eyes open EEG		Eyes closed EEG		
Tapping measure	Tapping hand	O_1/O_2	P_3/P_4	O_1/O_2	P_3/P_4	
Interval	Left	5	1	0	0	
	Right	0	0	0	1	
						Total = 7
q^2	Left	10	15	17	1	
	Right	2	17	5	12	
						Total = 79

[a] Each cell count represents the number of significant ($p < .05$) correlations out of 30 possible.

TABLE 19.4. Correlations between eyes open EEG spectral intensity ratios (left hemisphere/right hemisphere) and the q^2 measure of verbal-manual tapping ratios ((Tapping + Speech)/Tapping only).[a]

	Left-Hand Tapping					Right-Hand Tapping				
EEG Ratio	Theta	$Alpha_1$	$Alpha_2$	$Beta_1$	$Beta_2$	Theta	$Alpha_1$	$Alpha_2$	$Beta_1$	$Beta_2$
O_1/O_2	.07	.10	.44**	.45**	.39*	.04	-.11	.38*	.31	.24
P_3/P_4	.24	.49**	.47**	.36*	.33	.33	.51**	.45**	.43**	.44**

[a] Frequency bands (in c/s): Theta (4.3-6.9), $Alpha_1$ (7.7-9.5), $Alpha_2$ (10.3-12.0), $Beta_1$(12.9-15.5), $Beta_2$ (16.3-18.9).
* $p < .05$
** $p < .01$

significant correlations (7 out of 11) occur for the parietal leads. The fact that all significant correlations were positive in sign indicates a relatively greater EEG amplitude in the left versus right hemisphere, which is associated with greater disruption of tapping performance during speech. The strongest effect was observed in the alpha frequency bands, although effects were also seen in beta frequency bands as well (but never in theta).

DISCUSSION

The observation that 24% of our mentally retarded subjects were left-handed agrees reasonably well with published observations. Thus, Hécaen and DeAjuriaguerra (1964) reported an incidence of 17% left-handedness in mentally retarded persons compared to an 8% incidence in normal control children. Hicks and Barton (1975) reported a 13% incidence in a mildly and moderately retarded group, but this increased to 28% in a severely and profoundly retarded group. Bradshaw-McAnulty, Hicks, and Kinsbourne (1984) also found increasing left-hand preference with increasing severity of mental retardation. A reasonable explanation of this observation is the fact that pathological sinistrality is known to result from brain damage (Satz, 1972, 1973).

Figure 19.1B illustrates the modal response pattern observed in our sample of mentally retarded subjects. Thus, concurrent speaking equally disrupted left- and right-hand tapping performances. This effect is clearly seen in the group results of Table 19.1, in which left- and right-hand means and standard deviations are quite similar (only the overall difference between tapping alone and tapping plus speech was significantly different). These results are qualitatively similar to those reported for normal left-handed adults (Hicks, 1975), in whom it appears that the cerebral organization of speech is more varied and less well established (Rasmussen & Milner, 1975). These results suggest two conclusions: (a) the cerebral dominance for speech in our sample of mentally retarded subjects is poorly established, and (b) the Kinsbourne and McMurray paradigm of verbal-manual interference provides a useful assessment of cerebral dominance.

It appears that there may be certain experimental conditions in which variability measures are more sensitive to tapping disruption (e.g., Kee, Morris, Bathurst, & Hellige, 1986). However, in the present study, verbal-manual interference was reflected in both slower tapping rates and more variable performance. Hiscock, Kinsbourne, Samuels, and Krause (1987) reported a similar result in their study of finger tapping and memory encoding in children, but only tapping rate was sensitive to the effects of stimulus category, difficulty level, age, and hand. Nevertheless, they indicate that the failure of variability in the memory-encoding condition should not be generalized to all dual task conditions. Moreover, it should be noted that their measure of variability was the coefficient of variation (standard deviation divided by the mean), whereas in the present study we used the half mean square successive difference, or q^2 (von Neumann et al., 1941), which has quite different properties than the standard deviation. For example, consider the following two number sequences representing hypothetical tapping intervals (arbitrary units): 1, 1, 5, 5 and 1, 5, 1, 5. Each sequence results in identical interval means, standard deviations and histogram plots. Clearly, however, the second sequence represents a more variable tapping performance. This added dimension of variability is quantified by q^2, which is larger by a factor of 3 for the latter sequence. The robust nature of q^2 is also indicated in the results of Table 19.2 ,which tested for the equality of variances among the different experimental conditions. Larger t values were obtained for q^2 than for the interval standard deviation. Moreover, both left- and right-hand results were significant for q^2, with concurrent speech and tapping being significantly more variable.

Evidence for increased sensitivity of the q^2 measure is also seen in the results of Table 19.3. Thus, the number of significant correlations between the interval measure and EEG activity is no more than expected by chance alone. However, the number of significant correlations involving the q^2 measure far exceeds chance expectation. Moreover, the nonrandom basis of these findings is further enhanced by the fact that all significant correlation coefficients are positive in sign. The effect is strongest for the P_3/P_4 (eyes open) EEG.

Table 19.4 presents the actual correlations between q^2 tapping ratios and eyes open EEG spectral intensity ratios computed within conventional frequency bands. Overall, P_3/P_4 ratios show a greater number of significant correlations than O_1/O_2 ratios. This difference is most apparent for right-hand tapping, and in the Alpha_1 band for both left- and right-hand tapping. All significant correlations are positive in sign. Thus, parietal EEG, rather than occipital EEG, is more strongly correlated with the degree of time-shared interference. This finding may be due to the fact that parietal electrode placements are in closer proximity to brain centers involved in the control of speech, i.e., Wernicke's area. This parietal-occipital difference is most evident in the Alpha_1 (7.7-9.5 c/s) band. This may reflect the registration in parietal leads of the sensorimotor "mu" rhythm, an alpha-like rhythm recorded over motor regions during the voluntary suppression of movement (Sterman & Wyrwicka, 1967).

A positive correlation between q^2 tapping ratios (speech+ tapping)/(tapping alone) and EEG spectral intensity ratios (left hemisphere/right hemisphere) implies that larger tapping ratios (greater time-shared interference) is associated with relatively greater EEG amplitudes in the left hemisphere (an electrophysiological sign of left hemisphere *de*activation). In other words, subjects in whom the baseline (eyes open) EEG show relatively less left hemisphere dominance also show a greater propensity for disruption during time-shared motor performance. Since the EEG was recorded under baseline conditions, it might be assumed that the EEG ratios reflect a general level of CNS lateralization (an assessment of CNS maturation). Thus, large left/right ratios imply a less mature pattern, since one would expect relatively more left hemisphere activation in predominantly right-handed subjects. We know that children show greater time-shared interference than adults (Hiscock, 1982). In our sample, those subjects with less left hemisphere lateralization also showed greater interference, an effect that is analogous to the child-adult comparison.

In conclusion, the assessment of baseline hemispheric EEG patterns is seen to correlate significantly with interference during time-shared motor performance in a sample of institutionalized mentally retarded individuals. Thus, the quantitative assessment of the EEG may prove useful in predicting and understanding the mechanisms involved in motor performance in such individuals. It is possible that such assessments may prove helpful in prescribing sensorimotor remediation.

REFERENCES

Aird, R. B., & Gastaut, Y. (1959). Occipital and posterior electroencephalographic rhythms. *Electroencephalography and Clinical Neurophysiology, 11*, 637-656.

Bigham, H. B., Dustman, R. E., & Beck, E. C. (1970). Visual and somatosensory evoked responses from mongoloid and normal children. *Electroencephalography and Clinical Neurophysiology, 28*, 576-585.

Bradshaw-McAnulty, G., Hicks, R. E., & Kinsbourne, M. (1984). Pathological left-handedness and familial sinistrality in relation to degree of mental retardation. *Brain and Cognition, 3*, 349-356.

Brinkman, J., & Kuypers, H. G. J. (1972). Split-brain monkeys: Cerebral control of ipsilateral and contralateral arm, hand and finger movements. *Science, 176*, 536-539.

Brown, W. S., Marsh, J. T., & Smith, J. C. (1973). Contextual meaning effects on speech-evoked potentials. *Behavioral Biology, 9*, 755-761.

Brown, W. S., Marsh, J. T., & Smith, J. C. (1979). Principal component analysis of ERP differences related to the meaning of an ambiguous word. *Electroencephalography and Clinical Neurophysiology, 46*, 709-714.

Buchsbaum, M., & Fedio, P. (1970). Hemispheric differences in evoked potentials to verbal and nonverbal stimuli in the left and right visual fields. *Physiology and Behavior, 5*, 207-210.

Butler, S. R., & Glass, A. (1974). Asymmetries in the electroencephalogram associated with cerebral dominance. *Electroencephalography and Clinical Neurophysiology, 36*, 481-491.

Creutzfeldt, O., Grünewald, G., Simonova, O., & Schmitz, H. (1969). Changes of the basic rhythms of the EEG during the performance of mental and visuomotor tasks. In C. R. Evans & T. B. Mulholland (Eds.), *Attention in neurophysiology* (pp. 148-168). London: Butterworths.

Galin, D., & Ellis, R. R. (1975). Asymmetry in evoked potentials as an index of lateralized cognitive processes: Relations to EEG alpha Asymmetry. *Neuropsychologia, 13*, 45-50.

Galin, D., & Ornstein, R. (1972). Lateral specialization of cognitive mode: An EEG study. *Psychophysiology, 9*, 412-418.

Gliddon, J. B., Busk, J., & Galbraith, G. C. (1975). Visual evoked responses as a function of light intensity in Down's syndrome and nonretarded subjects. *Psychophysiology, 12*, 416-422.

Hécaen, H., & DeAjuriaguerra, J. (1964). *Left-handedness: Manual superiority and cerebral dominance*. New York: Grune & Stratton.

Hicks, R. E. (1975). Intrahemispheric response competition between vocal and unimanual performance in normal adult human males. *Journal of Comparative and Physiological Psychology, 89*, 50-60.

Hicks, R. E., & Barton, A. K. (1975). A note on left-handedness and severity of mental retardation. *Journal of Genetic Psychology, 127*, 323-324.

Hicks, R. E., & Kinsbourne, M. (1978). Human handedness. In M. Kinsbourne (Ed.), *Asymmetrical function of the brain* (pp. 523-549). New York: Cambridge University Press.

Hiscock, M. (1982). Verbal-manual time sharing in children as a function of task priority. *Brain and Cognition, 1*, 119-131.

Hiscock, M., Kinsbourne, M., Samuels, M., & Krause, A. E. (1987). Dual task performance in children: Generalized and lateralized effects of memory encoding upon the rate and variability of concurrent finger tapping. *Brain and Cognition, 6*, 24-40.

Kee, D. W., Morris, K., Bathurst, K., & Hellige, J. B. (1986). Lateralized interference in finger tapping: Comparisons of rate and variability measures under speed and consistency tapping instructions. *Brain and Cognition, 5*, 268-279.

Kinsbourne, M., & Cook, J. (1971). General and lateralized effects of concurrent verbalization on a unimanual skill. *Quarterly Journal of Experimental Psychology, 23*, 341-345.

Kinsbourne, M., & McMurray, J. (1975). The effect of cerebral dominance on time sharing between speaking and tapping by preschool children. *Child Development, 46*, 240-242.

Kirk, R. E. (1982). *Experimental design: Procedures for the behavioral sciences* (pp. 106-109). Monterey: Brooks/Cole Publishers.

Margerison, J. H., St. John-Loe, P., & Binnie, C. D. (1967). Electroencephalography. In P. H. Venables & I. Martin (Eds.), *A manual of psychophysiological methods* (pp. 351-402). Amsterdam: North Holland.

Rasmussen, T., & Milner, B. (1975). Clinical and surgical studies of the cerebral speech areas in man. In K. J. Zülch, O. Creutzfeldt, & G. C. Galbraith (Eds.), *Cerebral localization: An Otfrid Foerster symposium* (pp. 238-257). Heidelberg: Springer-Verlag.

Rhodes, L. E., Dustman, R. E., & Beck, E. C. (1969). The visual evoked response: A comparison of bright and dull children. *Electroencephalography and Clinical Neurophysiology, 27*, 364-372.

Robbins, K. I., & McAdam, D. (1974). Interhemispheric alpha asymmetry and imagery mode. *Brain and Language, 1*, 189-193.

Satz, P. (1972). Pathological left-handedness: An explanatory model. *Cortex, 8*, 121-135.

Satz, P. (1973). Left-handedness and early brain insult: An explanation. *Neuropsychologia, 11*, 115-117.

Sterman, M. B., & Wyrwicka, W. (1967). EEG correlates of sleep: Evidence for separate forebrain substrates. *Brain Research, 6*, 143-163.

Tingley, B. M., & Allen, G. D. (1975). Development of speech timing control in children. *Child Development, 46*, 186-194.

von Neumann, J., Kent, R. H., Bellison, H. R., & Hart, B. I. (1941). The mean square successive difference. *Annals of Mathematical Statistics, 12*, 153-162.

Walter, D. O. (1963). Spectral analysis for electroencephalograms: Mathematical determination of neurophysiological relationships from records of limited duration. *Experimental Neurology, 8*, 155-181.

Wood, C., Goff, W., & Day, R. (1971). Auditory evoked potentials during speech perception. *Science, 173*, 1248-1251.

Zangwill, O. L. (1967). Speech and the minor hemisphere. *Acta Neurologica et Psychiatrica Belgica, 7*, 1013-1020.

SECTION VI. NEURAL SUBSTRATES OF DEVELOPMENT

The three chapters of this section have several things in common--all are studies of the visual system, all involve electrophysiological recording techniques, albeit several different types: electroretinograms (ERGs), visual evoked potentials (VEPs), EEGs, and extracellular recordings. Of course, as the name of the section implies, all of the chapters concern developmental studies of animals (hamsters, kittens, rats, rabbits, and ferrets, among others) and of human infants. In addition, these studies all have a heavy reliance on experimental techniques, such as lesioning, and supplementary information concerning anatomical, neurophysiological, and biochemical analyses.

Rose's chapter is concerned with theory, research, and application in the field of developmental psychobiology, a field that involves the correlative and ultimately the causative relationship between the development of biological processes (neural, hormonal, etc.) and the maturation of behavioral processes (learned and instinctive). The field of developmental psychobiology has been enriched by Professor Lindsley's contributions of more than 25 publications over a period of more than 30 years, articles concerned with theory, application and research.

Professor Lindsley's initial interest in the field began in the mid-1930s when he became concerned about the longitudinal sequence of EEG during the development of human infants. An early study was of his son David's EEG patterns, recorded when the latter was still *in utero*. In the ensuing years, numerous papers were published dealing with the theme of brain-behavior relationships during development; these initially concentrated on the EEGs of normal and abnormal children, and later, in the sixties, the publications shifted to the study of neural-behavior functions in immature kittens. One of Lindsley's principal contributions in the early sixties was to suggest the possibility that "nonspecific" brainstem and thalamic systems, in comparison with "specific" sensory systems, not only might subserve different neural, and hence behavioral functions, but also might mature at different rates.

Rose, based on a suggestion by Professor Lindsley, has begun the difficult task of correlating neural mechanisms with behavioral maturation, based on behavioral data of the kitten. Also, the behavioral analysis of both normal and experimentally manipulated techniques (e.g., the effects of subcortical lesions on later visual behavior) was investigated. Rose also describes studies that illustrate his approach to developmental problems. These studies can be used to collect baseline, normative data with normal or experimentally manipulated neural systems, and they also suggest possible

DOI: 10.4324/9781003164647- 25

applications for human neurobehavioral assessment.

Chalupa and Rhoades' chapter concerns the role of environmental influences, such as visual deprivation and visual restriction, on the golden hamster's visual system and visual behavior. A second consideration is the possibility that there are significant species differences in visual systems, which may be an important factor in their modifiability. Most prior research has concerned the visual system of the cat and monkey, so species differences of the hamster and the rabbit could be of significance.

The chapter summarizes and discusses several earlier reports by Chalupa and Rhoades about environmental influences and species differences. The majority of visual cells in the hamster's superior colliculus exhibit directional selectivity and selective responding to different speeds of stimulus movement and to different size stimuli. Interestingly, it apparently is not necessary for the hamster or the rabbit, unlike the cat, to have visual experience for the induction or the maintenance of normal development of the superior colliculus. Thus, hamsters reared in total darkness from birth did not differ in directional selectivity, speed preference, or receptor field organization from hamsters reared normally. An intriguing finding was that rearing hamsters in stroboscopic illumination induced greater neurophysiological and behavioral effects than visual deprivation, such as rearing hamsters in total darkness from birth.

Woodruff-Pak and Gerrity examined rapid CNS changes occurring in the first four months of infant life. They tested the hypothesis that cortically mediated visual behaviors are absent during the first months of infancy. Previously, Lindsley had noted that towards the end of the third month of life, the EEG in the occipital areas changes dramatically. He suggested that the beginning of organized rhythmic activity in occipital areas may indicate visual behavior being brought under control of the cortex. Until the end of the third month of life, visual behavior may be mediated subcortically. In this study, 15 infants were tested at home seven times: at 2, 8, 11, 13, 15, and 17 weeks of age. Size and fixation preferences (different size stimuli and horizontal stripes vs. bull's eyes), and neurological development, as measured by the absence or presence of reflexes, were tested. Also, one infant's EEG was tested seven times over a 9-week period from 2 to 11 weeks of age.

The results showed a shift in the manner in which infants responded to visual stimuli between 8 and 11 weeks of age. This shift was indexed by the disappearance of primitive reflexes. All reflexes were normal or strong at the 2nd week of testing, and weak or absent by the 17th week, with the most dramatic losses in primitive reflexes occurring between the 5th and 8th weeks. There also was a high, significant correlation between the decrease in the strength of the neonatal reflexes and the increasing preference for stimulus curvature, i.e., the bull's eye stimulus. Fixation preferences as a function of stimulus pattern showed no difference for the earlier weeks, but by 11 weeks the infants demonstrated a statistically significant preference for the bull's eye. By comparison, the younger infants preferred the larger stimuli. It was concluded that younger infants differentiate the stimulus on the basis of its size while older infants differentiate the stimulus on the basis of qualitative pattern characteristics.

20 · THE INTERPLAY OF THEORY, RESEARCH AND APPLICATION IN DEVELOPMENTAL PSYCHOBIOLOGY

G. H. Rose

Bowdoin College

In its relatively brief history, developmental psychobiology, as a recognized research concept, has shown phenomenal growth. Its concern is the correlative, and ultimately causative relationship between the development of biological processes (neural, hormonal, etc.) on the one hand, and the emergence and maturation of behavioral processes (learned and reflexive or instinctive) on the other. This paper is a brief comment on theory, research and application in a field that owes a considerable debt to Donald Lindsley with regard to each of these topics.

Lindsley's first studies on the relationship between brain electrical activity and development appeared in the mid-1930s (Lindsley, 1936, 1938, 1939) and are recognized as classics in the field. These reports were the first to clearly document the longitudinal sequence of EEG changes during development in human infants. As is well known, many of the records were obtained over the years from son David (now a neurophysiologist!), beginning when the latter was still in utero. In the ensuing years, more than 25 of Lindsley's publications dealt with the theme of brain-behavior relationships during development, initially concentrating on the EEG of normal and abnormal children and later, in the sixties, shifting to an analysis of neural-behavioral functions in immature animal models (kittens). Many of these early human studies also focused on the application of the EEG in assessing behavioral disorders in children (see, for example, Lindsley & Cutts, 1940, 1941; Lindsley & Henry, 1942).

In the early sixties, there were very few normative studies in the literature dealing with neurophysiological development (EEG, etc.) in mammals. Partly as a result of Hebb's (1949) influential writings, most animal *behavioral* developmental studies were concerned with effects of "early experience" on later behavior (e.g., see Newton & Levine, 1968). Interdisciplinary studies which attempted to relate behavioral and physiological maturation in mammals were very few indeed.

Likewise at that time, there was a paucity of theory dealing with brain-behavior maturation. If we exclude the general position of Hebb, the ethologists, and the more specific debates of Coghill (1929), Kuo (1939), and Windle (1944) dealing with whether eventual complex behaviors began as

 DOI: 10.4324/9781003164647- 26

total patterns or as isolated reflexes, one of the few theoretical attempts was that of the Russian physiologist, Anokhin, who organized the available neurobehavioral data into a major conceptual scheme of the maturational process (Anokhin, 1964). His theory of systemogenesis (first introduced in 1948), which emphasized heterochronic maturation, focused attention on the need to consider differential development of neural subsystems to subserve particular behavioral repertoires, especially those during postnatal development. A marvelous example of emerging neural-behavioral interactions is seen in the behavior of the rock, an altricial bird. Immediately after hatching, it responds to the sound "kar-r," emitted by a parent, by opening its beak for food. An analysis of the inner ear of the newly hatched bird shows that only those receptor elements able to respond to the frequency components in the sound "kar-r" have matured. Anokhin's deterministic model, however, stresses only preprogrammed (genetic) systems while ignoring contemporaneous interactive influences from various environments.

Lindsley's principal theoretical contribution in the field of developmental psychobiology in the early sixties was to suggest that "nonspecific" brainstem and thalamic systems, in contrast to "specific" sensory systems, might not only subserve different neural and hence behavioral functions (Lindsley, 1958a, 1958b), but they might also manifest different developmental sequences (Lindsley, 1964; Lindsley & Rose, 1965; Rose & Lindsley, 1965). That is, systems that transmit modality specific information (e.g., primary visual pathway) might mature at different rates than those nonspecific systems, such as the reticular formation, with its supposed role in arousal and attention.

Thus an initial interest in human EEG development, followed by adult animal studies on the neural basis of arousal, attention and perception (see, for example, Lindsley, 1961), led Lindsley and his co-workers from a concentration on development per se to a consideration of its usefulness as a model in understanding the eventual complexity of adult neurobehavioral interactions. There was, of course, an awareness (Rose, 1975) that brain models based on *adult* animal studies cannot be uncritically applied to developing systems because, as Anokhin's work emphasizes, the functional relationships of parts of the model's systems continuously vary with age. Also, neural or behavioral events may appear only during certain stages of development with unique variations occurring in different species.

Our initial step was to confirm and extend the findings of Marty (1952), based on visual evoked potential (VEP) studies in kittens, which suggested the developmental differentiation of a dual visual system. Such a duality was confirmed by demonstrating the early development of a long latency negative visual evoked potential recorded over a wide cortical region, and the later development of a shorter latency positive-negative complex confined to the primary visual cortex (Rose & Lindsley, 1965, 1968). Figure 20.1 is a

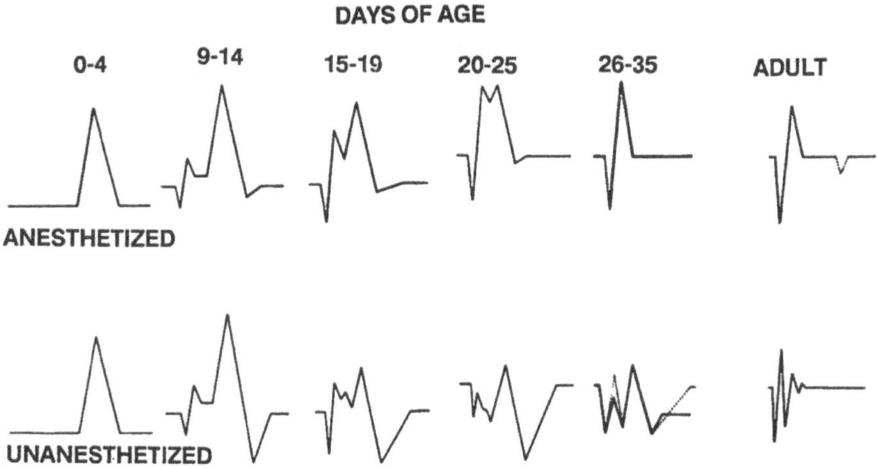

FIGURE 20.1. Schematic comparison of visual electrocortical response maturation obtained from anesthetized and unanesthetized kittens (Rose, Gruenau, & Spencer, 1972).

schematic representation of visual evoked potential changes during maturation as recorded from anesthetized and unanesthetized kittens. Rats show a similar sequence between 11 and 16 days of age (Rose, 1968), as do premature infants (30 to 40 weeks) (Umezaki & Morrell, 1970; Watanabe, Iwase, & Hara, 1973).

Available anatomical data led to suggestions that the initial negative wave was correlated with the early appearance of apical dendrites and axodendritic connections and reflected activity in the "nonspecific" system. In contrast, the positive-negative complex, which developed later, was felt to reflect activity in the primary (specific) visual system, and appeared to be correlated either with basilar dendritic development and axosomatic connections (Anokhin, 1964; Purpura, 1961), or the development of a new neuropil field in the fourth layer of the cortex (Scheibel & Scheibel, 1964). We were thus in a position to test the suggestion that:

> The development and maturation of the cortex and its integration with subcortical centers is in some measure reflected by the developing potentials as a function of age, with specific and nonspecific systems maturing at different times. We are going to propose ... that the first developing long-latency negative wave is associated with the nonspecific systems and that the later appearing positive and negative waves of short-latency are associated with the specific sensory system. (Lindsley, 1964, p. 5)

This basic hypothesis of the developmental differentiation of dual or multiple visual pathways, with differing functions, was supported by subsequent studies (Rose, 1971; Rose & Ellingson, 1970; Rose & Lindsley, 1965, 1968). The early appearing long latency negative wave is selectively affected by manipulations such as light adaptation, as well as lesions of the pretectal region (in particular, the brachium of the superior colliculus and the posterior lateral thalamic nucleus), suggesting the involvement of nonspecific or indirect pathways. In contrast, the later appearing, shorter latency, positive-negative complex, limited to the primary visual cortex, is relatively unaffected by light adaptation, but is reduced by lesions of the lateral geniculate body, suggesting the involvement of the direct geniculostriate system.

The concept of such a differential pathway system was further strengthened by follow-up studies comparing evoked potential development in unanesthetized and anesthetized kittens (Rose, Gruenau, & Spencer, 1972). These demonstrated the selective elimination, by barbiturate anesthesia, of those components felt to be mediated by the nonspecific system (see Figure 8, Rose et al., 1972).

The exact involvement and ontogenetic sequence of various subcortical and cortical mechanisms in the development of visual neural-behavioral relationships is still currently very much under investigation. The initial idea of a dual system of "specific and nonspecific components" is perhaps best replaced by that of multiple interactive visual systems (see also Graybiel, 1974; Jones, 1974). One should note, in a volume dedicated to Donald Lindsley, that these early concepts of *parallel* functional visual systems (involving large areas of the CNS), which were subsequently replaced by serial processing models (in restricted CNS regions), are a part of the *Zeitgeist* once again (see review by Stone & Bogdan, 1982). For that matter, so is the relationship between evoked potentials and behavior (see numerous papers in this volume; also Buchwald & Squires, 1982).

Although the early maturational studies demonstrated that the VEP sequence in Figure 20.1 is very similar in rats, rabbits, cats, dogs and monkeys, newer studies with ferrets (mustela putorius) demonstrate a range of variation among mammals (Rose, 1981a). Although a long latency negative wave is the initial response recorded, as in other species studied, the VEP onset occurs considerably later in maturation, at 30 days of age, and then follows a unique developmental pattern. The resultant VEP waveform and topography has the following different characteristics in this species (see Figure 20.2): (1) The VEP of the primary visual cortex of the deeply anesthetized adult ferret is never a biphasic positive-negative potential as seen in other mammals; an additional component is always present, resulting in a W-shaped wave. Since this waveform is similar to that of unanesthetized cats, serial VEP and EEG recordings under differing dose levels (sodium pentobarbital) verified that the response was indeed characteristic of appropriate (including deep) anesthetization. Furthermore, at these levels

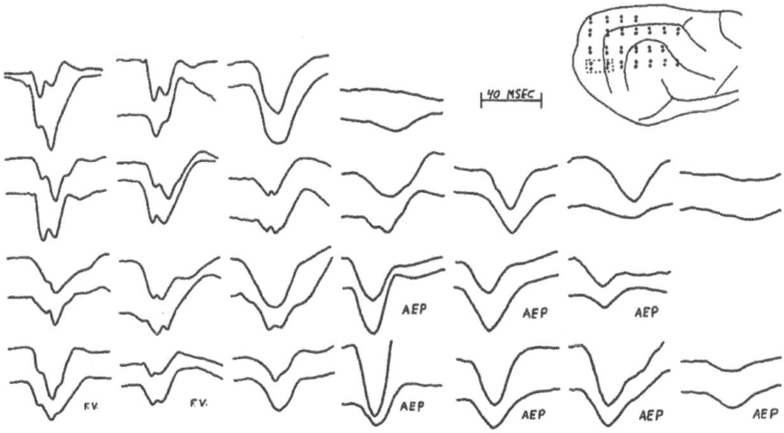

FIGURE 20.2. Topography of visual and auditory (AEP) potentials in the anesthetized ferret. F.V. = "fast visual" region indicated by dotted square on brain figure.

the auditory evoked response is a typical single biphasic wave. (2) Visual evoked potentials recorded from well-defined lower (ventral) regions of the posterior striate cortex (see dotted square in Figure 20.2) have an average initial peak latency of 20 msec in contrast to 35 msec recorded from the dorsal lateral gyrus, i.e., responses occurred 15 msec earlier in this region. (3) This difference is even greater in immature preparations; for example, a difference of as much as 40 msec is obtained at 35 days of age when peak latency at ventral and dorsal visual regions are approximately 95 and 135 msec, respectively.

Preliminary unit studies suggest a predominance of *y* cells in the "fast visual" region, typical of region 18 in other species. Further comparative neural and behavioral studies are underway in this species since the ferret may possess unique anatomical characteristics such as found in the tree shrew by Diamond and his associates (Diamond & Hall, 1969; Killackey, Snyder, & Diamond, 1971), thereby maximizing its research potential. The tree shrew has two separate visual pathways to the cortex (lateral geniculate to striate cortex; superior colliculus to pulvinar to temporal cortex) which subserve different visual-perceptual functions. It is of theoretical interest to determine what similar specific visual behavioral functions also exist in the ferret. Not only is this species unique from a developmental research prospectus, but from an evolutionary viewpoint it (unlike the tree shrew) is a good comparison species to the cat. That is, one can compare related carnivores with somewhat different visual behaviors, which presumably would be reflected in differing neural substrates.

These studies serve to emphasize Lindsley's insistence that an understanding of visual behavioral functioning is incomplete unless it also involves the comprehensive "visual system," a point often overlooked in current texts which focus mainly on neurons in the striate cortex when discussing the neural basis of visual perception.

Following Lindsley's suggestions, we began the difficult task of correlating neural mechanisms with behavioral maturation. Figure 20.3 illustrates some of the behavioral data available from the kitten. These results support the notion that in the younger kitten, with cloudy ocular media (Thorn, Gollender, & Erickson, 1976) and an immature visual system, visual input serves to arouse and alert the kitten rather than convey specific visual information. Later, when the media clears and the system matures, the animal is better able to respond to specific features of the visual environment.

We were initially interested in a behavioral analysis of both normal and experimentally manipulated systems which, in addition to elucidating the mechanisms involved, might also have future practical application. The studies of Norton and Lindsley (1971) and Atwell and Lindsley (1975), on the effects of subcortical lesions on later visual behaviors, reflected such an emphasis. In the sixties and seventies, as in the past, most animal behavioral studies were concerned with a limited set of phenomena (e.g., learning) in a limited number of species (e.g., rat and cat), and normative developmental studies, with some exceptions, reflected that emphasis. But by far the greatest concern was with the effects of varying "early experience" on subsequent brain-behavior maturation, again in a few species.

Due in part to the influence of ethologists and behavioral biologists and the appearance of publications, such as the series entitled, "Studies on the Development of Behavior and the Nervous System" (Gottlieb, 1976), the volume on *Biological Constraints* by Seligman and Hager (1972), etc., there emerged a more complete conceptualization of the interaction of biological and environmental influences on the one hand, and the role of normative baseline developmental data in interpretation of "manipulations" on the other. This has led to the formulation of a number of significant questions (Rose, 1981b). For example, is there a limit to the interpretation of "early experience" studies which are ignorant of normative developmental data, biological constraints, species differences, developmentally inappropriate measures, etc.? Or, under the term "early experience," why not include normal developmental strategies (e.g., light avoidance) which may vary with age and/or species? Do our experimental manipulations maintain, facilitate, or induce various processes, and are our experiments designed to reflect these options? What other systems (e.g., endocrine) are influenced by the early experience (e.g., light) which may influence the neural mechanisms of concern (e.g., visual)? Is there equal attention to the study of behaviors that

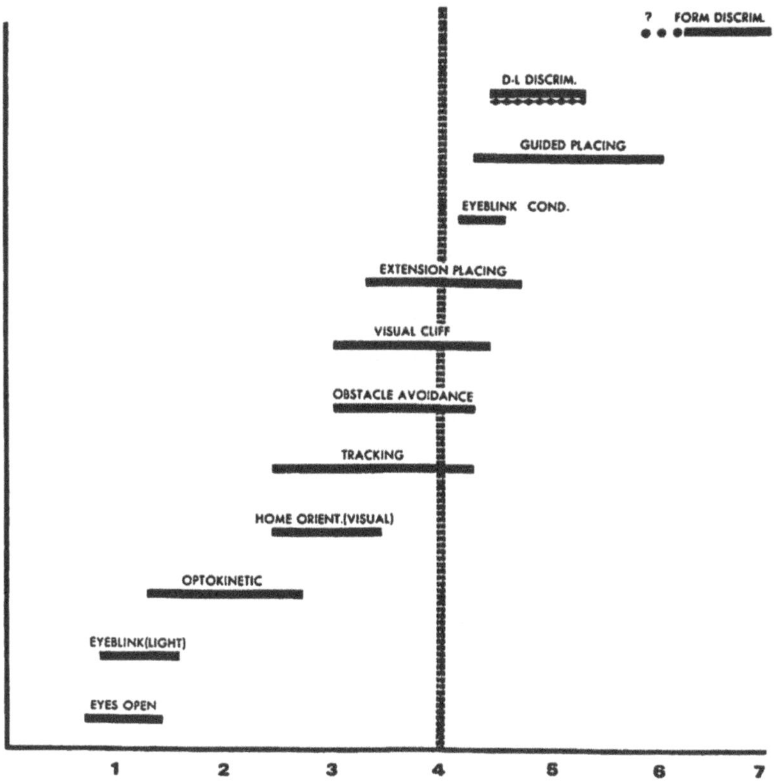

FIGURE 20.3. Maturation of visual function in kittens (Rose & Collins, 1975b).

"fit" various diverse species in their habitats, as there is to those species and behaviors that "fit" available techniques and apparatus?

Concerned with these issues and aware of the difficult and unique conceptual and methodological problems in working with immature animals (Rose, 1975, 1981b), we embarked on an early behavioral assessment program, admittedly following somewhat traditional methods, whereas our current thinking increasingly involves a more holistic/dialectical approach.

The following four studies briefly illustrate the various past approaches. They reflect the kinds of baseline normative behavioral data felt to be useful for correlation with normal or experimentally manipulated (e.g., lesioned) neural systems. In addition, they also suggest applications for human neurobehavioral assessment.

We chose an overall experimental paradigm for testing visual behavior development which consisted of a two-by-two factorial design, with two levels representing the principal problems of concern, namely, (a) the degree to which initial visual-perceptual abilities were dependent upon, or constrained by, early learning proficiencies in contrast to reflexive or instinctive behavior, and (b) the degree to which the immature animals' responses were of consequence in determining subsequent stimulation, e.g., response dependent or independent.

An example of a *response-independent learning* paradigm is classical conditioning. Developmental eyeblink conditioning studies (Norman, Collins, & Rose, 1975; Rose, 1975) established that, in kittens and rabbits, a somatic CS conditioned earliest, followed by auditory, and then visual stimuli; the latter, in the case of kittens, was effective in producing stable conditioning by approximately 34 days of age. Recall that the visual EP waveform reached maturity between four and five weeks of age (Figures 20.1 & 20.2).

The specific question of interest was the degree to which the earliest indications of this type of associative learning was constrained by motor immaturity. This was tested in a follow-up "latent learning" study: an experimental group of kittens were trained with visual CS for seven days (100 trials/day) beginning at 21 days of age, while a control group of litter mates received random presentations of both CS and US. Conditioning was not evident at 29 days of age. After two days of rest, both groups received further conditioning trials. The animals which had previous training showed immediate conditioning (at 31 days of age), whereas the former control groups gradually conditioned after 200-300 trials (i.e., by 34 days of age) as in the initial study (see Figure 5.7 in Rose, 1981b).

One conclusion is that the association between the CS and UCS was actually established *prior* to the age when CNS motor connections were sufficient to *express* the conditioned response; i.e., the eventual motor responses gave the impression that learning occurred later in development. These results also suggest a similar paradigm to be used with very young infants where behavioral testing is limited by motor immaturity.

For the *response-dependent learning* paradigm we used another procedure to circumvent the poor motor abilities of young kittens, this time an unorthodox feedback operant technique (for a more conventional approach, see Rose & Collins, 1975b). In this study (Rose, Norman, Naifeh, & Collins, 1975) instead of reinforcing a selected *motor* response to light flash, we directly reinforced a specified change in the visually evoked cortical potential (VEP). That is, eight kittens between one and one-half and three weeks of age were trained to increase (or decrease) the very predominate long latency negative (N2) wave of the cortical VEP. The tone functioned as a discriminative stimulus (S^D) signaling the onset of a trial, light flash as the CS, and mild electrical shock as a negative reinforcer in a shock-avoidance procedure. Specific criterion changes within a specified time-amplitude window, as analyzed by an on-line computer, determined whether reinforcement would be given or withheld. The kitten could use any strategy (e.g., changes in orientation to the light, etc.) to produce the desired result since we were interested in learning, not in a CNS feedback technique per se.

The results demonstrated that dramatic conditioned shifts (see Figures 2 and 3 in Rose et al., 1975) in the N2 wave, of 100% to over 500%, could be obtained in the experimental animals as early as 14 days of age. Yoked controls (e.g., light flash only; non-contingent shock) showed a maximum amplitude increase (or decrease) of only 35% from baseline.

These results represent some of the earliest ages at which operant conditioning, utilizing the visual modality, has been demonstrated. What is puzzling is that controls indicate that the results are not correlated with obvious overt changes such as head or eye movement, pupillary changes, etc. Further physiological-behavioral analysis might detect whether, in fact, general state alterations (attentional, emotional, etc.) involving various physiological systems (endocrine, etc.), in addition to discrete neural events, have been conditioned, and are reflected in the VEP alterations.

The final two examples of early infant animal assessment procedures are ones that minimize learning and study instead reflexive or instinctive behavior. In the *response-dependent* situation we devised a method, dubbed RAP (*R*esponse *A*djusting *P*rocedures), to study the infant animal's control over stimulus preferences (Rose & Collins, 1975a). Briefly, instead of measuring a predetermined response to specific stimuli, as in most orientation and learning studies, the animal's developmentally and species appropriate responses (e.g., crawling, etc.) are utilized to select preferred stimulation from a stimulus array (e.g., light of different intensities).

In a study of the development of species-specific light preferences (Rose, 1981c), a situation was arranged so that an infant rat's position (e.g., left, middle, right) in a long narrow enclosure triggered different lighting conditions (e.g., dark, medium, bright). Testing began at 7, 8, 9, and 10 days of age. Regardless of when testing began, infant rats first showed a highly reliable preference for dark over medium or bright light by 10 days of age (there were appropriate controls for position, odors, etc.). This is 5 to 6 days prior to eye opening but coincidental with the onset of visual evoked cortical potentials in rats (Rose, 1968). At 10 days of age and thereafter, bright light onset inhibited forward movement causing a retreat into the dark. Furthermore, such a preference was maintained up to 35 days of age in spite of considerable age-related differences in activity levels which showed a dramatic peak at eye opening (15-16 days of age). It would be useful to repeat these baseline studies in related species, and also under field conditions, to determine their generality or species relatedness.

Finally, as an example of a response-independent assessment technique, one with potential human clinical application, Rose and Collins (1975a, 1978) studied the early responsiveness of the newborn kitten to light as measured by behavioral (blink and squint, see Figure 20.4A) and EMG activity from eyelid

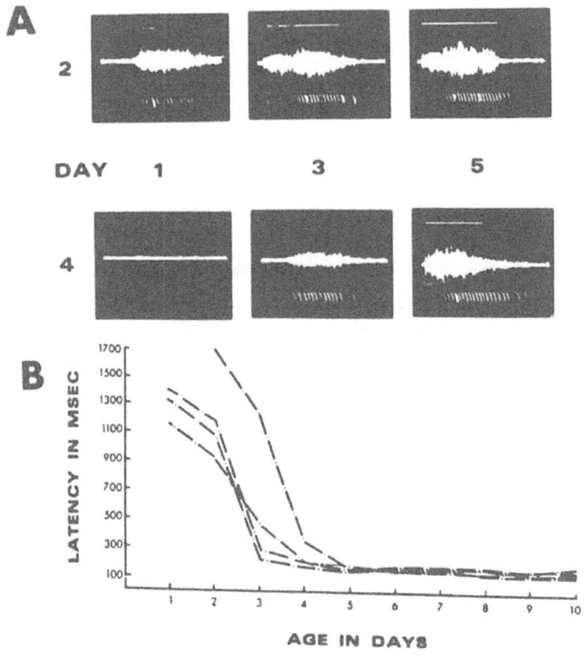

FIGURE 20.4. Responses to light in newborn kittens. A. EMG response from eyelid muscles (orbicularis oculi), from normal kitten (#2) and runt (#4). B. Latency of EMG response as a function of age. Kittens #2 and #4 are same as in B.

muscles (orbicularis oculi, see Figure 20.4B).

At birth, a threshold flash duration of 500 msec evoked a 1200- to 1500-msec EMG and behavioral responses. A dramatic change (see Figure 20.4C) in light duration threshold and consequent EMG activity was seen over the next several days resulting in a 125- to 150-msec EMG to a 15-msec light flash at 6 days of age in normal kittens. The occasional kitten that did not show this pattern usually was the runt of the litter (#4 in Figure 20.4) and had later difficulties often resulting in death.

By determining threshold responses to varying wave lengths of light in the dark-adapted kitten, we have been able to obtain scotopic spectral threshold curves using only the EMG as the response measure as shown in Figure 20.5. Even on the first day of life, there appears a tendency for the curves to peak at 500 nm; this is quite evident by day 2, and by day 4 the curves appear stable. The form of the curves on day 6 describes a very close approximation to the rhodopsin absorption curves (Wald, 1945) and to the

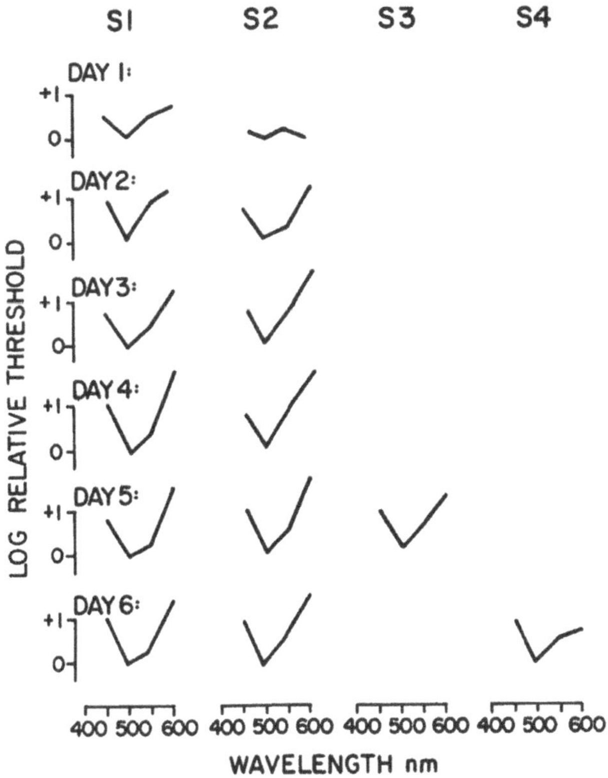

FIGURE 20.5. Scotopic spectral threshold curves (log relative threshold) in four kittens; two were tested daily and one each at 5 and 6 days of age. The data are standardized relative to the threshold at 500 nm.

spectral sensitivity curves for the adult cat (Dodt & Walther, 1958). The primary deviation is a tendency for the curves to fall off more sharply at 450 nm than would be predicted based on rhodopsin absorption. This is possibly due to the fact that the lens and pupil are covered with a vascular tunica during this period (Thorn et al., 1976). The hemoglobin absorption of this network and that of the closed eyelid itself would be expected to produce this sharpening effect due to the strong absorption bands of hemoproteins found between 400 and 450 nm (Crawford & Marc, 1976).

The possible application of the visual blink response technique in human infant assessment is suggested for the following reasons: (1) This reflex is currently part of the Brazelton newborn infant neurobehavioral examination (Brazelton, 1973). (2) Adult animal (Tokunaga et al., 1958) evidence indicates that the resultant EMG activity involves two types of reflexive responses: a short latency monosynaptic reflex mediated by the sensory nucleus of the fifth nerve, and a longer latency polysynaptic reflex involving thalamic nuclei and possibly the reticular formation. These responses have been differentiated in the adult by lesions and anesthetics, which once again suggests developmental differentiation in kittens and possibly human infants as well. (3) Finally, the procedure may present a method for obtaining photopic and scotopic spectral sensitivity curves very early in human development.

I have briefly reviewed some of our developmental evoked potential and behavioral studies which were considerably influenced by Lindsley's research and theoretical orientation. As is also evident in Lindsley's work, studies of gross electrical activity (evoked potentials, EEG, etc.) in animals should also be accompanied by anatomical, neurophysiological, and biochemical analyses at the cellular level. Obviously, such analyses are difficult, if not impossible, with humans. However, we must not lose sight of the continued value of evoked potential studies in humans (see others in this volume) and animals (see also Bullock, 1981), both because of the increased sophistication of statistical analyses of the phenomena, and because evoked potentials, as an expression of many single neurons, reflect the multiparameter complexity of even the simplest behavioral activity. We also need additional neurobehavioral studies, both within and across species, that reflect the interplay between developmental, evolutionary, and environmental factors. The multifaceted approach, so championed by Donald Lindsley, can provide the foundations for understanding both normal and aberrant neurobehavioral events in immature species, including the human infant.

REFERENCES

Anokhin, P. K. (1964). Systemogenesis as a general regulator of brain development. In W. A. Himwich (Ed.), *The developing brain, progress in brain research* (Vol. 9, pp. 58-86). Amsterdam: Elsevier.

Atwell, C. W., & Lindsley, D. B. (1975). Development of visually evoked responses and visually guided behavior in kittens: Effects of superior colliculus and lateral geniculate lesions. *Developmental Psychobiology, 8,* 465-478.

Brazelton, T. B. (1973). *Neonatal Behavioral Assessment Scale.* Philadelphia: Lippincott Co.

Buchwald, J. S., & Squires, N. S. (1982). Endogenous auditory potentials in the cat: A P300 model. In C. W. Woody (Ed.), *Conditioning* (pp. 503-515). New York: Plenum.

Bullock, T. H. (1981). Neuroethology deserves more study of evoked responses. *Neuroscience, 6,* 1203-1215.

Coghill, G. E. (1929). *Anatomy and the problem of behavior.* Cambridge, England: Cambridge University Press.

Crawford, M., & Marc, R. (1976). Light transmission of cat and monkey eyelids. *Vision Research, 16,* 323-324.

Diamond, I. T., & Hall, W. C. (1969). Evolution of neocortex. *Science, 164,* 251-262.

Dodt, E., & Walther, J. (1958). Der shotopsiche dominatar der katz im sichtbaren und ultravioletten spektralbreich. *Pflügers Archiv für Die Gesampte Physiologie des Menschen und der Tierre, 266,* 166-174.

Gottlieb, G. (1976). The roles of experience in the development of behavior and the nervous system. In G. Gottlieb (Ed.), *Neural and behavioral specificity* (Vol. 3, pp. 25-54). New York: Academic Press.

Graybiel, A. M. (1974). Studies on the anatomical organization of posterior association cortex. In F. O. Schmitt & F. G. Worden (Eds.), *The neurosciences: Third study program* (pp. 205-214). Cambridge, MA: MIT Press.

Hebb, D. O. (1949). *The organization of behavior.* New York: Wiley.

Jones, E. G. (1974). The anatomy of extrageniculate visual mechanisms. In F. O. Schmitt & F. G. Worden (Eds.), *The neurosciences: Third study program* (pp. 215-228). Cambridge, MA: MIT Press.

Killackey, H., Snyder, M., & Diamond, I. T. (1971). Function of striate and temporal cortex in the tree shrew. *Journal of Comparative and Physiological Psychology Monographs, 74,* 1-29.

Kuo, Z. Y. (1939). Total pattern or local reflexes? *Psychological Review, 46,* 93-122.

Lindsley, D. B. (1936). Brain potentials in children and adults. *Science, 84,* 354.

Lindsley, D. B. (1938). Electrical potentials of the brain in children and adults. *Journal of General Psychology, 19,* 285-306.

Lindsley, D. B. (1939). A longitudinal study of the occipital alpha rhythm in normal children: Frequency and amplitude standards. *Journal of Genetic Psychology, 55,* 197-213.

Lindsley, D. B. (1958a). Psychophysiology and perception. In R. Patton (Ed.), *The description and analysis of behavior* (pp. 48-91). University of Pittsburgh 10th Annual Conference on Current Trends in Psychology. Pittsburgh: University of Pittsburgh Press.

Lindsley, D. B. (1958b). The reticular system and perceptual discrimination. In H. H. Jasper et al. (Eds.), *International symposium on reticular formation of the brain* (pp. 513-534). Boston: Little, Brown.

Lindsley, D. B. (1961). The reticular activating system and perceptual integration. In D. E. Sheer (Ed.), *Electrical stimulation of the brain* (pp. 331-349). Austin, TX: University of Texas Press.

Lindsley, D. B. (1964). Brain development and behavior: Historical introduction. In W. A. Himwich & H. E. Himwich (Eds.), *The developing brain: Progress in brain research* (Vol. 9, pp. 1-5). Amsterdam: Elsevier.

Lindsley, D. B., & Cutts, K. K. (1940). Electroencephalograms of "constitutionally inferior" and behavior problem children: Comparison with those of normal children and adults. *Archives of Neurology and Psychiatry, 44,* 1199-1212.

Lindsley, D. B., & Cutts, K. K. (1941). Clinical and electroencephalographic changes in a child recovering from encephalitis. *Archives of Neurology and Psychiatry, 44,* 1199-1212.

Lindsley, D. B., & Henry, C. E. (1942). The effect of drugs on behavior and the electroencephalograms of children with behavior disorders. *Psychosomatic Medicine, 4,* 140-149.

Lindsley, D. B., & Rose, G. H. (1965). Development of visual evoked responses in kittens. *Federation Proceedings, 24,* 274.

Marty, R. (1952). *Dèveloppement post-natal des responses sensorielles du cortex cerebral chez le chat et le lapin.* Thèse Doctoral Science. Paris: Masson.

Newton, G., & Levine, S. (1968). *Early experience and behavior: The psychobiology of development.* Springfield, IL: C. C. Thomas.

Norman, R. J., Collins, J. P., & Rose, G. H. (1975). *Development of classical conditioning in kittens and rabbit pups.* Paper presented at the meeting of the International Society for Development Psychobiology, New York.

Norton, T. T., & Lindsley, D. B. (1971). Visual behavior after bilateral superior colliculus lesions in kittens and cats. *Federation Proceedings, 30,* 615.

Purpura, D. P. (1961). Structure and function of cortical synaptic organizations activated by corticipetal afferents in newborn cat. In M. A. B. Brazier (Ed.), *Brain and behavior* (Vol. 1, pp. 95-138)). Washington, DC: American Institute of Biological Sciences.

Rose, G. H. (1968). The development of visually evoked electrocortical responses in the rat. *Developmental Psychobiology, 1,* 35-40.

Rose, G. H. (1971). The relationship of electrophysiological and behavioral indices of visual development in mammals. In M. B. Sterman, D. J. McGinty, & A. M. Adinolfi (Eds.), *Neural ontogeny and behavior* (pp. 145-183). New York: Academic Press.

Rose, G. H. (1975). CNS maturation and behavioral development. In N. A. Buchwald (Ed.), *Brain mechanisms in mental retardation* (pp. 171-178). New York: Academic Press.

Rose, G. H. (1981a). Cortical evoked potential and single unit activity to visual and auditory stimuli in mature and immature ferrets (mustela putorius). *Neuroscience Abstracts, 7,* 542.

Rose, G. H. (1981b). Animal studies in developmental psychobiology: Method, theory, and human implications. In S. Friedman & M. Sigman (Eds.), *Preterm birth and psychological development* (pp. 73-105). New York: Academic Press.
Rose, G. H. (1981c). A response adjustment procedure for assessing stimulus preference in preweanlings. *Developmental Psychobiology, 14,* 559-563.
Rose, G. H., & Collins, J. P. (1975a). *A developmental RAP: Assessing early visual-perceptual abilities in animals using response adjusting procedures.* Paper presented at the meeting of the International Society for Developmental Psychobiology, New York.
Rose, G. H., & Collins, J. P. (1975b). Light-dark discrimination and reversal learning in early post-natal kittens. *Developmental Psychobiology, 8,* 511-518.
Rose, G. H., & Collins, J. P. (1978). *Threshold responses to white light and hue (spectral sensitivity) in newborn kittens.* Paper presented at meeting of the Eastern Psychological Association, Washington.
Rose, G. H., & Ellingson, R. J. (1970). Ontogenesis of evoked responses. In W. A. Himwich (Ed.), *Developmental neurobiology* (pp. 393-440). Springfield, IL: Charles C. Thomas.
Rose, G. H., Gruenau, S. P., & Spencer, J. W. (1972). Maturation of visual electrocortical responses in unanesthetized kittens: Effects of barbiturate anesthesia. *Electroencephalography Clinical Neurophysiology, 33,* 141-158.
Rose, G. H., & Lindsley, D. B. (1965). Visually evoked electrocortical responses in kittens: Development of specific and nonspecific systems. *Science, 148,* 1244-1246.
Rose, G. H., & Lindsley, D. B. (1968). Development of visually evoked potentials in kittens: Specific and nonspecific responses. *Journal of Neurophysiology, 31,* 607-623.
Rose, G. H., Norman, R. J., Naifeh, K., & Collins, J. (1975). Plasticity of visual evoked potentials in kittens demonstrated by operant conditioning. *Physiology and Behavior, 14,* 557-561.
Scheibel, M. E., & Scheibel, A. B. (1964). Some neural substrates of postnatal development. In M. Hoffman & L. Hoffman (Eds.), *Review of child development research* (Vol. 1, pp. 481-519). New York: Russell Sage.
Seligman, M., & Hager, J. (1972). *Biological boundaries of learning.* New York: Appleton.
Stone, J., & Bogdan, D. (1982). Parallel processing of information in visual pathways: A general principle of sensory coding? *Trends in Neurosciences, 15,* 441-446.
Thorn, F., Gollender, M., & Erickson, P. (1976). The development of the kitten's visual system. *Vision Research, 16,* 1145-1149.

Tokunaga, A., Oka, A., Murao, H., Okumura, T., Hirata, T., Miyashita, Y., & Yoshitatsu, S. (1958). An experimental study on facial reflex by evoked electromyography. *Medical Journal of Osaka University, 9*, 397-411.

Umezaki, H., & Morrell, F. (1970). Developmental study of photic evoked responses in premature infants. *Electroencephalography and Clinical Neurophysiology, 28*, 55-63.

Wald, G. (1945). Human vision and the spectrum. *Science, 101*, 653-658.

Watanabe, K., Iwase, K., & Hara, K. (1973). Visual evoked responses during sleep and wakefulness in pre-term infants. *Electroencephalography and Clinical Neurophysiology, 34*, 571-577.

Windle, W. F. (1944). Genesis of somatic motor function in mammalian embryos: A synthesizing article. *Physiological Zoology, 17*, 247-260.

21 · ENVIRONMENTAL INFLUENCES UPON THE DEVELOPMENT OF THE GOLDEN HAMSTER'S VISUAL SYSTEM

L. M. Chalupa and R. W. Rhoades

University of California at Davis

In the last two decades a great deal of attention has been devoted to the role of the environment in the development of the mammalian visual system. In general, three types of questions have been examined: (1) what is the functional organization of the visual system in neonatal (i.e. visually inexperienced) organisms; (2) to what extent can normal organization be modified by either binocular or monocular deprivation; and (3) how can the developing visual system be changed by various types of visual restriction paradigms?

Hubel and Wiesel (Hubel & Wiesel, 1963, 1970; Wiesel & Hubel, 1963, 1965) first applied single-cell recording techniques to deal systematically with the first two problems, while the work of Hirsch and Spinelli (1970) and Blakemore and Cooper (1970) extended these methods to the study of visual restriction. It should be noted that the differentiation between visual deprivation and restriction is primarily heuristic, since lid-suturing is usually employed for visual deprivation, and while this deprives the animal of pattern vision, it does permit considerable diffuse visual input through the closed eyelids. While a number of important controversies remain to be resolved, major strides have been made in providing answers to each of the questions listed above (for reviews see Barlow, 1975; Daniels & Pettigrew, 1976; Hubel & Wiesel, 1977; Movshon & Van Sluyters, 1981; Sherman & Spear, 1982).

For the most part neurophysiological studies in this area have been concerned with the visual system of the cat and monkey. However, the work of Chow (Chow & Spear, 1974; Grobstein & Chow, 1975) and others (Mize & Murphy, 1973) on the development of the rabbit's visual system suggests that species differences may be an important factor in the modifiability of the mammalian visual system. Partly for this reason we decided to investigate the possible effects of environmental influences upon the visual system of the golden hamster. Our choice of this species was also guided by the fact that Schneider and his colleagues have demonstrated a remarkable degree of anatomical reorganization in the hamster's visual system following various types of neonatal lesions (Schneider, 1973, 1977). These neurophysiological experiments have been concerned primarily with the hamster's superior colliculus. First, we employed conventional extracellular recording

 DOI: 10.4324/9781003164647- 27

procedures to examine the response properties of single neurons in the normal hamster's optic tectum. Results of these studies have been reported in detail elsewhere (Chalupa & Rhoades, 1977; Rhoades & Chalupa, 1976, 1977), and only the major findings will be summarized here.

Neurons in the superficial laminae of the superior colliculus (*stratum zonale, stratum griseum superficiale*, and *stratum opticum*) respond exclusively to visual stimuli and they have relatively small receptive fields. Visual cells in the deeper laminae (those ventral to the *stratum opticum*) generally have very large receptive fields and show rapid habituation with repeated stimulation. In addition, many cells in the deeper layers can be activated only by somatosensory stimuli. Auditory neurons are also present, but these appear to be confined to the caudal portion of the colliculus. As has been found in the cat (Gordon, 1973; Stein, Magãlhaes-Castro & Kruger, 1976) and mouse (Dräger & Hubel, 1975, 1976) there is clear spatial correspondence in the hamster's tectum between visual and extra-visual receptive fields (Chalupa & Rhoades, 1977; Finlay, Schneps, Wilson, & Schneider, 1978; Tiao & Blakemore, 1976).

The majority of visual cells in the hamster's colliculus exhibit directional selectivity and we have employed two criteria in classifying these neurons: (1) a statistical measure in which responses to stimulus movement in opposing directions are compared using a simple *t* test and (2) the traditional "null" (Barlow & Levick, 1965) criterion where a cell is considered as selective if no response or suppression of spontaneous activity is observed with stimulus movement in the non-preferred direction. In the normal hamster approximately 60% of the tectal cells are directionally selective according to the statistical criterion and about 28% are null cells (see Figure 21.1A, B, C). It should be noted that the null cells are included in the statistical category.

We have also classified visual cells in the superior colliculus according to their response to different speeds of moving stimuli. About 35% of the cells respond only to slowly moving stimuli (less than 50°/sec). Another 50% also respond to slow speeds, but these are more broadly tuned since they will also respond reliably to rapidly moved stimuli. The remaining 15% respond to fast movement (greater than 50°/sec). An example of each cell type is shown in Figure 21.1D, E, F.

While most visual cells in the hamster's colliculus respond best to stimulus movement, many neurons will respond, albeit not optimally, to flashed stationary spots of light. Typically these cells have homogeneous receptive fields, in that they respond in a characteristic fashion (e.g., with light onset and offset) to stimuli placed anywhere in the activating region of the receptive field. The activating region is defined as that portion of the receptive field which when stimulated elicits action potentials in a given cell. For the cells that do respond reliably to flashed spots of light, we have examined the effects of increasing stimulus size upon neuronal responsivity.

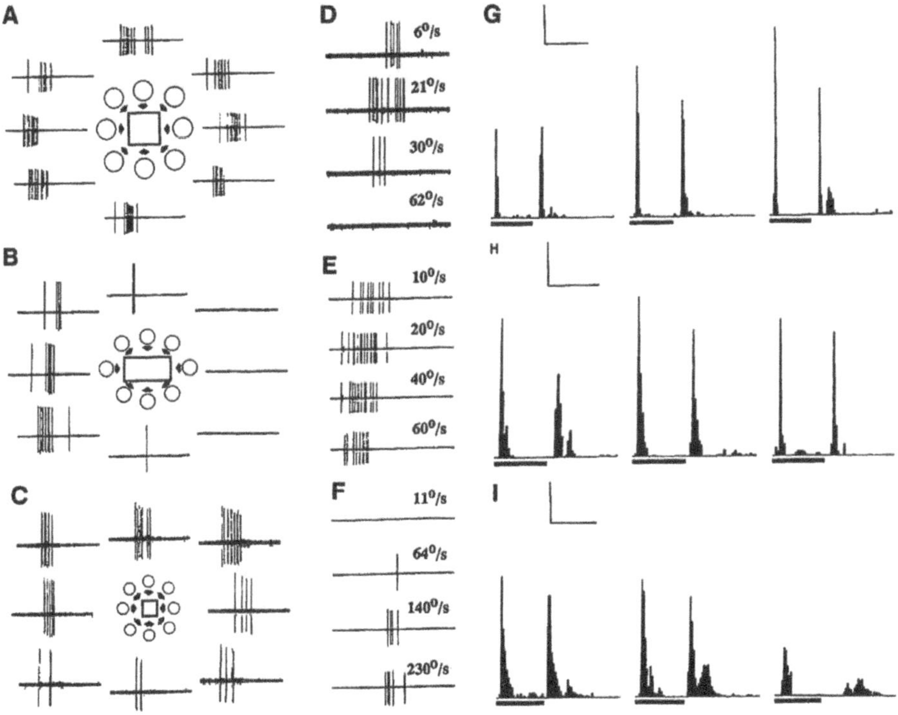

FIGURE 21.1. In A, responses of a non-directionally selective cell. The square denotes the activating region of the receptive field, and the circles spots of light which were swept across the field in each direction indicated by the arrows. Typical responses elicited by each direction of stimulus movement are also shown. Note that in this case there was no appreciable difference in the number of action potentials elicited by any pair of opposing directions of movement. In B, a cell judged as directionally selective according to the null criterion. This neuron responded to movement in the nasal (leftward) direction, but not in the temporal direction. In C, a statistically selective unit. This cell had no spontaneous activity and it responded reliably to each direction tested; however, note a preference for stimuli moved in the upper temporal direction (upper right). Oscilloscope traces are 5 sec long and the calibration is 10^{0} of visual angle. In D, tracings illustrating the responses of a cell which only responded to slowly moving stimuli (< 50^{0}/sec.). In E, a neuron which was broadly tuned for stimulus speed, but responded primarily to slowly moving spots. In F, a cell which responds to stimulus velocities in excess of 50^{0}/sec. In G, H, and I post-stimulus time histograms (PSTHs) illustrating the responses of three cells to spots of increasing size. In each row the first PSTH depicts the response to a flashed spot of light which was considerably smaller than the activating region of the receptive field; the next histogram was obtained using a stimulus approximately equal in size to this region, and the PSTH on the extreme right was obtained with stimuli much larger than the activating region. Thus, the cell in G exhibits response summation both within and beyond the borders of the activating region, that in H was not appreciably affected by stimulus size, and the neuron in I had its responses suppressed by the large stimulus. All PSTHs are based on 50 stimulus presentations and the bin width is 20 msec. The bar indicates the amount of time the stimulus was on (500 msec) and the calibrations are 25 impulses/bin and 500 msec.

Our strategy here has been to employ at least three spot sizes for each cell tested, including stimuli considerably smaller than the receptive field activating region, spots approximately equal to this area, and also stimuli considerably larger than the activating region. In all cases the stimulus was positioned on the center of the receptive field.

The results of this procedure enabled us to classify cells into three major types: (a) neurons whose responses are not affected by the size of the stimulus (19%); (b) cells that exhibit clear response summation within and beyond the border of the activating region (6%); and (c) neurons that show either complete suppression or partial suppression (75%) when the stimulus exceeds the dimensions of the cell's activating region. An example of each response pattern is shown in Figure 21.1 (G, H, I). The visual response properties described above for normal hamsters are summarized in Figure 21.2 (top).

With the above results as a basis for comparison, we next asked whether or not visual deprivation during development alters the functional organization of the superior colliculus (Rhoades & Chalupa, 1978). To this end, hamsters were raised from birth to adulthood (about 3 months of age) in total darkness. Each electrophysiological experiment was carried out on the day the animal was removed from darkness. We found that directional selectivity, speed preferences, and receptive field organization of superior collicular neurons were essentially the same in the visually deprived hamsters as in normals. These findings are depicted in the middle portion of Figure 21.2. However, there was an indication of some subtle effects, in that the latencies of "on" discharges to flashed spots were significantly longer in the visually deprived animals than in normal hamsters. These findings contrast markedly with the results obtained in the cat where visual deprivation (lid-suturing) has been reported (Flandrin & Jeannerod, 1977; Hoffmann & Sherman, 1975; Sterling & Wickelgren, 1970) to reduce the number of neurons in the superior colliculus which could be activated by visual stimuli, and for those cells which were visually responsive there was an additional reduction in binocularity and directional selectivity, as well as a shift in the distribution of speed preferences towards cells that respond to lower stimulus velocities. On the other hand, in the rabbit, visual deprivation has been found to produce no apparent effects on the functional organization of the superior colliculus (Chow & Spear, 1974). Thus in the hamster, as in the rabbit, visual experience is apparently not necessary for either the induction or the maintenance of relatively normal development of the superior colliculus.

Given these findings, we thought that it would be of some interest to determine whether or not visual restriction could modify the neuronal response properties of the hamster's tectum. We chose to raise animals from birth to adulthood in a stroboscopic environment which permits experience

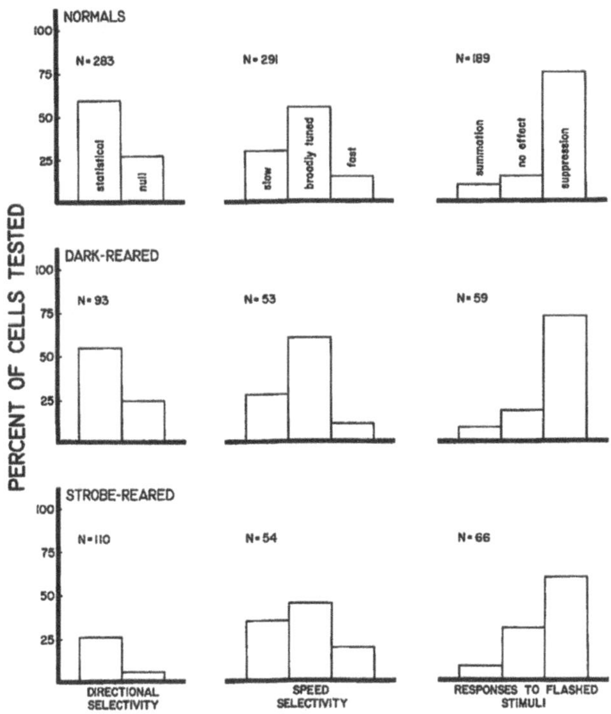

FIGURE 21.2. A summary of data for three collicular response properties which were quantitatively tested: directional selectivity (left), speed preference (center) and the effect of increasing spot size (right) in the three groups of hamsters, normals (top), dark-reared (middle), and strobe-reared (bottom). The number of cells tested is indicated above each distribution.

with visual patterns but abolishes the perception of moving visual stimuli. This paradigm has been shown to change the functional organization of the visual cortex (Cynader, Berman, & Hein, 1973, 1976; Cynader & Chernenko, 1976; Olson & Pettigrew, 1974), as well as that of the superior colliculus (Flandrin, Kennedy, & Amblard, 1976) in the cat.

In contrast to the lack of major effects in dark-reared hamsters, we observed that rearing animals in a stroboscopic environment did induce several changes in the receptive-field characteristics of superior collicular neurons. Of these, the most clear-cut was a decrease in the incidence of directionally selective cells (Chalupa & Rhoades, 1978a). In addition, the distribution of speed preferences was also altered, in that more cells in the strobe-reared animals responded only to slow velocities, and less were broadly tuned with regard to the speed of moving stimuli. Further, there was also an increase in the number of cells whose responses were not affected by increasing the size of stationary flashed stimuli. These results are shown in Figure 21.2 (bottom).

It should be noted that we did not find any increased responsivity in the visually restricted animals to stroboscopic stimulation. This was the case even when a strobe rate (2/sec) identical to that used in the rearing environment was employed as a stimulus (Chalupa & Rhoades, 1978b). Some of these findings dealing with visual restriction in the cat have been interpreted as indicating that during development the visual system is able to incorporate or selectively adapt to certain features of the environment (Cynader, Berman, & Hein, 1973; Hirsch & Spinelli, 1970; Pettigrew & Freeman, 1973; Van Sluyters & Blakemore, 1973). However, the results of restriction experiments in the cat are difficult to interpret since in this species visual deprivation also has marked consequences in visual cortex and the superior colliculus. Our findings in the hamster suggest that an aberrant visual environment (i.e., strobe-rearing) can disrupt normal visual functional organization. Furthermore, this disruption can be obtained in a visual system which develops relatively normally without visual experience.

The neurophysiological findings described above raised questions concerning the possible behavioral concomitants of dark and strobe rearing. Initially we examined the ability of the visually deprived and the strobe-reared animals to orient to and follow visual stimuli. These rather informal tests did not reveal any significant differences between these two groups of hamsters and normal animals (Chalupa & Rhoades, 1978b; Rhoades & Chalupa, 1978). We next employed a more rigorous procedure in which normal, dark-reared, and strobe-reared animals were trained on a five-choice visual pattern discrimination (horizontal versus vertical stripes). Prior to the pattern discrimination, all animals were trained to criterion on a brightness discrimination task using the same training procedure and apparatus. No differences were evident among the three groups on the brightness discrimination. However, on the pattern task we found that the dark-reared hamsters took a significantly greater number of trials to attain criterion performance than normals. Further, the animals reared in the stroboscopic environment were, as a group, significantly impaired as compared to the visually deprived hamsters. All three groups of animals were trained without experimenter's knowledge of their rearing histories. These findings are discussed in greater detail elsewhere (Chalupa, Morrow, & Rhoades, 1978).

These behavioral results correlate rather nicely with our neurophysiological findings. It should be pointed out, however, that no claim can be made that the behavioral impairment was due to dysfunction in the superior colliculus. Indeed, it is possible that other areas of the visual system were also modified to a lesser or greater extent by visual deprivation and restriction.

Perhaps the most intriguing finding was that stroboscopic illumination during development induced greater neurophysiological as well as behavioral effects than visual deprivation. Why should this be the case and what is the mechanism by which this effect is mediated? At present the answers to these questions are unknown, although it is likely that the disruption induced by strobe-rearing is not due to changes at the retina. This is indicated by the fact that the size of visual receptive fields was no greater in the strobe-reared animals than in visually deprived and normal hamsters. We also ascertained that the spatial resolution of collicular cells was not appreciably different in the restricted hamsters than in the other two groups.

Our work provides a demonstration of the functional and behavioral effects of visual restriction in a mammalian species other than the cat. It should be noted, however, that similar results to those we obtained in the hamster have been reported recently in the rabbit's visual system (Pearson, Berman, & Murphy, 1981; Pearson & Murphy, 1983). It is our belief that a comparative approach to the problem of visual system plasticity will serve as a useful means to further elucidate the factors involved in the environmental influences upon the developing brain.

ACKNOWLEDGMENT

A. Leslie Morrow ran the animals in the behavioral study. Supported in part by Chancellor's Patent Fund and MH29548 from NIMH.

REFERENCES

Barlow, H. B. (1975). Visual experience and cortical development. *Nature (London)*, *258*, 199-204.

Barlow, H. B., & Levick, W. R. (1965). The mechanism of directionally selective units in rabbit's retina. *Journal of Physiology (London)*, *178*, 477-504.

Blakemore, C., & Cooper, G. F. (1970). Development of the brain depends on the visual environment. *Nature (London)*, *228*, 477-478.

Chalupa, L. M., Morrow, A. L., & Rhoades, R. W. (1978). Behavioral consequences of visual deprivation and restriction in the golden hamster. *Experimental Neurology*, *61*, 442-454.

Chalupa, L. M., & Rhoades, R. W. (1977). Responses of visual, somatosensory and auditory neurons in the golden hamster's superior colliculus. *Journal of Physiology (London)*, *270*, 595-626.

Chalupa, L. M., & Rhoades, R. W. (1978a). Directional selectivity in hamster superior colliculus is modified by strobe-rearing but not by dark-rearing. *Science*, *199*, 998-1001.

Chalupa, L. M., & Rhoades, R. W. (1978b). Modification of visual response properties in the superior colliculus of the golden hamster following stroboscopic rearing. *Journal of Physiology (London), 274*, 571-592.

Chow, K. L., & Spear, P. D. (1974). Morphological and functional effects of visual deprivation of the rabbit visual system. *Experimental Neurology, 420*, 429-447.

Cynader, M., Berman, N., & Hein, A. (1973). Cats reared in stroboscopic illumination: Effects on receptive fields in visual cortex. *Proceedings of the National Academy of Sciences U.S.A., 70*, 1353-1354.

Cynader, M., Berman, N., & Hein, A. (1976). Recovery of function in cat visual cortex following prolonged deprivation. *Experimental Brain Research, 25*, 139-156.

Cynader, M., & Chernenko, G. (1976). Abolition of directional selectivity in the visual cortex of the cat. *Science, 193*, 504-505.

Daniels, J. D., & Pettigrew, J. D. (1976). Development of neuronal responses in the visual system of cats. In G. Gottlieb (Ed.), *Neural and behavioral specificity* (pp. 195-232). New York: Academic Press.

Dräger, U. C., & Hubel, D. H. (1975). Responses to visual stimulation and relationship between visual, auditory, and somatosensory inputs in mouse superior colliculus. *Journal of Neurophysiology, 38*, 690-713.

Dräger , U. C., & Hubel, D. H. (1976). Topography of visual and somatosensory projections to mouse superior colliculus. *Journal of Neurophysiology, 39*, 91-101.

Finlay, B. L., Schneps, S. E., Wilson, K. G., & Schneider, G. E. (1978). Topography of visual and somatosensory projections to the superior colliculus of the golden hamster. *Brain Research, 142*, 223-235.

Flandrin, J. M., Kennedy, H., & Amblard, B. (1976). Effects of stroboscopic rearing on the binocularity and directionality of cat superior colliculus neurons. *Brain Research, 101*, 576-581.

Flandrin, J. M., & Jeannerod, M. (1977). Lack of recovery in collicular neurons from the effects of early deprivation or neonatal cortical lesion in kitten. *Brain Research, 120*, 362-366.

Gordon, B. G. (1973). Receptive fields in deep layers of cat superior colliculus. *Journal of Neurophysiology, 36*, 157-178.

Grobstein, P., & Chow, K. L. (1975). Receptive field development and individual experience. *Science, 190*, 352-358.

Hirsch, H. V. B., & Spinelli, D. N. (1970). Visual experience modifies distribution of horizontally and vertically oriented receptive fields in cats. *Science, 168*, 869-871.

Hoffmann, K.-P., & Sherman, S. M. (1975). Effects of early binocular deprivation on visual input to cat superior colliculus. *Journal of Neurophysiology, 38*, 1049-1059.

Hubel, D. H., & Wiesel, T. N. (1963). Receptive fields of cells in striate cortex of very young, visually inexperienced kittens. *Journal of Neurophysiology, 26,* 994-1002.

Hubel, D. H., & Wiesel, T. N. (1970). The period of susceptibility to the physiological effects of unilateral eye closure in kittens. *Journal of Physiology (London), 206,* 419-436.

Hubel, D. H., & Wiesel, T. N. (1977). Functional architecture of macaque monkey visual cortex. *Proceedings of the Royal Society, London, 198,* 1-59.

Mize, R. R., & Murphy, E. H. (1973). Selective visual experience fails to modify receptive field properties of rabbit striate cortex neurons. *Science, 180,* 320-323.

Movshon, J. A., & Van Sluyters, R. C. (1981). Visual neural development. *Annual Review of Psychology, 32,* 477-522.

Olson, C. R., & Pettigrew, J. D. (1974). Single units in visual cortex of kittens reared in stroboscopic illumination. *Brain Research, 70,* 189-204.

Pearson, H. E., Berman, N., & Murphy, E. H. (1981). Stroboscopic rearing reduces direction selectivity in rabbit visual cortex. *Developmental Brain Research, 1,* 127-131.

Pearson, H. E., & Murphy, E. H. (1983). Effects of stroboscopic rearing on the response properties and laminar distribution of single units in the rabbit superior colliculus. *Developmental Brain Research, 9,* 241-250.

Pettigrew, J. D., & Freeman, R. D.(1973). Visual experience without lines: Effect on developing cortical neurons. *Science, 182,* 599-601.

Rhoades, R. W., & Chalupa, L. M. (1976). Directional selectivity in the superior colliculus of the golden hamster. *Brain Research, 118,* 334-338.

Rhoades, R. W., & Chalupa, L. M. (1977). Differential effects of stimulus size on "on" and "off" responses of superior collicular neurons. *Experimental Neurology, 57,* 57-66.

Rhoades, R. W., & Chalupa, L. M. (1978). Receptive field characteristics of superior colliculus neurons and visually guided behavior in dark-reared hamsters. *Journal of Comparative Neurology, 177,* 17-32.

Schneider, G. E. (1973). Early lesions of the superior colliculus: Factors affecting the formation of abnormal retinal projections. *Brain, Behavior and Evolution, 8,* 73-109.

Schneider, G. E. (1977). Growth of abnormal connections following focal brain lesions: Constraining factors and functional effects. In W. H. Sweet, S. Obrador, & J. G. Martin-Rodriquez (Eds.). *Neurosurgical treatment in psychiatry, pain and epilepsy* (pp. 5-26). Baltimore: University Park Press.

Sherman, S. M., & Spear, P. D. (1982). Organization of visual pathways in normal and visually deprived cats. *Physiological Reviews, 62,* 738-855.

Stein, B. E., Magãlhaes-Castro, B., & Kruger, L. (1976). Relationship between visual and tactile representations in cat superior colliculus. *Journal of Neurophysiology, 39*, 401-419.

Sterling, P., & Wickelgren, B. G. (1970). Function of the projection from visual cortex to the superior colliculus. *Brain, Behavior and Evolution, 3*, 210-218.

Tiao, Y.-C., & Blakemore, C. (1976). Functional organization in the superior colliculus of the golden hamster. *Journal of Comparative Neurology, 168*, 459-482.

Van Sluyters, R. C., & Blakemore, C. (1973). Experimental creation of unusual neuronal properties in visual cortex of kittens. *Nature (London), 246*, 506-508.

Wiesel, T. N., & Hubel, D. H. (1963). Single-cell responses in striate cortex of kittens deprived of vision in one eye. *Journal of Neurophysiology, 26*, 1003-1017.

Wiesel, T. N., & Hubel, D. H. (1965). Comparison of the effects of unilateral and bilateral eye closure on cortical unit responses in kittens. *Journal of Neurophysiology, 28*, 1029-1040.

22· NERVOUS SYSTEM DEVELOPMENT AND PATTERN PREFERENCE IN INFANTS

D. S. Woodruff-Pak and K. M. Gerrity

Temple University

The purpose of this study was to examine behavioral correlates of the rapid central nervous system changes occurring in the first 4 months of infant life. This research originates directly from Lindsley's (1936, 1938, 1939) early developmental work on the human EEG and is designed to test the hypothesis that cortically mediated visual behaviors are absent in the first few months of infancy. Lindsley was one of the first to note that toward the end of the third month after birth the EEG in occipital areas changes dramatically. The occipital EEG of the neonate is relatively flat and arrhythmic, but during the third month a prominent 50 μV rhythm of 3-4 Hz emerges. Lindsley suggested that the beginning of organized rhythmic activity in occipital areas may signify the onset of integrated cortical activity in these regions. It may be only at this period that visual behavior comes under the control of the cortex. Thus, until the third month visual behavior may be mediated primarily at a subcortical level.

While systematic attempts to test this hypothesis by measuring infant behavioral capacity before and after the onset of rhythmic occipital EEG activity have not been undertaken, investigations of neonatal EEG have continued to suggest the functional significance of the onset of rhythmic activity in posterior regions (Dreyfus-Brisac, Samson, Blanc, & Monod, 1958; Ellingson, 1967). Research on neurological development of the infant cortex at the cellular level indicated that cortical cell myelination and dendritic branching show little progress during the first month of life, but undergo marked maturational change during the second and third months (Conel, 1939, 1941, 1947). This burst of neurological development immediately precedes the onset of occipital alpha activity. On the other hand, Scheibel and Scheibel's (1964a, 1964b, 1971) work on the development of alpha-like activity in kittens suggests that alpha onset is linked in time to the development of apical dendrites of pyramidal cells which increase brainstem-cortex interconnections. By 32 to 33 weeks gestation in humans, Purpura (1975, 1976) has found dendritic spines on both apical and basilar dendrites. This evidence suggests that the neuronal prerequisites for alpha may be present before three months of age.

The development of the capacity to function as a cortically integrated

 DOI: 10.4324/9781003164647- 28

organism, as signaled by the emergence of organized rhythms in the occipital lobes, can be used as a perspective from which to examine behavioral development (Woodruff, 1978). Given the infant's dramatic increase in neurological potential around the age of 3 months, this should be a critical period in which to observe behavioral changes. The onset of occipital rhythmic activity corresponds with a time of change in neonatal reflex patterns. At approximately 3 to 4 months, many reflexes drop out of the infant's behavioral repertoire (Fiorentino, 1972; Scheibel & Scheibel, 1964a, 1964b; Taft & Cohen, 1967) due to increasing cortical inhibition of lower centers. Reflex arcs exist below the cortical level, and before integration occurs between subcortical and cortical centers, stimulation of the infant elicits an involuntary response. As maturing cortical centers become integrated with subcortical areas, primitive reflex behavior is inhibited, and voluntary responses emerge. The infant thus progresses to a neurologically more mature, integrated state. This disappearance of primitive reflexes is a clear behavioral sign indicating the development of the central nervous system.

Lindsley and Wicke (1974) inferred the lack of cortical functional integration in infant behavior prior to 3 months from the reflex behavior of infants born without a cerebral cortex. Although these infants do not live more than 3 months, while alive they exhibit the same reflex development as a normal newborn of the same age. This suggests the lack of cortical integration in the normal neonate.

The inference that visual behavior is mediated at a subcortical level in the first months of life was also made by Bronson (1974) based on his interpretation of the neuroanatomical, neurophysiological, and behavioral literature on early infancy. Bronson devised a model of neonatal visual development involving visual pathways through the superior colliculus and pulvinar as well as through the lateral geniculate to areas 17, 18, and 19. Bronson argued that the pathways through the superior colliculus and pulvinar comprise a secondary visual system and develop earlier and are myelinated sooner than the geniculo-striate pathways (comprising the "primary visual system"). Examining the behavioral data in infancy, he concluded that the discriminations which infants have been shown to be capable of performing in the first month of life could be mediated by the secondary system, and he believed that the primary system was not functional until the second month of life.

Haith and Campos (1977) challenged Bronson's (1974) model proposing instead that all parts of the visual system are equally immature. They argued, for example, that event-related potential (ERP) research indicates that the primary visual system functions in the neonate and that foveal vision appears to be functional in the neonate. Bronson (1978) responded to these arguments by suggesting that presence of an ERP does not necessarily signify functional capacity, and by presenting evidence that the secondary system can carry out some foveal processing. He stipulated additionally that since he presented a stage theory, critical tests of the theory should be based on longitudinal rather than cross-sectional data.

There is considerable evidence available from studies of neurophysiological development and the development of EEG, behavioral reflexes, and visual behavior that dramatic changes occur in the second and third months of life. However, relationships among these variables in the developing individual have not been examined. A major reason that such studies have not been undertaken in infancy is that few techniques for the accurate measurement of infant behavior had been developed until the last decade. We undertook the present *longitudinal* study using some of these recently developed techniques to study the relationship between visual development and development of the nervous system. Behaviors which could be mediated subcortically as well as behaviors requiring cortical control were included to test the hypothesis that infant behavioral capacity changes around the time that cortical and subcortical structures become integrated.

To study visual discrimination and preference, we used a behavioral paradigm devised by Ruff and Turkewitz (1975) to measure qualitative and quantitative aspects of infants' visual response. We hypothesized that during the first months of infancy, preference would occur as a function of quantitative dimensions of the stimuli (size) which could be mediated at a subcortical level. When primitive reflex measures showed a diminution, thus signifying integration between cortical and subcortical structures, visual preference should shift to qualitative stimulus dimensions (curved vs. straight elements) mediated at a cortical level.

METHOD

Subjects. Fifteen infants, 10 females and 5 males, participated in the behavioral study for the first 4 months of their lives. Following the Lindsley tradition of longitudinal EEG testing in his own children, a 16th subject, the first author's son (JTP) was tested in the behavioral paradigm and also had his EEG measured during the first 3 months of his life. All infants were full-term (mean gestational age of 40.5 weeks within a range of 39-42 weeks) and had an Apgar of 9 or above.

Apparatus. Stimuli were presented in a fiberboard table-top viewing apparatus, composed of three panels. The infant faced the rear panel and was surrounded on the front and both sides to block out interfering visual stimuli. The mother was seated before the table on which the white viewing screen was placed with her back toward the rear panel with her infant held at her shoulder. This placed the infant on eye level with the stimuli at a distance of approximately 12 inches. The openings for the stimuli were 11 x 11 inches, and the distance between these openings was 3 inches. Fixations were observed from behind the stand through a 1/4-inch hole between the stimuli, and fixation time was recorded by the experimenter. The stimuli could not be seen by the experimenter when they were in place, so the experimenter was blind as to which stimuli were being fixated. A cardboard flap on the inside of the apparatus covered the stimuli until they were presented. Fixations were timed by a Lafayette clock-counter, and presentations were timed using a pre-recorded cassette and timed signals to indicate the beginning and end of each trial.

Stimuli. The stimuli were the same as those used by Ruff and Turkewitz (1975) and were five different sizes of each of a bull's-eye and striped configuration. The bull's-eye was composed of 12 curvilinear segments and the horizontal stripes were composed of 12 straight segments arranged in two columns. Size was varied by designing bull's-eyes and stripes of five different total areas. A bull's-eye and horizontal stripe pattern of the same size were equated for number of segments, length of segments, and overall black/white ratio. Contour was the same for stimuli of the same size. Each stimulus presentation included two stimuli, a bull's-eye and a horizontal stripe stimulus. All five sizes of the bull's-eye pattern were paired with the intermediate-size striped pattern, and all five sizes of the striped pattern were paired with the intermediate-size bull's-eye pattern. Hence, there were ten possible stimulus pairs. Since the bull's-eye/stripes pair of intermediate size was common to both series, there was a total of nine different pairs presented.

Procedure. Each infant was tested at home seven times: at 2, 8, 11, 13, 15, and 17 weeks of age. The mother was asked to sit with her back to the apparatus and hold her infant so that he or she could look over the mother's shoulder. The nine stimulus pairs were shown in the same random order to all subjects and then shown again in a different random order with the positions of the stimuli reversed. Thus, the two stimulus types appeared equally on the left and the right throughout the 18 trials, appearing four times on the left in one order and five times in the other order. Each pair was presented twice to the infant for 5 seconds, so the total possible viewing time for each stimulus was 10 seconds. The experimenter observed the direction of the infant's gaze through the small viewing hole and recorded the direction and duration of each fixation. Fixations were recorded on the basis of the observer's judgment of the direction of gaze and not by corneal reflection. Neither the experimenter nor the mother could see the stimuli. Interrater reliability for this procedure was found by Ruff and Birch (1974) to be .95.

Neurological Assessment. On the same day that visual behavior was assessed, the strength of seven reflexes was measured. These reflexes are described in Table 22.1. One experimenter (KMG) was trained by a pediatric neurologist to elicit the reflexes, and they were scored on a 4-point scale devised by Parmelee (1971) and Sigman, Kapp, Parmelee, and Jeffrey (1973). On the Parmelee scale a reflex is scored as absent (0), weak (1), normal (2), or strong (3). Since there were seven reflexes tested, a total score indicating the greatest possible immaturity was 21.

Electroencephalographic Measurement. In one male infant (JTP), EEG as well as visual and reflex behavior was measured. EEG assessments were made at 2, 5, 7, 8, 9, 10, and 11 weeks of age, and behavioral assessments were made at 2, 5, 7, 8, 11, and 13 weeks. Grass silver-silver chloride cup electrodes were attached with Grass electrode cream to sites measured as O_1, O_2, and C_z according to the 10-20 system and referenced to linked earlobes. EEG was recorded on a Beckman type R dynograph with inputs to a Vetter Model A, FM tape recorder.

TABLE 22.1. Reflexes evaluated over the first 4 months of life.

Reflex	Description
1. Rooting	Stroking the corner of the mouth and moving the finger laterally across the cheek causes the baby to move his tongue, mouth, and head towards the stimulated side.
2. Sucking	Elicited by placing a finger or a nipple in the infant's mouth.
3. Palmar-mental	Produced by pressure on both palms; infant opens mouth and closes eyes.
4. Grasping	Infant's automatic grasp is used to pull him to sit by placing a forefinger in each palm (evaluation of motor maturity).
5. Tongue retrusion	Infant responds by pushing out tongue whenever hard object is placed in mouth.
6. Stepping reflex	In standing position, infant held under arms and inclined forward, takes rhythmical steps characterized by heel strike.
7. Withdrawal reflex	Legs extended, soles of feet stimulated results in extension of toes followed by pulling legs to torso.

RESULTS

As anticipated, neurological development, as measured by reflexes, proceeded rapidly in the first four months of life. All reflexes were normal or strong at the 2-week testing and weak or absent by the 17-week testing. This result is shown in Figure 22.1, which presents reflex scores over the four-month period for 15 infants. The most dramatic change in the nervous system as assessed by primitive reflexes occurred between the fifth and eighth week. Twelve of the 15 infants showed the greatest maturational change in reflexes at this point in their development. The remaining three showed greatest neurological score change between the 8- and 11-week testings.

Changes in visual responding as a function of age were examined in a 2x4x7 analysis of variance testing the effects of stimulus pattern (bull's-eye vs. horizontal stripes), stimulus size (large size 5 and 4 vs. small size 1 and 2), and age on fixation time. All three effects were statistically significant as was the age x size interaction and the age x pattern interaction. Infants fixated more at older than at younger ages ($F(6,98)=2.97$, $p<.01$), they preferred larger stimuli ($F(3,98)=29.0$, $p<.01$), particularly when they were younger ($F(18,294)=4.41$, $p<.01$), and they preferred bull's-eyes to horizontal stripes ($F(1,98)=73.0$, $p<.01$), only when they were older ($F(6,98)=6.09$, $p<.01$).

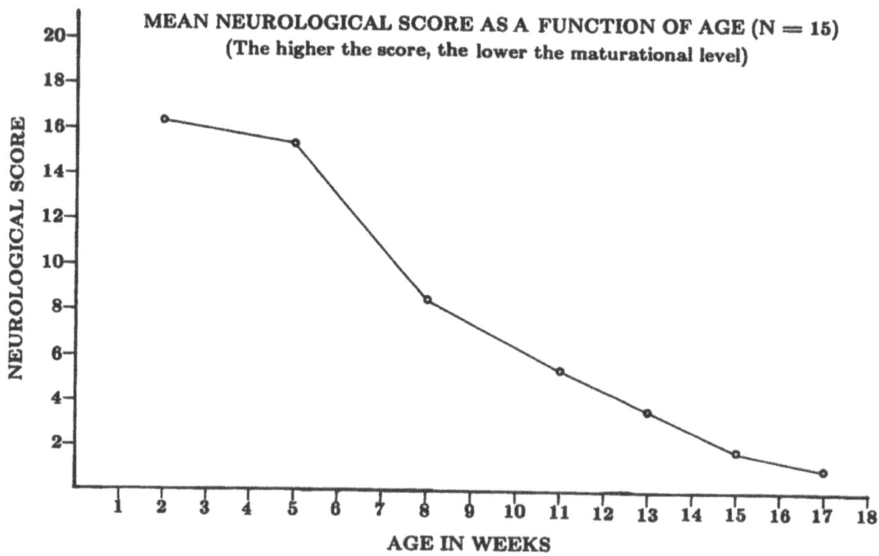

FIGURE 22.1. Mean neurological scores as a function of age. (The higher the score, the lower the maturational level.)

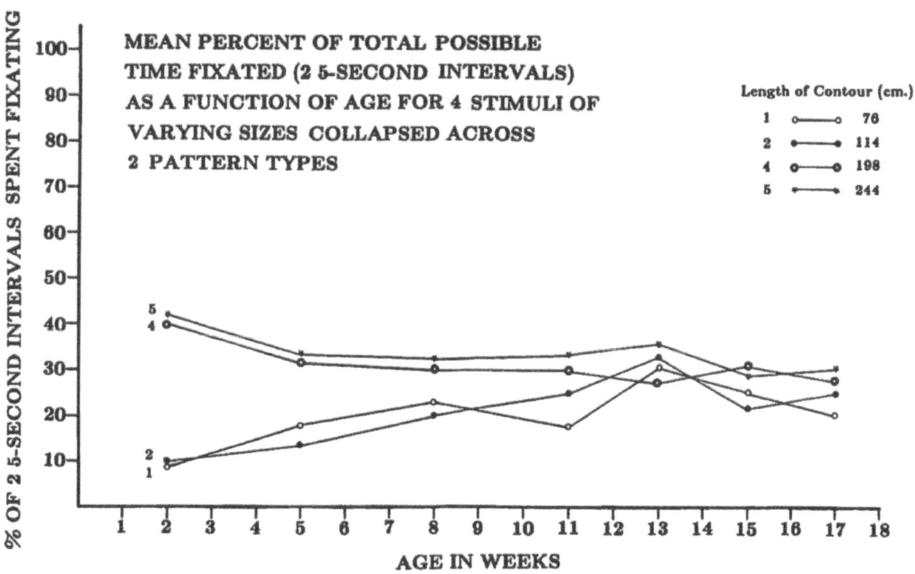

FIGURE 22.2. Mean percent of total possible time fixated (two 5-second intervals) as a function of age for four stimuli of varying sizes collapsed across two pattern types.

The effect of age and stimulus size on fixation is presented in Figure 22.2. Young infants clearly have a preference for larger stimuli as they fixated the two largest stimuli four times as long at the 2-week testing and twice as long at the 5-week testing as they fixated the two smallest stimuli.

The intermediate stimulus, not shown in Figure 22.2, was fixated an intermediate length of time between fixation time for the largest and smallest stimuli. Post hoc comparisons using the Scheffe test indicated that the differences between the largest and smallest stimuli and between the second largest and second smallest stimuli were both statistically significant at the .01 level of confidence at the 2-week testing and at the .05 and .10 levels, respectively, at the 5-week testing. By the 8-week testing the infants were not preferentially viewing stimuli on the basis of size as the differences between fixation time were not statistically significant. At the 5-week testing the largest stimulus was preferred over the smallest stimulus at the .05 level, but there were no significant preferences on the basis of size after that session. Thus, the data suggest that size is a salient stimulus characteristic for young infants, but by the time they are 8 weeks of age their fixation preferences seem determined much less by the size of the stimulus. It was also at this 8-week testing that we observed the greatest change in nervous system maturation as indexed by primitive reflex score.

Fixation preferences as a function of stimulus pattern are shown in Figure 22.3. During the first two test sessions at 2 and 5 weeks there was virtually no difference in preference for horizontal stripes or bull's-eyes. These were the same two sessions in which there was a clear preference for stimuli on the basis of size. At 8 weeks, 5 of the 15 infants began to show some preference for bull's-eyes, but the effect was not statistically significant. By 11 weeks the preference for bull's-eyes was evident in 12 of the 15 infants, and post hoc Scheffe tests indicated that the effect was significant at the .05 level as it was for all of the subsequent testings. Throughout the period of the study infants showed about the same amount of looking time for the horizontal stripes, but they significantly increased the amount of time they fixated bull's-eyes at the age of 11 weeks and continued to fixate them longer for the rest of the study. Thus, infants did not appear to discriminate differences between patterned stimuli until 8 weeks of age, but by the time they were 11 weeks they were clearly showing a preference for curvature.

The correlations between the decrease in strength of the neonatal reflexes and the increasing preference for curvature were strikingly high. When the average reflex score for all subjects at each age of testing was correlated with the average preference for bull's-eye score at that age, the

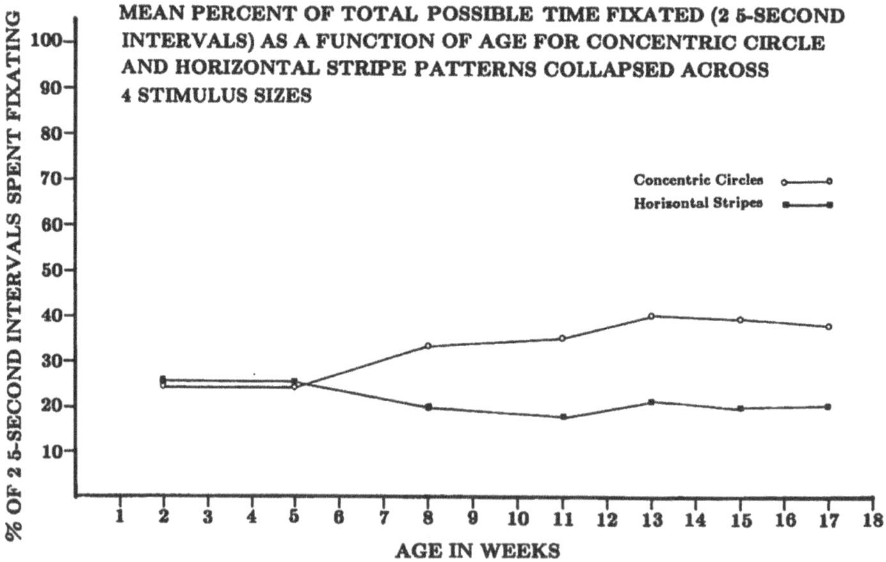

FIGURE 22.3. Mean percent of total possible time fixated (two 5-second intervals) as a function of age for concentric circle and horizontal stripe patterns collapsed across four stimulus sizes.

correlation was -.946. Neonatal reflex score was a powerful predictor of visual preference for pattern elements.

Neurological maturation in the subject (JTP) for whom there is longitudinal EEG data showed a pattern of neurological maturation similar to that in the other 15 subjects. He showed the greatest change in neurological score between the 5- and 8-week testings. Additionally, he preferred the largest over the smallest stimuli in the initial test sessions, and first showed a clear preference for bull's-eye stimuli at 8 weeks.

While JTP showed a preference for curvature at the age of 8 weeks, he did not have clear 50 μV 3-4 Hz rhythmic activity in his EEG at the 8- or 9-week testing. It was not until the 10-week testing that the occipital rhythm of 3-4 Hz began to be apparent. While the onset of the occipital EEG rhythm at 10 weeks was earlier than published reports of mean age at which the rhythm appears, it was still preceded by clear behavioral evidence of preference for curvature.

DISCUSSION

The results of this study suggest that there is a shift in the manner in which infants respond to visual stimuli between 8 and 11 weeks of age, and this shift is preceded by rapid changes in neurological maturation as indexed by the disappearance of primitive reflexes. The greatest decline in reflex strength took place between 5 and 8 weeks, while the shift in visual behavior became apparent at 8 weeks and was clear by 11 weeks. These results add to a growing body of literature (Fantz & Fagan, 1975; Karmel, Hoffman, & Fegy, 1974; Harter, Deaton, & Odom, 1977; Ruff & Turkewitz, 1975; Schanel-Klitsch & Woodruff, 1985) demonstrating that infants younger than 9 to 10 weeks discriminate stimuli on a different basis from infants 11 weeks or older. Younger infants differentiate on the basis of the size of the stimulus while older infants differentiate on the basis of qualitative characteristics of pattern elements. Young infants treat stimuli of the same pattern but of different size as different stimuli, while older infants attending to pattern characteristics of stimuli respond to preferred pattern elements regardless of size.

Ruff and Turkewitz (1975) argued that such results could not be explained simply on the basis of acuity differences between young and older infants. The stimuli used by Ruff and Turkewitz were those used in the present investigation, and the segment width of even the smallest segment was well above the threshold of the youngest infant. Since preferential fixation occurred even between two stimuli a great deal above acuity threshold (e.g., between the largest and the intermediate stimulus), it is unlikely that acuity is the cause of the preference shift.

The explanation favored by Ruff and Turkewitz and by the present investigators involves changes in the central nervous system. Lindsley (1938) and Lindsley and Wicke (1974) suggested that subcortical and cortical structures are not integrated at birth and that the infant functions on a subcortical level until the third month of life. Bronson (1974) suggested that the secondary visual system, involving the superior colliculus and pulvinar and characterized by poor foveal vision and greater sensitivity to peripheral stimulation, may predominate in the control of infant visual behavior during the first two months of life. The primary visual system is relatively immature until the end of the second month when it supersedes the secondary system. Functional capacity in this geniculo-striate system provides the neurological basis for pattern vision which emerges at this time. In his model, Bronson argued that patterns of myelinogenesis and neuronal growth in the infant cortex follow the order in which information is processed. Subcortical structures such as the superior colliculus and lateral geniculate mature earlier than the cortex. Thus, visual behaviors that can be mediated at a subcortical level appear in young infants, while behaviors requiring cortical control emerge only after the cortex is functional.

While Bronson suggested histological measures to index the onset of function in the cortex, Lindsley suggested that the onset of rhythmic occipital EEG might signify the integration of cortical and subcortical structures and hence the onset of functional cortical capacity. Bronson's criteria for cortical control cannot be measured in normal human infants while Lindsley's can. Longitudinal data on one subject suggested that the geniculo-striate system is functionally involved in behavior before the onset of organized rhythmic activity in that system. Stimulus preference based on pattern discrimination occurred 2 weeks before the appearance of the occipital rhythm. Subsequent cross-sectional research (Klitsch & Woodruff, 1980) with 32 infants aged 1, 2, 3, and 4 months has supported our initial result that organized rhythmic EEG does not usually precede preference for curved pattern elements. Regression analysis on these data predicted a behavioral shift from preference for size to preference for pattern elements at 1.7 months, while the prediction for the appearance of EEG alpha rhythm was 3.5 months. Spectral analysis of the EEGs of the infants indicated the presence of prominent activity in the 3-5 Hz bandwidth in some infants as early as one month, but the majority of infants exhibited such activity later. These data suggest that while the rhythmic activity may be associated with the onset of cortical function, it does not pace or precede that function.

The onset of the EEG alpha rhythm does not appear to signal the functional integration of subcortical and cortical visual structures. However, our behavioral data suggest a qualitative change in visual processing beginning at 5 weeks of life. In kittens, Rose and Lindsley (1968) and Atwell and Lindsley (1975) identified changes in ERP components with development which they experimentally demonstrated to be related to the primary and secondary visual systems (also see the chapter by Rose in the present volume). Secondary visual system components emerged first and were later integrated with components from the primary visual system. Ellingson (1964), Harter et al. (1977) and Karmel et al. (1974) have described developmental changes in human neonatal ERPs which show parallels in integration of ontogenetically earlier and later waveforms. Vaughan (1975) cautions that such simple formulations as specific and nonspecific systems will probably not suffice to explain the complex neurophysiological substrate of the visual ERP. However, careful examination of longitudinal visual ERP data coupled with assessment of visual behavioral capacity in the developing neonate may yield information about the functional integration of subcortical and cortical visual structures.

REFERENCES

Atwell, C. W., & Lindsley, D. B. (1975). Development of visually evoked responses and visually guided behavior in kittens: Effects of superior colliculus and lateral geniculate lesions. *Developmental Psychobiology, 8*, 465-478.

Bronson, G. (1974). The postnatal growth of visual capacity. *Child Development, 45*, 873-890.

Bronson, G. (1978, March). *The postnatal growth of visual capacity: Considerations and clarifications.* Paper presented at the International Conference on Infant Studies, Providence, RI.

Conel, J. L. (1939). *The postnatal development of the human cerebral cortex: Vol. I. The cortex of the newborn.* Cambridge, MA: Harvard University Press.

Conel, J. L. (1941). *The postnatal development of the human cerebral cortex: Vol. II. The cortex of the one month infant.* Cambridge, MA: Harvard University Press.

Conel, J. L. (1947). *The postnatal development of the human cerebral cortex: Vol. III, The cortex of the three month infant.* Cambridge, MA: Harvard University Press.

Dreyfus-Brisac, C., Samson, D., Blanc, C., & Monod, N. (1958). L'electroencephalogramme de l'enfant normal de moins de 3 ans. *Etudes Neo-Natales, 7*, 143-175.

Ellingson, R. J. (1964). Cerebral electrical responses to auditory and visual stimuli in the infant (human and sub-human studies). In P. Kellaway & I. Peterson (Eds.), *Neurologic and electroencephalographic correlative studies in infancy* (pp. 78-116). New York: Grune & Stratton.

Ellingson, R. J. (1967). The study of brain electrical activity in infants. In L. L. Lipsitt & C. Spiker (Eds.), *Advances in child development and behavior (Vol. 3*, pp. 53-97). New York: Academic Press.

Fantz, R. L., & Fagan, J. F. (1975). Visual attention to size and number of pattern details by term and preterm infants during the first six months. *Child Development, 46*, 3-18.

Fiorentino, M. R. (1972). *Normal and abnormal development: The influence of primary reflexes on motor development*. Springfield, IL.: Charles C. Thomas.

Haith, M. M., & Campos, J. J. (1977). Human infancy. In M. Rosenzweig & L. W. Porter (Eds.), *The Annual Review of Psychology, 28*, 251-294.

Harter, M. R., Deaton, F. K., & Odom, J. V. (1977). Maturation of evoked potentials and visual preference in 6 to 45-day-old infants: Effects of check size, visual acuity, and refractive error. *Electroencephalography and Clinical Neurophysiology, 42*, 595-607.

Karmel, B. Z., Hoffman, R. F., & Fegy, M. J. (1974). Processing of contour information by human infants evidenced by pattern-dependent evoked potentials. *Child Development, 45*, 39-48.

Klitsch, E. S., & Woodruff, D. S. (1980, September). *EEG alpha rhythm and cortical control of neonatal vision*. Paper presented at the 88th Annual Meeting of the American Psychological Association, Montreal.

Lindsley, D. B. (1936). Brain potentials in children and adults. *Science, 84*, 354.

Lindsley, D. B. (1938). Electrical potentials of the brain in children and adults. *Journal of General Psychology, 19*, 285-306.

Lindsley, D. B. (1939). A longitudinal study of the occipital alpha rhythm in normal children: Frequency and amplitude standards. *Journal of Genetic Psychology, 55*, 197-213.

Lindsley, D. B., & Wicke, J. D. (1974). The electroencephalogram: Autonomous electrical activity in man and animals. In R. F. Thompson & M. M. Patterson (Eds.), *Bioelectric recording techniques* (pp. 3-83). New York: Academic Press.

Parmelee, A. H. (1971). *Newborn neurological examination*. Unpublished manuscript. University of California at Los Angeles.

Purpura, D. P. (1975). Morphogenesis of visual cortex in the preterm infant. In M. A. B. Brazier (Ed.), *Growth and development of the brain* (pp. 33-49). New York: Raven Press.

Purpura, G. H. (1976). Structure-dysfunction relations in the visual cortex of preterm infants. In M. A. B. Brazier & F. Coriani (Eds.), *Brain dysfunction in infantile febrile convulsions* (pp. 223-240). New York: Raven Press.

Rose, G. H., & Lindsley, D. B. (1968). Development of visually evoked potentials in kittens: Specific and nonspecific responses. *Journal of Neurophysiology, 31*, 607-623.

Ruff, H. A., & Birch, H. G. (1974). Infant visual fixation: The effect of concentricity, curvilinearity, and number of directions. *Journal of Experimental Child Psychology, 17*, 460-473.

Ruff, H. A., & Turkewitz, G. (1975). Developmental changes in the effectiveness of stimulus intensity on infant visual attention. *Developmental Psychology, 11*, 706-710.

Schanel-Klitsch, E., & Woodruff, D. S. (1985). Compound pattern perception in early infancy. *Child Study Journal, 15*, 1-12.

Scheibel, M. E., & Scheibel, A. B. (1964a). Some neural substrates of postnatal development. In M. Hoffman & L. Hoffman (Eds.), *Review of child development research (Vol. 1*, pp. 481-519). New York: Russell Sage.

Scheibel, M. E., & Scheibel, A. B. (1964b). The developing brain. *Progress in Brain Research, 9*, 6-25.

Scheibel, M. E., & Scheibel, A. B. (1971). Selected structural-functional correlations in the post-natal brain. In M. B. Sterman, D. J. McGinty, & A. M. Adinolfe (Eds.), *Brain development and behavior* (pp. 1-21). New York: Academic Press.

Sigman, M., Kapp, C. B., Parmelee, A. H., & Jeffrey, W. (1973). Visual attention and neurological organization in neonates. *Child Development, 44*, 461-466.

Taft, L. T., & Cohen, H. J. (1967). Neonatal and infant reflexology. In J. Hellmuth (Ed.), *Exceptional infant: Vol. I. The normal infant* (pp. 79-120). New York: Brunner/Mazel.

Vaughan, H. G. (1975). Electrophysiological analysis of regional cortical maturation. *Biological Psychiatry, 10*, 513-526.

Woodruff, D. S. (1978). Brain electrical activity and behavior relationships over the life span. In P. B. Baltes (Ed.), *Life span development and behavior* (Vol. 1, pp. 111-179). New York: Academic Press.

SECTION VII. HIGHER COGNITIVE PROCESSES

The term "higher cognitive processes" as applied to the papers of this section extends to a wide range of topics including learning and conditioning, motivation and emotions, reading disability, language laterality and stuttering, the confidence of a subject in making a decision, and even the relationship between the contingent negative variation (CNV) and high social status. The common theme throughout this section is a continuous concern with brain-behavior relationships. Only one report is completely behavioral (Friedman et al.), but even in this report there is an attempt to provide information about brain-behavior relationships, specifically, in this case, of reading disability. All of the remaining studies measured both behavioral and electrophysiological responses and investigated their relationships.

Thompson and Berry's report is an extensive, detailed review of Professor Lindsley's contributions to physiological psychology in general and to the areas of learning and motivation in particular. The review begins with a survey of Professor Lindsley's psychophysiological contributions in the late forties and early fifties, including the 1949 publication by Lindsley, Bowden and Magoun reporting on the behavioral effects of lesions in the reticular formation. In 1951, Professor Lindsley published his important activation theory of emotion in Stevens' influential *Handbook of Experimental Psychology*. The Thompson and Berry review continues to survey Professor Lindsley's research on learning and motivation into the late 1970s with his neurophysiological research on the topics of hippocampal EEG and hypothalamic systems. Hippocampal EEG data in cats collected by Coleman and Lindsley during lever-press learning for rewards provide strong, supplementary support of the general notion of an inverted U-shaped function relating activation to learning. Furthermore, the data show that, as an index of a behavioral state of arousal or activation, hippocampal EEG correlates well with attentional processes throughout the course of appetitive learning in cats. In rabbits, the data actually predict the responsiveness of hippocampal neurons to conditioning stimuli and the behavioral rate of classical aversive conditioning. Thompson and Berry provide interesting hippocampal data of their own, giving the first demonstration of how a purely neurophysiological measure taken prior to the beginning of training can predict the subsequent behavioral rate of learning.

The review also describes how, when medial hypothalamic stimulation is given to well-trained animals performing an operant lever-pressing task, the animals stop, look around, and appear to be attending to other cues. This behavior is accompanied by a pronounced theta rhythm. Earlier Schlag-Rey and Lindsley examined the effects of medial and lateral hypothalamic stimulation on photic evoked potentials in visual and association cortex of the cat, and they cautiously concluded that central neural processes may be attenuated by such stimulation.

The Simonov paper, "Need-Information Interaction of Brain Structures," concerns the role of four main brain structures in the genesis of emotional states and in the organization of purposeful behavior. The four brain structures are the frontal neocortex, the hippocampus, the hypothalamus, and the amygdala. The frontal neocortex is responsible for behavioral orientation to high probability events, while the hippocampus provides reactions to events of low probability. The hypothalamus orients behavior towards the satisfaction of a dominating need, and the amygdala creates a balance or dynamic coexistence between competing needs.

Simonov has an even higher-order organization than these four brain structures. He combines the hippocampus-amygdala structures into an "emotional system," corresponding to the role of emotions in the formation of behavior--its orientation toward satisfaction of the actual need considering the available situation (amygdala) and the broadening of the range of meaningful external stimuli (hippocampus). The frontal neocortex and hypothalamus brain structures are regarded as the cerebral substratum of the "will." It is the "will" that stabilizes behavior oriented towards satisfaction of initially dominant needs, in spite of an unfavorable forecast and competing motives.

Friedman and his colleagues present a theoretical and empirical analysis of reading disability (RD) in the framework of current research in neuropsychology and cognitive psychology. Current theories of RD postulate slower or less complete differentiation of cerebral functions, especially of the left hemisphere. Initially, these researchers investigated different processing skills associated with left and right hemispheric functioning and reading disability. The RD subjects performed more poorly on left hemisphere tests of verbal skill but were equal to normal controls on right hemisphere, non-verbal tasks. An analysis of the strategies actually employed in reading and spelling showed an orderly, fixed sequence of different strategies. There also is evidence for clinical subgroups of RD that can be related to the different types of strategy classification. Other important components of RD include the report that RD children are generally inattentive and distractible, which in turn has been related to either deficient left hemisphere functioning or to incomplete dominance. Friedman and co-workers discuss how cognitive style may be an important factor in the development of an RD child. Using the Witkin Rod and Frame Test of Field Articulation, they found that, as predicted, RD subjects tended to be much more field dependent than normal controls. The report concludes that although this research began with simplistic notions about brain-behavior relationships, the empirical research revealed numerous factors that are relevant to a better understanding of the identification, diagnosis and treatment of reading disability.

In several previous studies, Brown and his colleagues have reported that in normal right-handed males, differences in the wave forms of EEG event-related potentials (ERPs) produced by the noun and verb meanings of a homophone (e.g., "fire") are greater for responses from left hemisphere placements than homologous right hemisphere placements. In the present study, they utilize this asymmetry of noun/verb ERP correlation as a noninvasive measure of the hemispheric asymmetry of language processing. Language dominance was studied in three groups: (1) normal right-handed males, (2) normal right-handed females, and (3) male right-handed stutterers.

The results of Brown and co-workers' investigation lead to three conclusions. First, the interrelationship between hemispheric dominance, sex, and stuttering is not simply a one-dimensional conceptualization. For example, in addition to the more commonly studied left/right dimension, the anterior/posterior dimension of cortical language processing should be evaluated. Second, the relationship between the ERP index of laterality, sex, and stuttering indicates an association among these variables that is not necessarily causal. Finally, these data emphasize the usefulness of EEG and ERP methods, in general, and how they can be used to study language processing, in particular.

Rohrbaugh and co-workers obtained event-related potentials (ERPs) and 3-point confidence ratings for a line discrimination task, in which a horizontal line was presented briefly followed 2 seconds later by another horizontal line in the same location. Based on random selection on 50% of the trials, the comparison stimulus was the same length as the previously exposed standard line. On the other 50% of the trials, the comparison line was longer than the standard line. Task difficulty was manipulated by varying, in separate sessions, the length of the comparison lines: 20% longer in the Hard condition, 25% longer in the Medium condition, and 30% longer in the Easy condition. Also recorded, as an index of preparatory activity, was the contingent negative variation (CNV). The performance data indicated that confidence ratings were used by the subjects in a consistent and meaningful manner. Large, systematic, and significant effects were associated with confidence in the P300 measure. The P300 reflects the quality of the associated evaluation or decision processes, particularly the confidence with which the decisions are made. In comparison, the CNV measure showed no significant changes related to difficulty, correctness, or confidence.

The thesis of the Maltzman and Pendery investigation is that there are constraints on the laws of classical conditioning, namely, the laws do not apply in the same manner to all response systems. Laboratory experiments in classical conditioning are studying aspects of attention as reflected by variations in the galvanic skin response (GSR) index of the orienting response (OR). Conditions for the occurrence of an OR are due either to novelty and stimulus change or to the establishment of the significance of a stimulus. Stimulus significance may be biologically determined, obtained through "discovery" during the course of a subject's problem solving activity, or imparted by the experimenter's instructions. Maltzman and Pendery's results demonstrate that both "discovery" and experimenter instructions lead to the verbal conditioning of a GSR measure of the OR, similar to more classical conditioning in the laboratory. Also, their results suggest the importance of distinguishing between voluntary and involuntary types of ORs.

Barchas and her colleagues present data suggesting that the CNV correlates with a subject's self-evaluation of his own competence. The specific hypothesis is that the manipulation of a subject into a high competency state relative to another individual (an actor) produces a heightened CNV. Previous studies have shown that subjects manipulated into a high competence state react faster than subjects who are manipulated into a low competency state. CNV research also indicates that shorter latencies are associated with greater magnitude CNVs. Therefore, it was postulated that the high competence manipulation would produce an increase in CNV magnitude.

The results obtained by Barchas support the hypothesis that for all subjects who accepted and believed all aspects of the high competence manipulation, and therefore were strongly manipulated, the amplitude of the CNV increased in the post manipulation session. This was in comparison with a second group of subjects who were only weakly manipulated. In that group, only one subject showed an increase in CNV amplitude after manipulation. However, there was a marked, but not significant, pre-manipulation difference in the CNV magnitude between the two groups, which could account for the CNV magnitude differences. Nonetheless, CNV amplitude holds promise as a possible central nervous system correlate of complex social behavior.

23 · LEARNING AND MOTIVATION, A HISTORICAL OVERVIEW

R. F. Thompson and S. D. Berry

University of Southern California
and
Miami University

It is a great pleasure for us as physiological psychologists to participate in honoring the world's most eminent physiological psychologist--Donald Lindsley. Don was the pioneering figure in the development of electrophysiological recording in the context of behavior, one of the major research areas in physiological psychology.

Our title sounds more forbidding than we intended. We give a brief overview of our own field of interest--brain substrates of learning and memory--in the context of Donald Lindsley's contributions to the field. In our view, his activation theory of emotion, published in 1951, has resulted in genuine progress on one of the major problems in the field.

A dominant tradition in animal learning, and among many who study brain mechanisms of learning, has been the association of stimuli--the formation of S-R bonds. Indeed, for many scientists today, the basic hypothesis of association has not much changed from Hume's exposition of causality as an illusion resulting from the association of events, stimuli and ideas.

There has always been something missing in this conceptualization of learning. It is a mechanical, almost lifeless, approach to a process--learning--that in its very nature is biological. This missing element was supplied by Donald Lindsley in his extremely important and influential chapter on emotion in Stevens' 1951 *Handbook of Experimental Psychology*.

The first author was a beginning graduate student in psychology at the University of Wisconsin in 1952--where, among other things, we had to commit the *Handbook* to memory. I will never forget the impact of Don's chapter. His activation theory of emotion was a major departure from previous ideas. The notion of activation or arousal had no physiological existence before Don's chapter. It is a different way of looking at the motivational aspects of behavior, different in very important ways from traditional views of motivation, particularly in the context of learning.

The association of ideas tradition in learning was expressed most starkly by Watson (1930) in his S-R theory of behavior. Motivation had no real place in Watson's scheme--indeed, biology had no place. For Watson, the human infant was a tabula rasa of a peculiarly non-biological sort.

Karl Lashley was dominated by Watson's ideas. At the beginning of his

DOI: 10.4324/9781003164647- 30

search for the engram, his notion of the memory trace was very Watsonian. Each specific new stimulus-response connection that was formed from a specific learning experience was coded by a new connection formed in the brain. All one had to do was to train many animals in a learning task and take out different portions of the brain. This in itself, by the way, was a major conceptual advance and one of the starting points of modern physiological psychology. Franz and Lashley formalized the lesion-behavior approach--still one of the major techniques used to study brain and behavior.

Localization of function--the notion that each psychological "trait" or function has a specific locus of representation in the brain--was perhaps the major intellectual issue concerning brain organization at the time Lashley began his work. An extreme form of localization was popular early in the 19th century with F. Gall's phrenology. The neurologist P. Flourens moved away from that position in the 1840s by arguing for a lack of specific localization of function within the cerebrum. However, P. Broca's discovery of a speech center in the early 1860s began to move the pendulum back (see Boring, 1950). The critical and classic study on localization was that of G. Fritsch and E. Hitzig (1870). They stimulated the cerebral cortex electrically and defined the motor area (region which yielded limb movements when stimulated). Not only was it localized to a particular region of the cortex, but there was additional organization and localization within it. The subsequent work of D. Ferrier, H. Munk and F. L. Goltz, in the last three decades of the 19th century, identified the general locations of the visual and auditory regions of the cerebral cortex. Localization of function was winning the day.

As noted above, in Watson's behaviorism the learning of a particular response was held to involve the formation of a particular set of connections--a series set. Consequently, Lashley argued, it should be possible to localize the place in the cerebral cortex where that learned change in brain organization was stored--the engram. (It was generally believed at the time, consistent with Pavlov's dictum, see Pavlov, 1927, that learning was coded in the cerebral cortex.) Thus, behaviorism and localization of function were beautifully consistent--they supported the notion of an elaborate and complex switchboard on which specific and localized changes occurred when specific habits were learned.

Lashley set about systematically to find the places in the brain where learning occurred--the engrams--in an extensive series of studies culminating in his 1929 monograph *Brain Mechanisms and Intelligence*. In this research, he used mazes differing in difficulty, and made lesions of varying sizes in different regions of the cerebral cortex of the rat. The results profoundly altered Lashley's view of brain organization: the locus of the lesions was unimportant; the size was critically important, particularly for the most difficult maze. These findings, which led to Lashley's two theoretical notions of *equipotentiality* and *mass action*, had a very dampening effect on the young field concerned with brain substrates of learning and memory.[1]

Donald Hebb revitalized the field of brain mechanisms of learning in his 1949 book, *The Organization of Behavior.* Throughout the book, he grappled with the problem of motivation. As he states, there are two aspects of the problem: (1) Why is an animal active at all? (2) Why does this activity take particular forms, as in food seeking or learning particular tasks? To state it more specifically in the context of learning, why do organisms learn and why do they sometimes learn much better than at other times? The same is of course true of perception and performance--they vary enormously with other conditions. Why?

Physiological and neural mechanisms of motivation and emotion have been a particular province of physiological psychology since the 1930s. In recent years, the fields of "motivation" and "emotion" have tended to go separate ways. However, motivation and emotion have a common historical origin. In the 17th and 18th centuries, instinct doctrine served as the explanation for why organisms were driven to behave (at least infrahuman organisms, which did not possess souls). Darwin's emphasis on the role of adaptive behavior in evolutionary survival resulted in the extension of instinct doctrine to human behavior. Major sources of impetus for this were Sigmund Freud's and William McDougall's notions of instinctive human motivation. Watson rebelled violently against the notion of instinct and eventually rejected it out of hand, together with all biological mechanisms of motivation. As Lashley put it (1938), he "threw out the baby with the bathwater."

Experimental work on brain substrates of motivation and emotion began with the studies of J. P. Karplus and A. Kreidl (1910) on effects of stimulating the hypothalamus. In 1928, Bard showed that the hypothalamus was responsible for "sham rage." In the 1930s, S. W. Ranson and his associates at Northwestern University, particularly H. W. Magoun and W. R. Ingram, published a classic series of papers on the hypothalamus and its role in emotional behavior (see Ranson & Magoun, 1939). In the same period W. R. Hess and his collaborators in Switzerland were studying effects of stimulating the hypothalamus in freely moving cats (Hess, 1928).

It is against this backdrop that the modern field of the psychobiology of motivation developed. Karl Lashley was again a prime mover. His paper on "The experimental analysis of instinctive behavior" in 1938 was the key. He argued that motivated behavior varies, is not simply a chain of instinctive or reflex acts, is not dependent upon any one stimulus, and involves some

[1]Now, it appears that the memory traces for at least one basic category of associative learning--learning of discrete behavioral responses to deal with aversive stimuli--are embedded in a highly localized serial circuit. But the essential circuit includes the cerebellum, particularly the interpositus nucleus, and not the cerebral cortex (see D. A. McCormick & R. F. Thompson, 1984).

central state. His conclusions, that "physiologically, all drives are no more than expressions of the activity of specific mechanisms" and that hormones "activate some central mechanism which maintains excitability and activity," have a very modern ring.

Lashley's general notion of a central mechanism that maintains activity was developed by Frank Beach, in an important series of papers in the 1940s (see Beach, 1947), and by Morgan in the first edition of his important text, *Physiological Psychology* (1943), into a central excitatory mechanism and ultimately a central theory of drive. This was the state of the field when Hebb's book was published in 1949. The central excitatory mechanism was a largely theoretical notion with no solid physiological substrate. It was in the same year that Lindsley, Bowden, and Magoun (1949) and Moruzzi and Magoun (1949) published their epochal papers on the reticular activating system. Lindsley developed these discoveries into his activation theory of emotion. This theory supplied the critical links between motivation and learning. It accounts for why organisms are active and why performance, learning and perception are so variable. In his handbook chapter in 1951, he presents the theory in characteristically diffident and modest form:

> As far as it may be considered a theory, the conception to be described here may be labeled an "activation theory." It is based largely upon recent findings concerning the electroencephalogram and particularly the interaction of the cerebral cortex and subcortical structures. The activation theory is not solely an explanatory concept for emotional behavior but relates also the phenomena of sleep-wakefulness, to EEG manifestations of cortical activity, and to certain types of abnormal behavior revealed in neurologic and psychiatric syndromes. (Lindsley, 1951, pp. 504-505)

He presents the theory in some 4 pages, and although he does not spell out the implications for learning and perception, they are clearly there, particularly in regard to motivation. Hebb (1955) and Stellar (1954) incorporated Lindsley's activation theory into a general central theory of motivation.

A general characterization of motivation at the behavioral level has always presented difficulties. Judson Brown (1961) developed perhaps the most comprehensive account of two major alternative behavioral theories, drive theory and "associative" theory. Drive theory, originally developed by Clark Hull (1943), has been a dominant view in American psychology. The basic notion, familiar to everyone, is that learning--the formation of engrams or in Hull's terms the increase in habit strength--occurs in association with drive reduction. The hungry rat only learns when he eats. This has always been somewhat unsatisfactory, as Hebb (1955) has indicated. It is possible that if Hull had used primates or even carnivores rather than rats in his work, he would not have placed so much weight on primary need or drive reduction as the necessary motivating condition for learning. It has always seemed particularly remote to human learning. Lindsley's activation theory would seem to provide a more useful conceptual framework for the motivational conditions necessary for learning.

Figure 23.1 is the familiar inverted U function relating activation or arousal to performance and learning, developed from Lindsley's activation theory by Duffy (1962), Hebb (1955), and Malmo (1959). There are two basic notions embedded in this figure, both of which were first proposed by Don Lindsley. One is of course that learning, perception and performance vary with the degree of activation or arousal of the organism in an inverted U form. The second is that there is such a thing as "state"--i.e., state of activation or arousal--and that this state can be measured by physiological indices like the EEG.

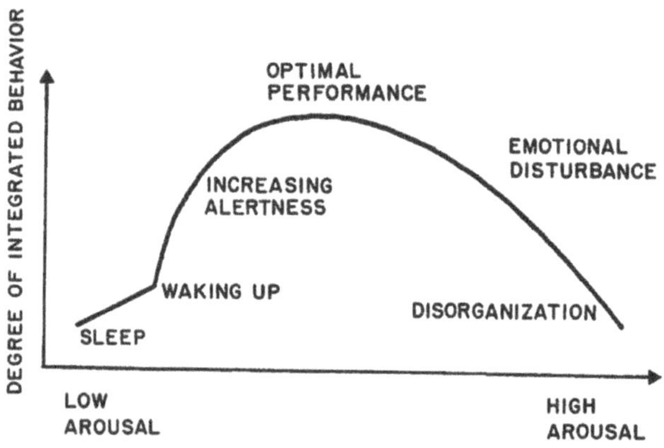

FIGURE 23.1. Hypothetical relationship between arousal level or "state" and task performance (based on Hebb, 1966).

In the current field concerned with the neurobiology of learning, these concepts have assumed dominant importance. To take a very simple example, in our own earlier work on habituation of spinal reflexes, Spencer, Groves, and I were unable to avoid a notion of state or level of activation. Even in the spinal cord, the temporal course and degree of simple habituation of a specific reflex path is modulated by the state of excitability of the spinal cord. There is no other way to account for the data. Examples of habituation of the hindlimb flexor response of acute spinal cat to repeated cutaneous stimulation are shown in Figure 23.2 for four different levels of stimulus intensity. For the weakest stimulus, there is essentially pure habituation (solid lines and dots are experimental data). However, as stimulus intensity is increased, there is a more pronounced and prolonged

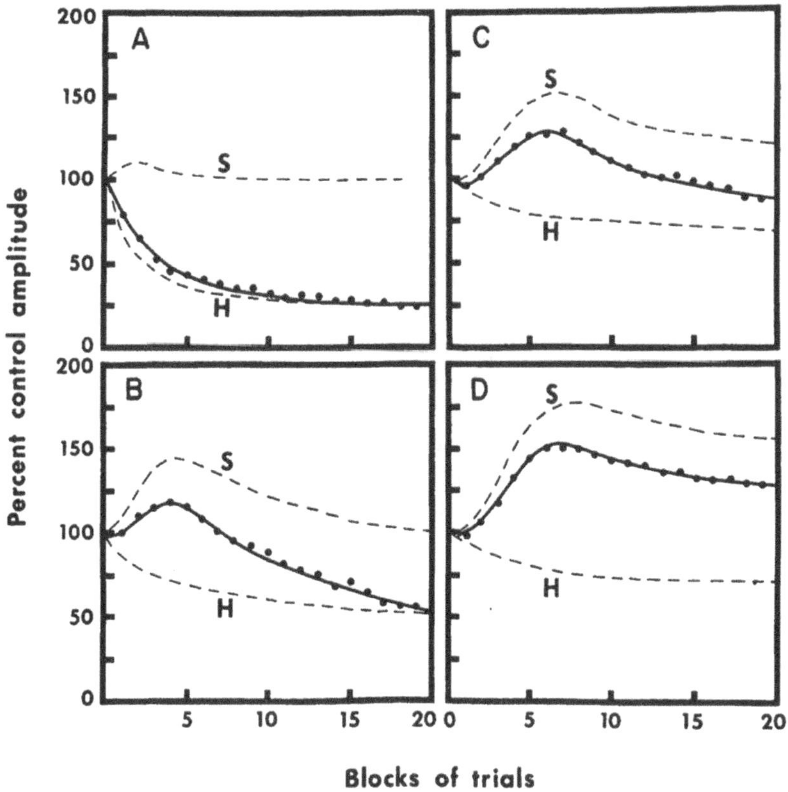

FIGURE 23.2. Habituation of hindlimb flexor response of acute spinal cat to cutaneous stimulation at different levels of stimulus activity. Solid lines and dots are data points; dashed lines are hypothetical habituation and sensitization processes. A-D: increasing stimulus intensity (from Groves & Thompson, 1970).

increment in response. A process of sensitization or arousal of "state" develops, schematized by the dashed line labeled "S."

We were able to show that this sensitization of state is an independent process superimposed upon habituation both in the spinal cord and in responses of intact animals (Groves & Thompson, 1970; Thompson & Spencer, 1966). This sensitization process does not disrupt habituation per se, but is rather a separate process, closely analogous to Lindsley's notion of arousal or activation.

The modern field of memory consolidation--studying the effects of drugs, neurotransmitters, electroconvulsive shock, etc., on learning and consolidation of memory--has come to the same position. As an example, a theory was proposed by Kety (1976) concerning the role of biogenic amines, particularly norepinephrine, in learning. He suggested that norepinephrine released in affective states could act to promote selective growth and persistence of neural circuits that lead to reward or relief. The view is growing that such hypothetical actions of biogenic amines may not be so specific but rather more general and nonspecific--they act as modulators, as in Lindsley's activation theory.

James McGaugh, Paul Gold, and their colleagues have developed this notion very clearly in their work (see Gold & McGaugh, 1975). An example is shown in Figure 23.3. The left portion of the figure shows results of an experiment where different groups of animals were given different doses of epinephrine following training on a one-trial passive avoidance task. The animals were pretrained to lick from a water spout and then given weak footshock. Later, they were tested for retention. The results shown take the form of the inverted U function. The most effective dose for facilitation of retention was 0.1 mg. Interestingly, if strong footshock was used, then 0.1 mg of epinephrine was disruptive--a lower dose facilitated retention. This is a

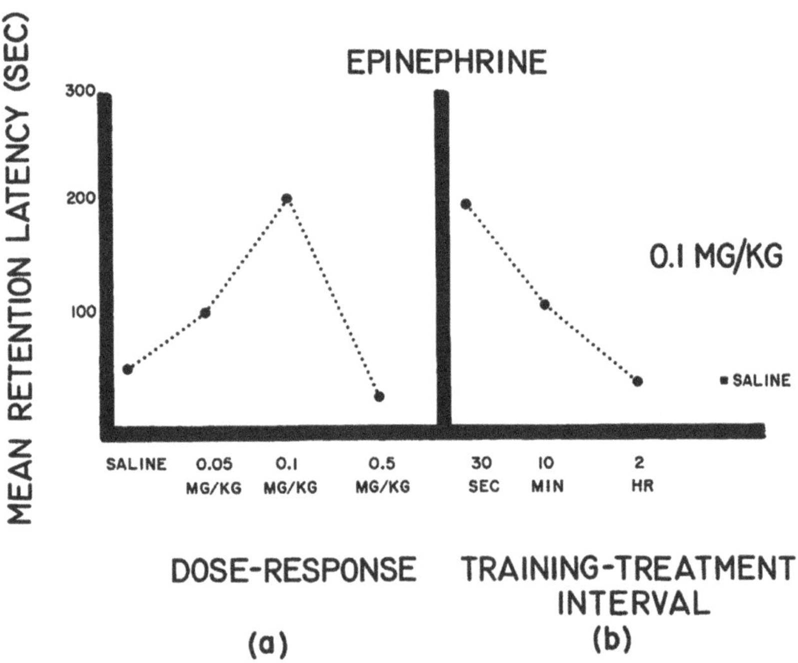

FIGURE 23.3. Relationship between treatment with epinephrine and facilitation of retention in a passive avoidance task. The left side shows the relationship between dosage level and latency. Note the inverted U form of the relationship. The right side illustrates the effects of delaying epinephrine administration after training (from Gold & McGaugh, 1975).

very nice example of the interaction among two sources of activation or arousal--peripheral stress and a brain chemical. This interaction is predicted from the Lindsley inverted U model of activation.

Complementary results are cited by Kety, McGaugh, and others concerning endogenous brain levels of substances such as norepinephrine, ACTH and vasopressin--results of many such studies can be placed on the inverted U function. To quote Gold and McGaugh:

> Many studies now show that post-trial administration of several drugs will facilitate memory for a wide variety of tasks (McGaugh, 1973). All effective drugs are known to increase arousal level. With all drugs tested, there is an inverted U dose-response curve for memory facilitation. The dose-response curve for a particular drug on different tasks is not constant. In different tasks, or with different motivational levels, the peak dose for memory facilitation shifts. Our hypothesis (author's note: their basic hypothesis was Lindsley's activation theory) predicts that the most effective dose is one that maximizes the nonspecific physiological consequences of training; *the most common general mechanism may be arousal level.* (Our italics.) Traditional behavioral studies suggest that the influence of arousal level itself on learned performance follows an inverted U function. Thus, high doses of memory facilitation treatments may be disruptive because the arousal level is so high that appropriate memory storage processes are blocked. Alternatively, high doses of these drugs may produce additional alterations in neural activity, such as brain seizures, which disrupt memory storage processes. Low doses, of course, are ineffective simply because they do not add sufficiently to the arousal processes. (Gold & McGaugh, 1975, p. 367)

Thus in the context of learning, motivation is level of activation. Too little activation results in low activity and poor learning. Moderate amounts lead to optimal levels of electrical activity of the brain (EEG arousal), optimal brain levels of neurochemicals and hormones, and optimal levels of behavioral arousal, perception and learning. Too much activation disrupts all these processes. One of the beauties of Lindsley's activation theory of emotion is that, unlike the mechanistic drive reduction approach, it accounts very well for the range of variations seen in human learning and performance.

Basic drives and drive reduction fit very well into activation theory. They are biological imperatives. When an animal is hungry, thirsty or sexually stimulated he becomes activated and aroused--apparently to optimal levels under relatively normal conditions. It is this activation and arousal, and not simply the increased tissue needs per se, that results in increased activity levels, exploration, improved perception and learning. The mammalian brain has developed a general mechanism to deal with particular sources of need and drive. Lindsley's work indicating the very powerful controls exerted by the ascending reticular system and hypothalamus on levels of cortical and hippocampal electrophysiological arousal is beautifully consistent with this view.

Perhaps the major challenge to activation theory has come from the field of human psychophysiology. John Lacey, in a classic paper in 1967, argued against the notion of a unified central state of arousal. His major point was very simple--there is at best an imperfect correlation among various peripheral physiological measures of arousal.

In our view, this need not imply the absence of a unified *central* state of arousal. In our earlier analysis of habituation, the "dual-process theory" (Groves & Thompson, 1970), we argued that any stimulus that elicits a response has two effects--it elicits the response over a specific stimulus-response pathway which tends with repetition to habituate, and it produces some degree of more general arousal or sensitization of the state of the organism (see Figure 23.2). In these terms, the various peripheral physiological responses that occur, for example, in the orienting response, are examples of the activities of particular stimulus-response pathways.

Each peripheral system, e.g., cardiovascular or sweat glands, has its own considerable complexities. They vary across species and exhibit different components even within the same species. The various components of the different peripheral response measures often do not behave in a manner consistent with each other, particularly in response to a repeated stimulus. It would be very surprising if they did. The peripheral systems measured in studies of orienting do not exist for the purpose of exhibiting orienting. This is most obvious in the case of the cardiovascular system--its function being to supply blood (oxygen, nutrients, etc.) to the body and most particularly to the brain.

The orienting reflex is by definition a phasic response to a stimulus. The peripheral responses occur and are over in a matter of seconds. However, the reflex is really superimposed upon the pre-existing state of the organism, including the states of the various peripheral systems. More importantly, the stimulus also produces tonic effects, particularly on the central state. Increased cortical EEG arousal may persist for many minutes. The cortical EEG arousal response--desynchronization--is itself a rather complicated response in a very complex piece of neural tissue--the cerebral cortex. It may have several generators, there are certainly several cortical projection systems involved, and pacemaker mechanisms exist at least in the thalamus and in the cortex itself. Superimposed upon it are a variety of specific sensory projection systems and feature-analytic processes. EEG arousal occurs in the context of a columnar organization and over assemblies of columns in the cerebral cortex.

It is not surprising the EEG arousal waxes and wanes, its areal distribution over the cortex fluctuates, and so on. Lindsley (1951) developed the notion of a central state of activation or arousal using cortical EEG desynchronizing as the measure. It serves as a tonic measure of state, in contrast to peripheral measures. It is an imperfect measure. Nonetheless, it shows a much more unified character than any peripheral response measure.

Is there any necessity for the concept of a more-or-less unified central state of arousal? It seems at the very least to have heuristic value. This is perhaps most evident with infrahuman laboratory animals. They may come to an experimental situation in any state from sleep to extreme arousal, with urination and defecation. The inverted U function relating arousal and performance is a genuine reality in animal studies. A difficulty with human subjects in a laboratory setting is that they may vary only over a narrow and restricted range of arousal. The deleterious effect of restriction of the range on predictive correlation is well known.

In infrahuman animals there appears to be a better index of central alerting or arousal than cortical EEG desynchrony--hippocampal theta--as recent work in Lindsley's lab has indicated very clearly. The hippocampus seems to be a particularly appropriate region for analysis of levels of activation or arousal and their influence on behavioral processes such as attention and learning, thanks to the pioneering studies by Adey (1966), Grastyan, Lissak, Madarasz, and Donhoffer (1959), and Green and Arduini (1954) on hippocampal EEG. In many mammals, the slow wave activity of the hippocampus is dominated by theta rhythm, a high amplitude, synchronous EEG pattern ranging from 3 to 12 Hz. The size and clarity of theta in different species appear to be inversely related to the development of the neocortex (Parmeggiani, 1967). These differences, the apparent difference in dominant theta frequency (approximately 7-12 Hz in rats, 3-7 Hz in rabbits, 4-8 Hz in cats), and the different behavioral correlates of theta between species have given rise to a controversial literature on the functional significance of theta. It is generally accepted that theta occurs during REM sleep, but the waking behavioral correlates are more difficult to explain (see Bennett, French, & Burnett, 1978; Winson, 1972, 1975, for reviews of species differences). In general, researchers working with rabbits or cats tend to interpret theta in terms of learning, attention, or arousal, while those working with rats note clear correlations between theta and movement patterns (for reviews, see chapters by Bennett; Black; Lindsley & Wilson; Vanderwolf, Kramis, Gillespie, & Bland; and Winson in Isaacson & Pribram, 1975; and Pribram & McGuinness, 1975).

Recent pharmacological evidence may help resolve this discrepancy. It has been demonstrated that, in both rat and rabbit, there are two different types of theta or rhythmic slow activity (RSA) (Kolb & Whishaw, 1977; Kramis, Vanderwolf, & Bland, 1975; Vanderwolf, 1975; Whishaw, 1976; Whishaw, Bland, Robinson, & Vanderwolf, 1976; Whishaw & Schallert, 1977). One type has a relatively high frequency (6-12 Hz) and correlates with "voluntary movement" (e.g., locomotion). This type is resistant to anticholinergics such as atropine sulfate, but is inhibited by urethane. The second type is of lower frequency (4-8 Hz), correlates with immobility and sensory stimulation, and is resistant to urethane but blocked by atropine. Perhaps the conflicting literature is in part due to the relative dominance of the first type of theta in rats and the second type in rabbits and cats.

Lindsley's work in this area has centered around the control of hippocampal EEG and behavioral processes by brainstem and hypothalamic systems (Anchell & Lindsley, 1972; Coleman & Lindsley, 1975, 1977; Lindsley & Wilson, 1975; Macadar, Chalupa, & Lindsley, 1974). These elegant studies demonstrated that, at the hypothalamic level, there are two contrasting systems involved in the control of theta: a medial system which, when stimulated at 100 Hz, elicited hippocampal theta (synchrony), and a more lateral system which desynchronized hippocampal EEG. Mapping the posterior hypothalamus with a combination of stimulation, lesions and cryogenic blockade, they found the medial system to correspond roughly to the region of the dorsal longitudinal fasciculus of Schutz, while the lateral system was in the region of the medial forebrain bundle. In an effort to delineate the brainstem contributions to these systems, they performed a systematic analysis, using electrical stimulation, of brainstem nuclei, especially the raphé nuclei, nuclei reticularis pontis oralis and caudalis, nucleus locus coeruleus, the periaqueductal gray substance, and the midbrain and pontine tegmental fields. They discovered five areas in which theta could be elicited: nucleus locus coeruleus, nucleus reticularis pontis oralis, ventrolateral periaqueductal gray, nucleus reticularis gigantocellularis of the pontine tegmentum, and a region of the mesencephalic tegmentum. In addition, two desynchronizing areas were found: the raphé nuclei and nucleus reticularis pontis caudalis.

Stimulation of the theta synchronizing areas in the awake, behaving cat results in cessation of ongoing behavior, orienting, and scanning the environment in a manner interpreted as an alteration of attentional processes. Stimulation of the lateral desynchronization system results in immobility and intense fixation of visual gaze--behaviors which do not often occur spontaneously in the testing situation. Even during asymptotic performance of an operant task, stimulation of the medial system resulted in attention shifts lasting for many minutes. In contrast, lateral stimulation arrested behavior for a much shorter time. These effects were interpreted as operation of an attentional mechanism, particularly one involved with shifts of attention.

Such interpretations are consistent with their observations of endogenous EEG frequency changes during the course of training (Coleman & Lindsley, 1977). Briefly, much theta was present during the initial stages of acquisition, when the task required close attention and shifts of attention between stimuli. Later, when performance was more automatic, theta activity decreased, and desynchrony became more apparent.

The behavioral correlations of non-movement-related hippocampal theta have been interpreted by many other authors to be a reflection of an arousal or attentional process (Adey, 1977; Bennett, 1975; Bennett et al., 1978; Berry & Thompson, 1978; Green & Arduini, 1954; Kemp & Kaada, 1975; Klemm, 1976; Lopes da Silva & Arnolds, 1978; Pribram & McGuinness, 1975; Radil-Weiss & Hlavicka, 1978). In part because of this relationship between hippocampal activity and processes such as arousal, that are closely related to learning, we have turned to the hippocampus in our analyses of neuronal substrates of learning. Our results corroborate Lindsley's observations and support his general notions of hippocampal function.

We have adopted the preparation developed by Gormezano (1966), classical conditioning of the rabbit nictitating membrane response, as a model system for analysis of brain mechanisms of learning (Thompson, Berger, Berry, & Hoehler, 1980; Thompson et al., 1976). In brief, conditioning animals are given paired tone CS (85 dB, 1 KHz, 350 msec) and airpuff UCS (210 g/cm^2 source, 100 msec overlapping last 100 msec of CS tone stimulus). The intertrial interval is varied around a mean of 60 sec. Every 9th trial is a tone-alone test trial; 117 trials are typically given per day. Control animals are given a pseudorandom sequence of unpaired tone and airpuff stimuli, using a variable interval with a mean of 30 sec. Animals are held in a restraining device in a sound-shielded enclosure. For hippocampal EEG recording, chronic microelectrodes are implanted dorsally in the hippocampus one week prior to training (See Berry & Thompson, 1978, for further details). For multiple unit recordings, described later, a microelectrode is chronically implanted in the pyramidal cell layer of CA1 and CA3 or the granule cell layer of the dentate. For single unit recording, a micromanipulator is chronically implanted and a microelectrode inserted for each recording session (see Berger & Thompson, 1978a, 1978b).

In this situation, we have observed a striking increase in hippocampal unit activity over the course of learning. Unit patterns that develop in the hippocampus are compared with the behavioral nictitating membrane (NM) response in Figure 23.4. Examples are shown for a paired conditioning animal and an unpaired control animal at the beginning and the end of training. The hippocampal unit poststimulus histogram and averaged NM response for the first block of eight trials are given for one animal in Figure 23.4A and for the same animal after learning criterion was reached in Figure 23.4B--over the first block of eight trials there is a large increase in unit activity in the UCS period that precedes and closely parallels the behavioral NM response form. Over training, this hippocampal unit response increases and moves into the CS period as behavioral learning develops.

The unpaired control animal hippocampal data are completely different. The eight trial hippocampal unit activity and averaged NM are shown for

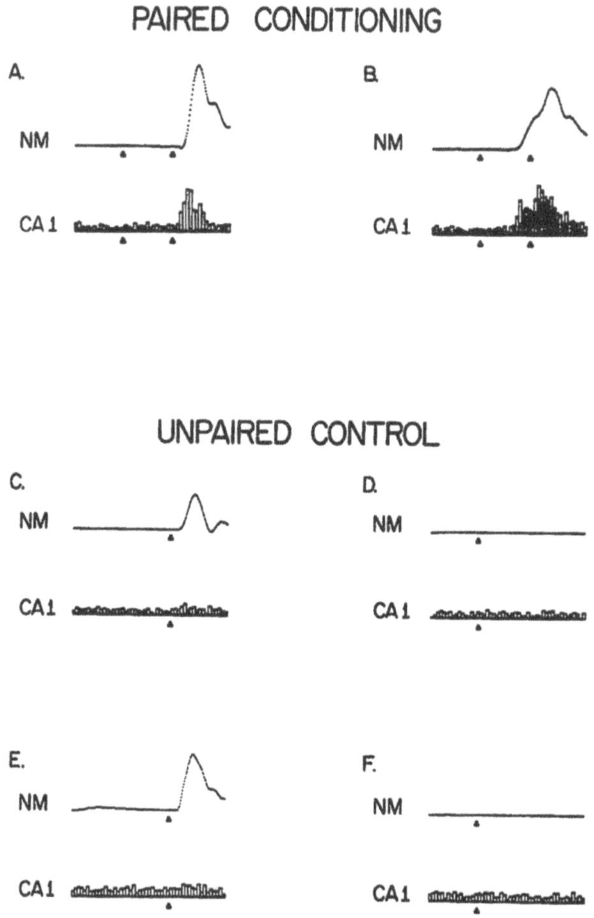

FIGURE 23.4. Upper trace: Average nictitating membrane (NM) response for one block of eight trials. Lower trace: Hippocampal unit post-stimulus histogram for one block of eight trials. (A) First block of eight paired conditioning trials, day 1. (B) Last block of eight paired conditioning trials, day 1, after conditioning has occurred. First cursor indicates tone outset; second cursor indicates airpuff onset. (C) First block of eight paired UCS-alone trials, day 1. (E) Last block of eight paired UCS-alone trials, day 2. Cursor indicates airpuff onset. (D) First block of eight unpaired CS-alone trials, day 2. Cursor indicates tone onset. Total trace length is 750 msec (redrawn from Berger, Alger, & Thompson, 1976).

airpuff alone trials at the beginning and end of unpaired training for a control animal in Figure 23.4C and E. Although there is a clear reflex NM response, there is little associated unit activity in the hippocampus. There is essentially no NM response or evoked hippocampal activity in tone alone trials (Figure 23.4D and F). The hippocampal unit responses illustrated in Figure 23.4 are closely paralleled for all animals in both conditioning and control groups (see Berger & Thompson, 1978a).

In Figure 23.5 are shown the standard scores of unit activity for both paired (N=21) and unpaired (N=12) groups across all blocks of training trials. For both UCS and CS periods, unit activity in the hippocampus for conditioned animals increases and remains high over all 26 blocks of paired trials (334 trials total, solid lines). In contrast, standard scores for animals given control training remain low across blocks of unpaired trials (broken lines) in both the CS and UCS periods.

Because hippocampal unit activity appeared so highly developed at the end of the first block of paired trials (see Figure 23.4), an individual trial analysis for the first eight pairings was completed for all animals. A robust NM response to the airpuff was usually present from the first trial, yet the

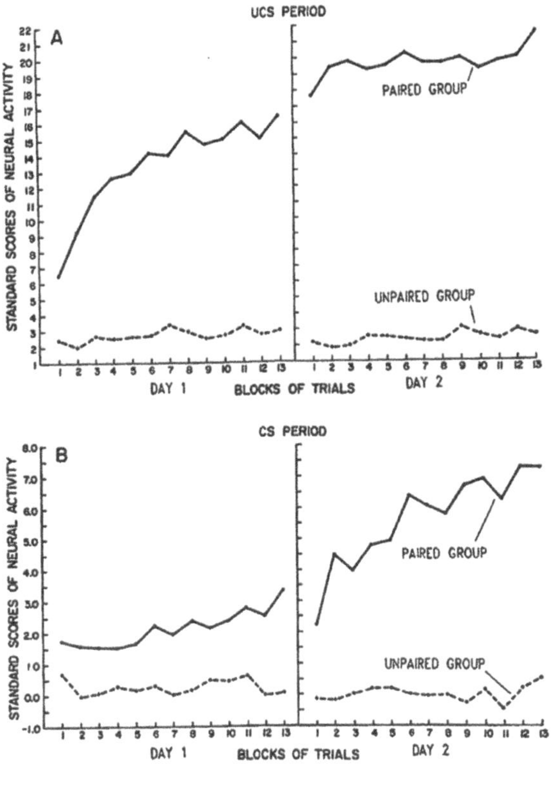

FIGURE 23.5. Comparison of standard scores of unit activity across blocks of training between animals given paired training and the unpaired control group. (A) changes during the airpuff (UCS) period; (B) changes during the tone (CS) period (from Berger & Thompson, 1978a).

hippocampal unit response did not appear to develop in conditioned animals until later within the first block. The paired and unpaired group have the same low level of hippocampal activity in trial 1; however, conditioning animals increase and control animals decrease over the first eight trials of training such that the two groups diverge significantly (Berger & Thompson, 1978a).

To summarize the results of our current studies, under conditions of paired training where behavioral learning will occur, unit activity in the hippocampus increases rapidly, initially in the UCS period, forms a temporal "model" of the behavioral response and precedes it in time. As hippocampal activity begins to occur in the CS period, behavioral learning begins to occur. This increased unit activity in the hippocampus does not develop in unpaired control animals. In paired animals the hippocampal activity begins to develop in the first few trials of training and may well be the earliest sign of learning in the brain (Berger & Thompson, 1978a). Further, the growth of the hippocampal response is completely predictive of subsequent behavioral learning. If the hippocampal response does not develop, the animal will not learn (Hoehler & Thompson, 1980). This learned hippocampal response is projected over a major limbic pathway: hippocampus to lateral septum (Berger & Thompson, 1977). A single unit analysis has established that the hippocampal response is generated largely, if not entirely, by pyramidal neurons (Berger & Thompson, 1978b).

Perhaps the most surprising aspect of the hippocampal response is that it actually forms a temporal model of the behavioral response--see Figure 23.4. To determine how good a model it is, we computed a "real-time" correlation coefficient for each eight-trial block for each of 21 conditioning animals between the NM response amplitude and the number of unit discharges of hippocampal neurons (Berger, Laham, & Thompson, 1980). Thus, in the example shown in Figure 23.4A, the two measures were taken every 3 msec from the beginning of the PreCS period to the end of the UCS period and a product-moment correlation coefficient was computed. This was done for each block of each animal and then averaged across animals. The result--the mean correlation between brain and behavior over the blocks of training--is plotted in Figure 23.6.

As can be seen, the correlation is initially modest, but grows over training as the unit activity comes to form a good model of behavior (r=.63). Note that this real-time correlation could not be perfect because the hippocampal response occurs earlier in time than the behavioral response. To determine the maximum possible correlation we time-shifted the hippocampal response so it and the NM had the same onset point and compared only the active phase of the response. At the end of training this time-adjusted correlation is r=.88, a very good model indeed.

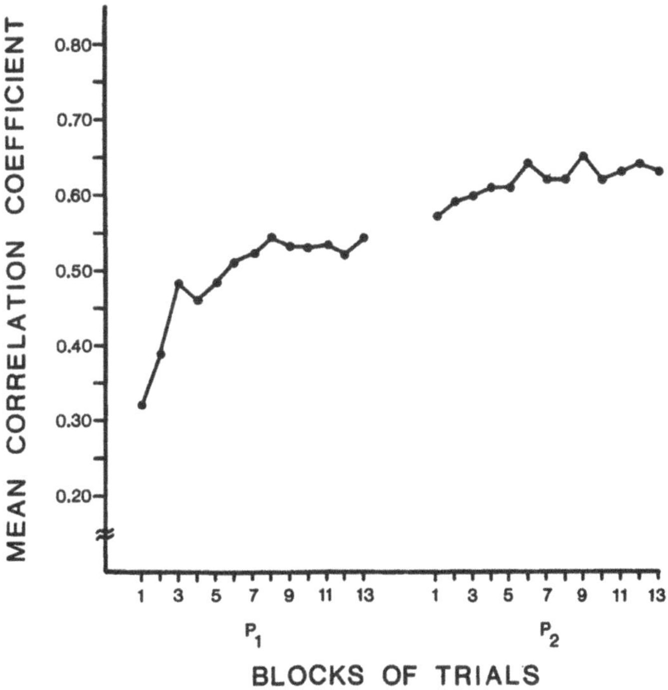

FIGURE 23.6. Correlation between NM response amplitude and number of unit discharges of hippocampal neurons across blocks of training trials. P_1 and P_2 are the first and second days of paired training, respectively (from Berger et al., 1980).

These data suggest that the hippocampus is processing information of current significance to the animal. Unit activity grows rapidly when there is a meaningful behavioral and biological contingency between a neutral warning stimulus and one that threatens damage. It is almost as though information of importance to the animal is being held in the hippocampus to act on the other regions of the brain, much as short-term memory is supposed to function in human information processing.

Given Lindsley's interpretation of theta as a substrate of arousal and attentive and exploratory processes, one would expect that this growth in unit activity and the concomitant learning would be facilitated by theta.

A very simple study of hippocampal EEG indicated that theta activity in the hippocampus is indeed related to learning (Berry & Thompson, 1978, 1979). In brief, 2-min time samples of spontaneous hippocampal EEG were recorded at the beginning and end of each day of paired conditioning training in 16 animals. Initially we simply compared the frequency spectrum of the 2-

min time sample of EEG taken just prior to the beginning of the training on the first day of training against number of trials to criterion over the several days of training. Examples of the frequency spectra from two animals are shown in Figure 23.7. Animal *a* has a preponderance of activity in the theta range whereas animal *b* has a more mixed pattern with greater higher frequency. As it happens, *a* is a good learner, and *b* is a poor learner.

In order to characterize the overall (2-22 Hz) EEG in terms of a low/high frequency dichotomy, a ratio was computed: the percentage of 8-22 Hz activity divided by the percentage of 2-8 Hz activity. Thus, the ratio for the histogram values in Figure 23.7a was 0.40 while that for 23.7b was 1.33. The correlation between this measure and trials to criterion was highly significant ($r=+.72$, $df=14$, $p<.01$). Note the clear linear trend of the correlation illustrated by the scatter plot and best fitting linear regression line in Figure 23.8.

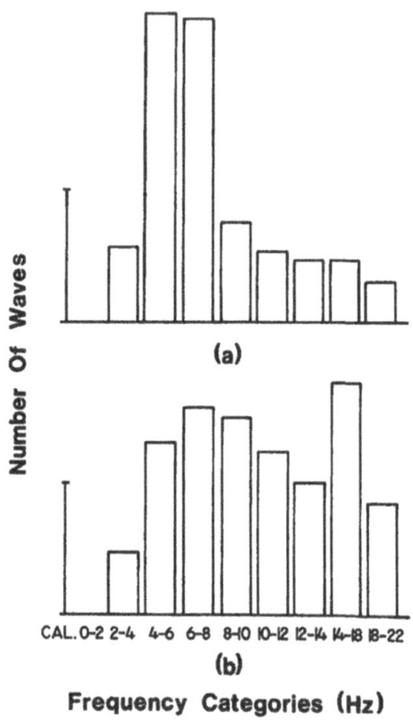

FIGURE 23.7. Samples of EEG frequency analysis, showing the amount of activity in each frequency category during a 2-min spontaneous EEG sample: (a) EEG pattern with a large proportion of low frequency (<8 Hz) activity, (b) EEG sample with a large proportion of higher frequency (>8 Hz) EEG (from Berry & Thompson, 1978).

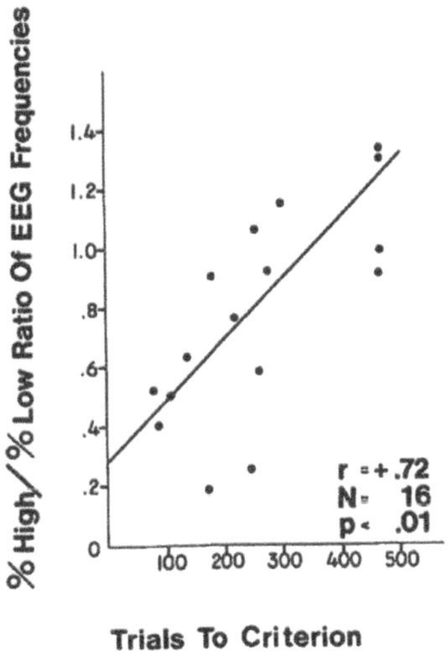

FIGURE 23.8. Scatterplot of EEG ratio (% high frequency/% low frequency) against trials to criterion illustrating the relationship between EEG frequencies and rate of NM conditioning (from Berry & Thompson, 1978).

Thus, a brief time sample of hippocampal EEG taken prior to the onset of training is highly predictive of subsequent learning rate, even over a period of days. A higher proportion of hippocampal theta (2-8 Hz) predicts faster rates of learning. To our knowledge, this is the first demonstration that a purely *neurophysiological* measure taken prior to the beginning of training can predict the subsequent *behavioral* rate of learning.

Prokasy (1972) has developed a most interesting mathematical model of behavioral learning for this particular paradigm (i.e., classical conditioning of rabbit NM). Prokasy's analysis indicates that learning occurs in two phases--an initial phase that extends from the beginning of training until the animal begins to give conditioned responses, and a second phase that extends from this point until the response is well-learned. Phase one is more variable and more likely to be influenced by "motivation," "arousal," and other conditions.

In part to determine if Prokasy's model could be extended to physiological measures, we compared the amount of *change* in the low/high EEG ratio over training against the number of trials of training required to give the fifth conditioned response for each of the 16 animals described above. The result, shown in Figure 23.9, is striking and would seem to provide "physiological" substantiation of Prokasy's model. The correlation between amount of change in EEG ratio and number of trials to the fifth CR is -.93, a highly significant value. The relationship is such that the greater the change in EEG frequency, the more rapid is learning. Since more theta is also predictive of fast learning (Figure 23.8), good learners would seem to

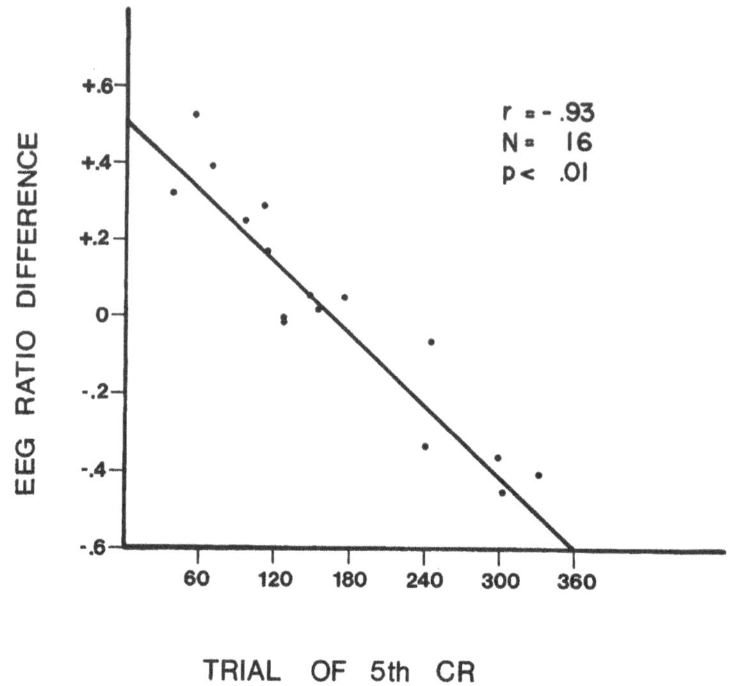

FIGURE 23.9. Correlation between the EEG frequency changes during learning and the number of trials before the fifth conditioned response (from Berry & Thompson, unpublished observations).

shift toward a desynchronized EEG as they learn. Slow learners exhibiting a desynchronized EEG initially would seem to shift more slowly toward theta as they learn. The individual records indicate that this is indeed the case. These results, incidentally, are in close agreement with the data of Coleman and Lindsley (1977) in their analysis of hippocampal EEG during lever press learning for reward in cats. The data provide strong additional support for the general notion of an inverted U function relating alerting and arousal to learning.

As mentioned above, when medial hypothalamic stimulation is given to a well-trained animal performing an operant lever-pressing task, the animal stops, looks around, and appears to be attending to many other cues--often for several minutes. This behavior is accompanied by prominent theta. This result raises the fascinating question of the extent to which behavioral processes can be influenced by manipulation of brain systems that influence theta. To quote Lindsley and Wilson (1975):

> If . . . the electrical activity of the hippocampus reflects its underlying functional state, then to be able to manipulate its electrical activity in contrasting ways by stimulating the medial and lateral hypothalamic systems should prove to be a powerful approach to the study of hippocampal function and its integrative relationships with other neural structures in the control and regulation of behavior. (p. 273)

Schlag-Rey and Lindsley (1975) examined the effects of medial and lateral hypothalamic stimulation on photic evoked potentials in visual and association cortex of the cat. They reported an attenuation of averaged evoked response amplitudes in both primary and secondary visual cortex during both types of stimulation, although the attenuation was somewhat greater with stimulation of the lateral system. They cautiously conclude that central neural processes may be attenuated by these hypothalamic stimulations.

Two additional aspects of their findings should be emphasized, however. First, it can be seen that, during stimulation of the medial system, the latency of evoked responses is shorter--information is getting to the cortex faster. Second, the attenuation of responses to the light flash during medial stimulation is not inconsistent with an attentional interpretation of theta synchronization. Activation of this system may result in the animal becoming much more attentive to stimuli other than the repeated light flash, as demonstrated in Lindsley's work on attention effects on the human evoked potential (Spong, Haider, & Lindsley, 1965).

In keeping with these findings of differences between medial and lateral systems, it is well known that the medial and lateral septal nuclei play quite different roles in the regulation of hippocampal activity (DeFrance, Kitai, & Shimono, 1973a, 1973b; McLennan & Miller, 1974). We have also found different patterns of change in multiple unit activity in medial and lateral septal regions during rabbit NM conditioning (Berger & Thompson, 1978b). Basically, we have shown that cells in the medial septal nucleus (MSN) display a response that can best be characterized as arousal in nature, while cells in the lateral septal nucleus (LSN) show learning-related changes which develop later than, but parallel, the increases in multiple unit activity recorded from hippocampus. These different relationships to NM conditioning suggest that, corresponding to the anatomy of these regions, the MSN is functionally afferent to the hippocampus, while the LSN is a major recipient of hippocampal efferents.

Additional data from our laboratory demonstrate the importance of the arousal-related MSN input during NM conditioning. Briefly, we have demonstrated (Berry & Thompson, 1979) that small, localized lesions of the medial septal nucleus, which significantly disrupt hippocampal theta, also retard the behavioral rate of learning. Importantly, such lesions also significantly attenuate the growth of hippocampal unit responses seen in normal rabbits. Figure 23.10 illustrates the form of behavioral disruption as reflected in the number of conditioned responses (CRs) across trial blocks. The delay in occurrence of the first few CRs until late in training in the lesion group parallels the deficit seen on this task following other forms of hippocampal disruption (Salafia, Romano, Tynan, & Host, 1977). Since EEG in the normal rabbit appears to correlate well with phase 1 in Prokasy's (1972) two-phase model, it is important to note that disruption of hippocampal EEG by septal lesions results in a behavioral deficit that can be characterized as an extension of phase 1, the period of training prior to the appearance of CRs. Also, recall that in Prokasy's model, phase 1 is more likely to be influenced by variables affecting motivation or arousal. These results are completely consistent with the hypothesized role of hippocampal theta in central alerting or arousal mechanisms, and raise the exciting possibility that treatments which enhance theta may optimize learning.

Others have stressed the correlation between hippocampal theta and voluntary movement (Vanderwolf, 1971, 1975) or higher-order motor functions such as spatial mapping (Black, 1975; Winson, 1978). The

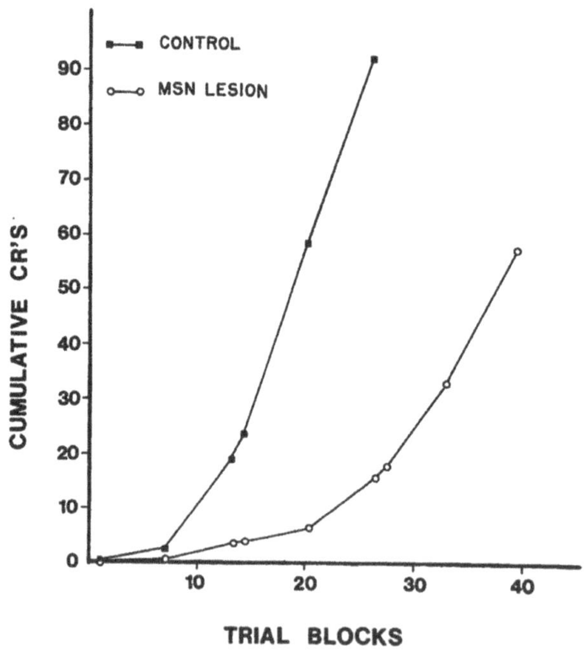

FIGURE 23.10. Mean cumulative conditioned responses across training blocks for animals in the control and MSN and MSN lesion groups (from Berry & Thompson, 1979).

demonstration of clear-cut spatial correlates of the firing pattern of somehippocampal neurons (O'Keefe, 1976; O'Keefe & Nadel, 1978; Olton, Branch, & Best, 1978), the almost invariant correlation between exploratory locomotion and theta in the rat (Vanderwolf, 1969) and the relationship between speed of locomotion and frequency of theta (McFarland, Teitelbaum, & Hedges, 1975) all point in the direction of a relationship between hippocampal function and processing of spatial information in the rat.

Indeed, a deficit in spatial memory has been demonstrated after disruption of theta through medial septal lesions (Winson, 1978). Although Winson's interpretation stresses the spatial nature of this deficit, and its relation to memory rather than acquisition, his results are very consistent with our finding of an impairment in acquisition of the conditioned NM response in rabbits. While not emphasized in his discussion, rats with MSN lesions tended to acquire his task more slowly, especially during the non-spatial, cue-assisted phase of his shaping procedure. Thus, although the spatial mapping hypothesis of hippocampal function is attractive for several reasons, it appears from the results of a variety of experiments in several species that low frequency hippocampal theta is clearly related to more general processes than movement per se or spatial learning per se. Equally clear correlations have been observed between theta and learning in nonspatial tasks in both cat and rabbit (Adey, 1966; Berry & Thompson, 1978; Coleman & Lindsley, 1977). Even in rats, Bennett et al. (1978) have shown that the amount of hippocampal theta during a DRL task is related to the relevance of an external cue ("attention to an environmental cue") rather than to motor aspects of behavior.

In addition, theta has been observed in behavioral states clearly not related to movement. For example, Lindsley and Wilson (1975) state: "We have observed theta rhythm if an animal appeared alert or attentive, even though completely immobile" (p. 274). The original characterizations of hippocampal theta (Green & Arduini, 1954) emphasized its relation to arousal processes as opposed to motor responses. However, as Schwartzbaum (1975) has suggested, behavioral processes such as movement and arousal are not necessarily (logically) orthogonal (i.e., ongoing behavior can perhaps limit the range of variation in certain types of arousal processes). One striking example of this sort of phenomenon is the finding that the amplitude of field potentials evoked in CA1 by stimulation of the angular bundle in rats is dependent upon the tonic behavioral state of the animal (Winson & Abzug, 1977). For example, smaller potentials occur during REM sleep and alertness; larger potentials are elicited during slow wave sleep. It is intriguing that the qualitative aspects of the behavioral states involved in these experiments are dichotomized along theta--non-theta lines. Thus alertness, or another behavioral state, can affect the basic reactivity of brain systems to afferent input. In our conditioning situation, some pre-existing state--highly correlated with hippocampal theta rhythm--limits both behavioral learning and the reactivity of hippocampal multiple unit activity to the conditioning stimuli.

Thus, the data show that, as an index of a behavioral state of arousal or activation, hippocampal EEG correlates well with attentional processes throughout the course of appetitive learning in cats and, in rabbits, actually *predicts* the responsiveness of hippocampal neurons to conditioning stimuli *and* the behavioral rate of classical aversive conditioning. All these results illustrate most clearly the profound and continuing impact of Donald Lindsley's work and ideas in the field of physiological psychology.

ACKNOWLEDGMENTS

Supported by research grants from NIMH (MH26530), NIH (NS12268), NSF (BMS75-00453), and the McKnight Foundation. We thank Cathy Berry for illustrations, Juta Keithe for photography, and Carol Hibbert for manuscript.

REFERENCES

Adey, W. R. (1966). Neurophysiological correlates of information transaction and storage in brain tissue. In E. Stellar & J. M. Sprague (Eds.), *Progress in physiological psychology* (Vol. I, pp. 3-43). New York: Academic Press.

Adey, W. R. (1977). The sensorium and the modulation of cerebral states: Tonic environmental influences on limbic and related systems. In B. M. Wenzel & H. P. Ziegler (Eds.), *Tonic functions of sensory systems* (pp. 396-420). (Annals of the New York Academy of Sciences, 290.) New York: New York Academy of Sciences.

Anchell, H., & Lindsley, D. B. (1972). Differentiation of two reticulo-hypothalamic systems regulating hippocampal activity. *Electroencephalography and Clinical Neurophysiology, 32*, 209-226.

Bard, P. (1928). A diencephalic mechanism for the expression of rage with special reference to the sympathetic nervous system. *American Journal of Physiology, 84*, 490-515.

Beach, F. A. (1947). A review of physiological and psychological studies of sexual behavior in mammals. *Physiological Review, 27*, 240-307.

Bennett, T. L. (1975). The electrical activity of the hippocampus and processes of attention. In R. L. Isaacson & K. H. Pribram (Eds.), *The hippocampus* (Vol. 2, pp. 71-99). New York: Plenum.

Bennett, T. L., French, J., & Burnett, K. N. (1978). Species differences in the behavioral correlates of hippocampal RSA. *Behavioral Biology, 22*, 161-177.

Berger, T. W., Alger, B. E., & Thompson, R. F. (1976). Neuronal substrate of classical conditioning in the hippocampus. *Science, 192*, 482-485.

Berger, T. W., Laham, R. I., & Thompson, R. F. (1980). Hippocampal unit-behavior correlations during classical conditioning. *Brain Research, 193*, 229-248.

Berger, T. W., & Thompson, R. F. (1977). Limbic system interrelations: Functional subdivision among hippocampal-septal connections. *Science, 197*, 587-589.

Berger, T. W., & Thompson, R. F. (1978a). Neuronal plasticity in the limbic system during classical conditioning of the rabbit nictitating membrane response. I. The hippocampus. *Brain Research, 145*, 323-346.

Berger, T. W., & Thompson, R. F. (1978b). Identification of pyramidal cells as the critical elements in hippocampal neuronal plasticity during learning. *Proceedings of the National Academy of Science*, 1978, *75*(3), 1572-1576.

Berry, S. D., & Thompson, R. F. (1978). Prediction of learning rate from the hippocampal electroencephalogram. *Science, 200*, 1298-1300.

Berry, S. D., & Thompson, R. F. (1979). Medial septal lesions retard classical conditioning of the nictitating membrane response in rabbits. *Science, 205*, 209-211.

Black, A. H. (1975). Hippocampal electrical activity and behavior. In R. L. Isaacson & K> H> Pribram (Eds.), *The hippocampus* (Vol. 2, pp. 129-167). New York: Plenum.

Boring, E. G. (1950). *A history of experimental psychology*. New York: Appleton-Century-Crofts.

Brown, J. S. (1961). *The motivation of behavior*. New York: McGraw-Hill.

Coleman, J. R., & Lindsley, D. B. (1977). Behavioral and hippocampal electrical changes during operant learning in cats and effects of stimulating two hypothalamic hippocampal systems. *Electroencephalography and Clinical Neurophysiology, 42*, 309-331.

Coleman, J. R., & Lindsley, D. B. (1975). Hippocampal electrical correlates of free behavior and behavior induced by stimulation of two hypothalamic-hippocampal systems in cat. *Experimental Neurology, 49*, 506-528.

DeFrance, J. F., Kitai, S. T., & Shimono, T. (1973a). Electrophysiological analysis of the hippocampal-septal projections: I. Response and topographical characteristics. *Experimental Brain Research, 17*, 447-462.

DeFrance, J. F., Kitai, S. T., & Shimono, T. (1973b). Electrophysiological analysis of the hippocampal-septal projections: II. Functional characteristics. *Experimental Brain Research, 17*, 463-476.

Duffy, E. (1962). *Activation and behavior*. New York: Wiley.

Fritsch, G., & Hitzig, E. (1870). Über die elektrische Erregbarkeit des Grosshirns. *Archiv für Anatomie und Wissenschaftliche Medizin (Leipzig)*, *37*, 300-332.

Gold, P. E., & McGaugh, J. L. (1975). A single-trace, two-process view of memory storage processes. In D. E. Deutsch & J. A. Deutsch (Eds.), *Short-term memory*. New York: Academic Press.

Gormezano, I. (1966). Classical conditioning. In J. B. Sidowski (Ed.), *Experimental methods and instrumentation in psychology* (pp. 385-420). New York: McGraw-Hill.

Grastyan, E., Lissak, K., Madarasz, L., & Donhoffer, H. (1959). Hippocampal electrical activity during the development of conditioned reflexes. *Electroencephalography and Clinical Neurophysiology*, *11*, 409-430.

Green, J. D., & Arduini, A. (1954). Hippocampal electrical activity and arousal. *Journal of Neurophysiology*, *17*, 533-557.

Groves, P. M., & Thompson, R. F. (1970). Habituation: A dual-process theory. *Psychological Review*, *77*, 419-450.

Hebb, D. O. (1949). *The organization of behavior*. New York: Wiley.

Hebb, D. O. (1955). Drives and the C.N.S. (conceptual nervous system). *Psychological Review*, *62*, 243-254.

Hebb, D. O. (1966). *A textbook of psychology* (2nd Ed.). Philadelphia: W. B. Saunders.

Hess, W. R. (1928). Stammganglien-Reizversuche (Verhandlungen der Deutschen Psychologische Gesellschaft, Sept., 1927), *Berichte über die Gesamte Physiologie*, *42*, 554.

Hoehler, F. K., & Thompson, R. F. (1980). Effect of the interstimulus (CS-UCS) interval on hippocampal unit activity during classical conditioning of the nictitating membrane response of the rabbit, *Oryctolagus cuniculus*. *Journal of Comparative and Physiological Psychology*, *94*, 201-215.

Hull, C. L. (1943). *Principles of behavior*. New York: Appleton-Century-Crofts.

Karplus, J. P., & Kreidl, A. (1910). Gehirn und Sympathicus. II. Ein sympathicuszentrum im zwischenhirn. *Pflügers Archiv für die Gesamte Physiologie des Menschen und der Tiere*, *135*, 401-416.

Kemp, I. R., & Kaada, B. R. (1975). The relation of hippocampal theta activity to arousal, attentive behavior and somato-motor movements in unrestrained cat. *Brain Research*, *95*, 323-342.

Kety, S. (1976). Biological concomitants of affective states and their possible role in memory processes. In M. R. Rosenzweig & E. L. Bennett (Eds.), *Neural mechanisms of learning and memory* (pp. 321-326). Cambridge, MA: MIT Press.

Klemm, W. R. (1976). Hippocampal EEG and information processing: A special role for theta rhythm. *Progress in Neurobiology, 7*, 197-214.

Kolb, B., & Whishaw, I. Q. (1977). Effects of brain lesions and atropine on hippocampal and neocortical electroencephalograms in the rat. *Experimental Neurology, 56*, 1-22.

Kramis, R., Vanderwolf, C. H., & Bland, B. H. (1975). Two types of hippocampal slow activity (RSA) in both the rabbit and rat: Relations to behavior and effects of atropine, diethylether and pentobarbital. *Experimental Neurology, 49*, 58-85.

Lacey, J. I. (1967). Somatic response patterning and stress: Some revisions of activation theory. In M. H. Appley & B. Trumbull (Eds.), *Psychological stress: Issues in research* (pp. 14-44). New York: Appleton.

Lashley, K. S. (1929). *Brain mechanisms and intelligence*. Chicago: University of Chicago Press.

Lashley, K. S. (1938). An experimental analysis of instinctive behavior. *Psychological Review, 45*, 445-471.

Lindsley, D. B. (1951). Emotion. In S. S. Stevens (Ed.), *Handbook of experimental psychology* (pp. 473-516). New York: Wiley.

Lindsley, D. B., Bowden, J., & Magoun, H. W. (1949). Effect upon the EEG of acute injury to the brain stem activity system. *Electroencephalography and Clinical Neurophysiology, 1*, 475-486.

Lindsley, D. B., & Wilson, C. L. (1975). Brainstem hypothalamic systems influencing hippocampal activity and behavior. In R. L. Isaacson & K. H. Pribram (Eds.), *The hippocampus* (Vol. 2, pp. 247-278). New York: Plenum.

Lopes da Silva, F. H., & Arnolds, D. E. A. T. (1978). Physiology of the hippocampus and related structures. *Annual Review of Physiology, 40*, 185-216.

Macadar, A. W., Chalupa, L. M., & Lindsley, D. B. (1974). Differentiation of brainstem loci which affect hippocampal and neocortical electrical activity. *Experimental Neurology, 43*, 499-514.

Malmo, R. B. (1959). Activation: A neuropsychological dimension. *Psychological Review, 66*, 367-386.

McCormick, D. A., & Thompson, R. F. (1984). Cerebellum: Essential involvement in the classically conditioned eyelid response. *Science, 223*, 296-299.

McFarland, W. L., Teitelbaum, H., & Hedges, E. K. (1975). Relationship between hippocampal theta activity and running speed in the rat. *Journal of Comparative Physiological Psychology, 88*, 324-328.

McGaugh, J. L. (1973). Drug facilitation of learning and memory. *Annual Review of Pharmacology, 13*, 229-241.

McLennan, H., & Miller, J. J. (1974). The hippocampal control of neuronal discharges in the septum of the rat. *Journal of Physiology, 237*, 607-624.

Morgan, C. T. (1943). *Physiological psychology*. New York: McGraw-Hill.

Moruzzi, G., & Magoun, H. W. (1949). Brainstem reticular formation and activation of the EEG. *Electroencephalography and Clinical Neurophysiology, 1*, 455-473.

O'Keefe, J. (1976). Place units in the hippocampus. *Experimental Neurology, 51*, 78-109.

O'Keefe, J., & Nadel, L. (1978). *The hippocampus as a cognitive map*. Oxford: Oxford University Press.

Olton, D., Branch, M., & Best, P. J. (1978). Spatial correlates of hippocampal unit activity. *Experimental Neurology, 58*, 387-409.

Parmeggiani, P. L. (1967). On the functional significance of the hippocampal theta rhythm. In W. R. Adey & T. Tokizane (Eds.), *Progress in brain research, 27*, 413-441. Amsterdam: Elsevier.

Pavlov, I. (1927). *Conditioned reflexes*. New York: Oxford University Press.

Pribram, K. H., & McGuinness, D. (1975). Arousal, activation, and effort in the control of attention. *Psychological Review, 82*, 116-149.

Prokasy, W. F. (1972). Developments with the two-phase model applied to human eyelid conditioning. In A. H. Black & W. F. Prokasy (Eds.), *Classical conditioning II: Current research and theory* (pp. 119-147). New York: Appleton-Century-Crofts.

Radil-Weiss, T., & Hlavicka, P. (1978). To the function of the reticulo-septo-hippocampal subsystem of the brain. *Recent Developments of Neurobiology in Hungary, 6*, 149-157.

Ranson, S. W., & Magoun, H. W. (1939). The hypothalamus. *Ergebnisse der Physiologie, 41*, 1-23.

Salafia, W. R., Romano, T., Tynan, T., & Host, K. G. (1977). Disruption of rabbit (*Oryctolagus cuniculus*) nictitating membrane conditioning by posttrial electrical stimulation of hippocampus. *Physiology and Behavior, 18*, 207-212.

Schlag-Rey, M., & Lindsley, D. B. (1975). Effect of medial and lateral hypothalamic stimulation on visual evoked potentials and hippocampal activity in alert cats. Cited in Lindsley and Wilson.

Schwartzbaum, J. S. (1975). Interrelationship among multiunit activity of the midbrain reticular formation and lateral geniculate nucleus, thalamocortical arousal and behavior in rats. *Journal of Comparative Physiological Psychology, 89*, 131-157.

Spong, P., Haider, M., & Lindsley, D. B. (1965). Selective attention and cortical evoked responses to visual and auditory stimuli. *Science, 148*, 395-397.

Stellar, E. (1954). The physiology of motivation. *Psychological Review, 61*, 5-22.

Stevens, S. S. (Eds.). (1951). *Handbook of experimental psychology*. New York: John Wiley.

Thompson, R. F., Berger, T. W., Berry S. D., & Hoehler, F. K. (1980). The search for the engram II. In D. McFadden (Ed.), *Brain mechanisms in behavior*. New York: Springer-Verlag.

Thompson, R. F., Berger, T. W., Cegavske, C. F., Patterson, M. M., Roemer, R. A., Teyler, T. J., & Young, R. A. (1976). The search for the engram. *American Psychologist, 31*, 209-227.

Thompson, R. F., & Spencer, W. A. (1966). Habituation: A model for the study of neuronal substrates of behavior. *Psychological Review, 173*, 16-43.

Vanderwolf, C. H. (1969). Hippocampal electrical activity and voluntary movement in the rat. *Electroencephalography and Clinical Neurophysiology, 26*, 407-418.

Vanderwolf, C. H. (1971). Limbic-diencephalic mechanisms of voluntary movement. *Psychological Review, 78*, 83-113.

Vanderwolf, C. H. (1975). Neocortical and hippocampal activation in relation to behavior: Effects of atropine, eserine, phenothiazines and amphetamine. *Journal of Comparative Psychology, 88*, 300-323.

Vanderwolf, C. H., Kramis, R., Gillespie, L. A., & Bland, B. H. (1975). Hippocampal rhythmic slow activity and neocortical low-voltage fast activity: Relations to behavior. In R. L. Isaacson & K. H. Pribram (Eds.), *The hippocampus* (Vol. 2, pp. 101-128). New York: Plenum.

Watson, J. B. (1930). *Behaviorism*. New York: Norton.

Whishaw, I. O. (1976). The effects of alcohol and atropine on EEG and behavior in the rabbit. *Psychopharmacology, 48*, 83-90.

Whishaw, I. O., Bland, B. H., Robinson, T. E., & Vanderwolf, C. H. (1976). Neurotransmitter blockade: The effects on two hippocampal RSA (theta) systems and neocortical desynchronization. *Brain Research Bulletin, 1*, 573-581.

Whishaw, I. O., & Schallert. T. (1977). Hippocampal RSA (theta), apnea, bradycardia and effects of atropine during underwater swimming in the rat. *Electroencephalography and Clinical Neurophysiology, 42*, 389-396.

Winson, J. (1972). Interspecies differences in the occurrence of theta. *Behavioral Biology, 7*, 479-487.

Winson, J. (1975). The theta mode of hippocampal function. In R. L. Isaacson & K> H> Pribram (Eds.), *The hippocampus* (Vol. 2, pp. 169-183). New York: Plenum.

Winson, J. (1978). Loss of hippocampal theta rhythm results in spatial memory deficit in the rat. *Science, 201*, 160-163.

Winson, J., & Abzug, C. (1977). Gating of neuronal transmission in the hippocampus: Efficacy of transmission varies with behavioral state. *Science, 196*, 1223-1225.

24 · NEED-INFORMATION INTERACTION OF BRAIN STRUCTURES

P. V. Simonov

Academy of Sciences of the USSR

Although concepts of "systemic," "integral" and "cortical-subcortical" organization of brain activity are generally accepted and repeated in hundreds of scientific papers, we do not have a general hypothesis emphasizing the actual anatomic composition of brain structures subserving these concepts in higher animals and humans. The present paper deals with the role of four main brain structures in the genesis of emotional states and in organizing purposeful behavior (see Simonov, 1986, for additional consideration of this topic). It is concluded that the frontal neocortex is responsible for behavioral orientation to high-probability events, while the hippocampus provides reactions to events of low probability. The hypothalamus orients behavior towards the satisfaction of a dominating need, while the amygdala creates a balance or dynamic coexistence between competing needs (motivations) and emotions.

According to Reflex Theory, the organism interacts with the environment through afferent, efferent, and central mechanisms; the latter transforming the external stimulus into the corresponding reaction. If we apply this theory to the composition and activity of the brain as a whole, we discover that our knowledge about central mechanisms estimating external stimuli and determining appropriate responses is extremely limited. Instead, we must integrate information from sensory physiology (including higher cortical representations), mechanisms of movement organization and of vegetative regulation, activating and deactivating systems controlling levels of arousal, and existing knowledge about brain structures underlying the strategies and tactics of behavior.

Textbooks usually contain references to frontal parts of the neocortex, limbic system, and the hypothalamus, but give little accurate information on the principal peculiarities of their functions, or the specificity of their "contribution" to the shaping of behavior. The functional characteristics of specific brain structures is purely descriptive and is difficult to translate into the language of neurophysiology. For example, Pribram (1961) and Luria (1966) suggested that analytic-synthetic activity of the posterior "internal system" of neocortex (tertiary areas of temporal, parietal and occipital lobes) mobilizes previously acquired experience on the basis of which the program of activity is being drawn, whereas the comparison of this program to the actual results of its fulfillment occurs in the frontal cortex. However, according to the same authors, under the influence of impulses from the hypothalamus, the state of tension that serves as the first impetus to search

DOI: 10.4324/9781003164647- 31

for a means to satisfy the actual need is evoked not in the posterior programming system of the neocortex, but in mediobasal areas of the frontal cortex.

As a further example, the hippocampus is involved in such a great variety of brain activities (forming inhibitions of conditioned reflexes, orienting reactions, emotions, memory, etc.) that its functional difference from other brain structures becomes elusive (Gray, 1982). As to the hypothalamus and nuclei of the amygdala, data obtained through direct stimulation and coagulation suggest that these two brain structures seem to duplicate the functions of each other, e.g., controlling hunger, thirst, sex, defense or attack (Fonberg, 1967).

We believe that progress in the physiological study of the brain structures discussed above has been limited by our indistinct ideas concerning the precise behaviors which can be "superimposed" on the available morphological data. According to the Information Theory of emotions (Simonov, 1964, 1986), emotion as the direct controller of behavior is determined by two factors: the presence of actual needs, and the possibility to satisfy these needs through interaction with the environment. The importance of environmental stimuli is determined by their relation to the available needs. All these stimuli may be divided into two major categories: those with high probability of reinforcing by directly satisfying a need, and those with low probability of reinforcement. Among actual needs we distinguish acute, dominating needs, and subdominant ones dynamically coexisting or competing with the dominant needs (see Figure 24.1).

Experimental data obtained in our laboratory demonstrate the importance of neural interactions in estimating signals coming from the environment, and in the choice of reactions. Thus, frontal neocortex (lateral area of prefrontal neocortex in dogs) supports orienting behavior to signals with high probabilities of reinforcement. After ablation the importance of environmental signals seems to be equalized and the animal reacts equally to high- and low-probability events (Mekhedova, 1971). Clinical observations are in agreement with these findings. Thus, lesions of temporal, parietal and occipital cortical areas do not impair probabilistic forecasting; but such disturbances are found in patients with frontal lobe defects (Bazhin et al., 1970).

Unlike the frontal areas of neocortex, the hippocampus is responsible for reactions to signals of low-probability events. This is illustrated in the results of Figure 24.2, which demonstrates that, following bilateral hippocampectomy, rats lose their ability to elaborate alimentary conditioned reflexes when the probability of reinforcement declines to 33 or 25 percent (Pigareva, 1977).

Errors in probabilistic forecasting are also typical of the "emotionally stressed" brain. In such conditions subjective estimations of objectively low

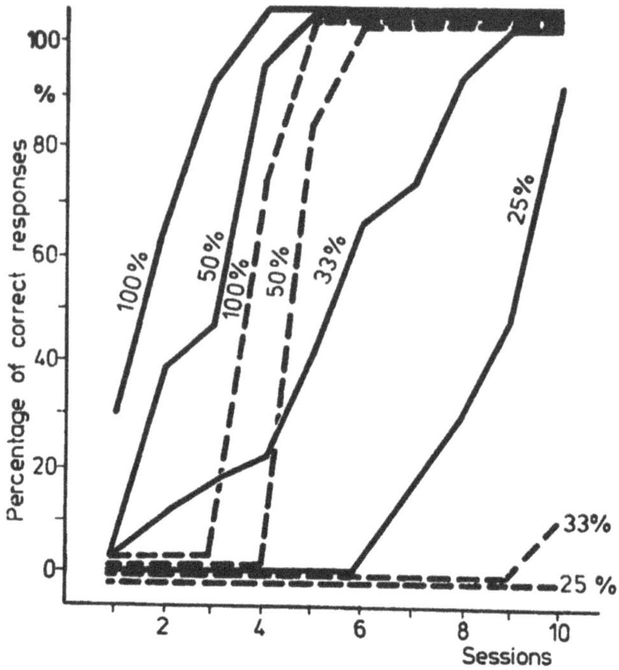

FIGURE 24.1. Elaboration of conditioned alimentary motor reflexes in normal (continuous line) and bilaterally hippocampectomized (dashed line) rats, with different probabilities of reinforcement.

probability events are increased. Such a reaction is expedient when there is a deficit in information on the necessary and sufficient means for satisfying a need. In other words, the hippocampus participates in the search for the missing information by broadening the set of engrams taken from memory and by lowering the criteria of "decision making" in comparing these engrams to the available stimuli.

Experiments were performed in amygdalectomized rats, using the paradigm of E. A. Asratyan, which involves a switching of alimentary and defensive conditioned reflexes (the same conditioned signal is reinforced in the morning by food, and in the evening by painful electrical stimuli). The results (see Table 24.1) show that such reflex elaboration is possible only by combining a weak pain stimulus and high alimentary excitation (three days food deprivation), or a strong electric current coupled with short (one day) alimentary deprivation (Pigareva, 1977).

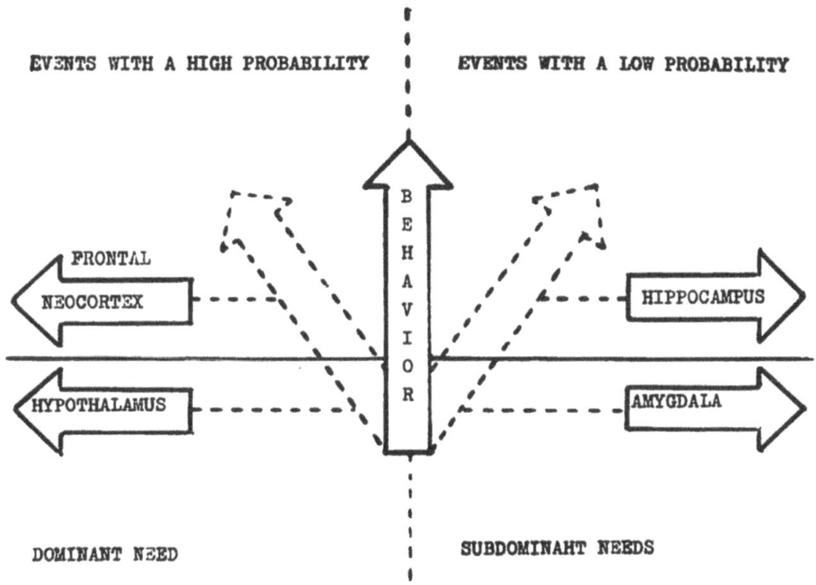

FIGURE 24.2. Schematic diagram of need-information interaction of four main brain structures. Frontal neocortex orients behavior to signals with high probability of their reinforcement by some factors, able to satisfy the organism's need. The hippocampus is necessary for reactions to signals with low probability of their reinforcement. Due to the interaction of hypothalamus and amygdala the dominant need of primary satisfaction is singled out taking into account subdominant needs, the available situation and previous experience.

These data correlate with the results of our own experiments on the rate of elaboration of a conditioned avoidance reaction to pain stimulation delivered to a second animal (Simonov, 1976). These experiments demonstrated that the consequences of amygdalectomy depend upon which of two competing motivations was relatively dominant in each animal prior to the operation: sensitivity to the "pain cry" of another rat receiving electric shock, or preference for the enclosed area containing the shock pedal rather than the open field of the experimental chamber.

Thus, if destruction of the hippocampus transforms the animal into an "automatic machine," reacting only to the signals of highly probable events and ignoring all other alternatives, lesions of the amygdala orient behavior toward satisfaction of dominant needs only. In the natural habitat such behavior would be deleterious. For example, an animal motivated by hunger would ignore stimuli signaling danger.

The results obtained in our laboratory conform to published results. Thus, in the experiments of White and Weingarten (1976), satiated

TABLE 24.1. Switching alimentary and defensive conditioned reflexes in control and amygdalectomized rats.[a]

	Duration of Food Deprivation			
	One Day		Three Days	
Shock Intensity	Control	Operated	Control	Operated
0.4	4(0)	5(0)	8(2)	8(5)
0.6	5(2)	5(0)	8(4)	8(4)
0.8	5(3)	5(0)	5(5)	5(0)
1.0	5(4)	5(0)	5(3)	5(0)
1.2	7(5)	7(4)	6(4)	6(1)
1.4	8(3)	8(5)	6(3)	6(0)

[a] In each column, the number of rats in the given group and the number of animals (in brackets) which mastered the "switching" paradigm. The switching criterion consisted of 100% performance on three consecutive experimental sessions.

amygdalectomized rats showed a higher degree of exploratory activity than the control group, the degree of alimentary activity of both groups being the same. However, the exploratory behavior of hungry amygdalectomized rats was inferior to the control group, although the alimentary activity of the operated animals was better than the controls. Amygdalectomy thus enhanced the type of behavior which was initiated by the dominant need. Gloor's (1960) idea of the amygdala as a part of the structural system determining the choice of behavior may be accepted, with the additional specification that the amygdala participates by "weighing" competing emotions evoked by competing needs.

The competition of emotions, rather than needs (motivations), is suggested by the following. It was shown that the basolateral amygdala deals with the previous experience of decreasing thirst; and not with "tissue thirst," or the detection of water-salt metabolism (Rolls & Rolls, 1973). In addition, lesions of the amygdala affect reactions caused by fear, and not by pain. Thus, such lesions in dogs impair classically conditioned defensive reflexes, but not instrumental ones, where the signs of fear disappear throughout an improvement of the conditioned defensive reaction. This permits the assumption that the amygdala is involved in the process of forming behavior at comparatively late stages, after the actualized needs have been compared to the prospects of their satisfaction and then transformed into corresponding emotional states. The amygdala thus belongs to the system of brain structures that implement the "interruptive function" of emotions (Simon, 1967) which orients behavior toward the primary satisfaction of the dominant need while taking into account the available situation and previous experience.

Lesions of the hypothalamus, on the other hand, often produce effects contrary to amygdalectomy. Thus, animals stop responding to "tissue thirst" and a decrease in blood glucose, but due to the amygdala continue to react to conditioned signals of food and water. Their previous experience, isolated from current needs, acquires a certain independence. In the experiments with avoidance reaction in rats during pain stimulation of a second animal, reactions were contrary to the consequences of amygdalectomy. Such rats usually "stick" in the doorway between the enclosed pedal area and the open field of the chamber.

We have attempted to generalize the experimental data derived from frontal neocortex, hippocampus, amygdala and hypothalamus concerning the shaping of goal-directed behaviors. This is presented schematically in Figure 24.1, which naturally oversimplifies the real complexity of brain interactions and fails to take into account many other functions of the designated brain structures. Nevertheless, it is because of its simplicity that this schematic representation is accessible to experimental verification.

All other brain structures play an executive or auxiliary role: sensory systems, mechanisms of movement organization (pyramidal and extrapyramidal), and regulation systems of arousal and vegetative functions as well. As for other structures of the limbic system, the septum is so closely interactive with the hippocampus that most authors prefer to speak about a single septo-hippocampal system. Central gray matter is specifically related with evaluation of stimulus aversion and reactions of avoidance, which can be observed during stimulation of medial hypothalamus (Sandner, Schmitt, & Karli, 1982). Non-specific thalamus is the place of convergence of sensory and motivational impulses (Casey & Keene, 1973), being limited primarily to transmissive functions. In other words, *just the four above-mentioned structures define to what outer stimuli, and by which behavioral reaction, the living organism responds at any given moment.* According to Mogenson and colleagues, connection of limbic and motor systems is realized through the ventral tegmentum and projection of nucleus accumbens to globus pallidum (Mogenson, Jones, & Yim Chi Yiu, 1980). The dorsocaudal globus pallidum is directly concerned with the preparation and realization of movements, while the ventrorostral globus pallidum is concerned with the transmission of motivational information to the motor system (Nishino, Ono, Muramoto, Fukuda, & Sasaki, 1985).

By the way, our diagram only seems simple. It will no longer be simple if we consider the most complex interrelations and intereffects of the four studied structures. The experimental study of sensory and nonsensory factors of perception showed that, in case of a strong need, the external stimulus may pick out from memory such an engram that corresponds to the given need and does not correspond to the real stimulus. For example, a hungry person starts to perceive the signals with vague meaning as associated with food. In other words, motivational level "hypothalamus-amygdala" and informational level "frontal cortex-hippocampus" really function as a single integrative complex.

The functions of the emotional system "hippocampus-amygdala" correspond to the role of emotions in the forming of behavior--its orientation toward satisfaction of the actual need considering the available situation (amygdala) and broadening the range of meaningful external stimuli (hippocampus). The system "frontal neocortex-hypothalamus" may be regarded as the cerebral substratum of will. Indeed, it is the will that stabilizes behavior oriented toward satisfaction of initially dominant needs in spite of an unfavorable forecast and competing motives (Simonov, 1975, 1986). From this point of view, "frontal cortex-hypothalamus" may be defined as a strategic system, and "hippocampus-amygdala" as a tactical system, that makes behavior more adaptive.

REFERENCES

Bazhin, B. F., Vanina, T. M., Malikova, I. V., Meerson, Y. A., Morkva, L. I., & Tonkonogiy, J. M. (1970). In B. F. Bazhin (Ed.), *Principles of probabilistic organization of behavior, recognition and medical diagnosis* (pp. 17-18). Leningrad (Russian).

Casey, K. L., & Keene, J. J. (1973). Unit analysis of the awake animal: Pain and self-stimulation. In M. I. Phillips (Ed.), *Brain unit activity during behavior* (pp. 115-119). Springfield: Thomas.

Fonberg, E. (1967). The motivational role of hypothalamus in animal behavior. *Acta Neurobiologie Experimentalis, 27*, 303-318.

Gloor, P. (1960). Amygdala. In J. F. Field & H. W. Magoun (Eds.), *Handbook of physiology*: Sec. 1. *Neurophysiology* (Vol. II., pp. 1395-1420). Washington, DC: American Physiological Society.

Gray, J. (1982). *The neurophysiology of anxiety: An inquiry into the functions of the septo-hippocampal system*. Oxford: Oxford University Press.

Luria, A. R. (1966). *Higher cortical functions in man*. New York: Basic Books.

Mekhedova, A. Y. (1971). The role of the frontal brain areas in the formation of conditioned reactions adequate to the probability and magnitude of their reinforcement. *Zhurnal visshej nervnoj deyatelnosti, 21*, 459-464 (Russian).

Mogenson, G., Jones, D., & Yim Chi Yiu (1980). From motivation to action: functional interface between the limbic system and the motor system. *Progress in Neurobiology, 14*, 69-97.

Nishino, H., Ono, T., Muramoto, K., Fukuda, M., Sasaki, K. (1985). Movement and non-movement related pallidal unit activity during bar press feeding behavior in the monkey. *Behavioral Brain Research, 15*, 27-42.

Pigareva, M. L. (1977). *Limbic mechanisms of switching over*. Moscow: Nauka (Russian).

Pribram, K. (1961). To the theory of physiological psychology. *Voprosi psychologii*, No. 2, 133-156 (Russian).

Rolls, B., & Rolls, E. (1973). Effects of lesions in the basolateral amygdala on fluid intake in the rat. *Journal of Comparative and Physiological Psychology, 83*, 240-247.

Sandner, G., Schmitt, P., & Karli, P. (1982). Effect of medial hypothalamic stimulation inducing both escape and approach on unit activity in rat mesencephalon. *Physiology and Behavior, 29*, 269-274.

Simon, H. A. (1967). Motivational and emotional controls of cognition. *Psychological Review, 74*, 29-39.

Simonov, P. V. (1964). On correlation between motor and vegetative components of conditioned defensive reflex in man. In E. A. Asratyan (Ed.), *Central and peripheral mechanisms of motor activity in animals and man* (pp. 65-66). Moscow: Nauka (Russian).

Simonov, P. V. (1975). *Higher nervous activity of man: Motivational-emotional aspects*. Moscow: Nauka (Russian).

Simonov, P. V. (1976). *Neurophysiological approach to analysis of intraspecific behavior*. Moscow: Nauka (Russian).

Simonov, P. V. (1986). *The emotional brain*. New York: Plenum.

White, N., & Weingarten, H. (1976). Effects of amygdaloid lesions on exploration by rats. *Physiology and Behavior, 17*, 73-79.

25 · CEREBRAL PROCESSING AND COGNITIVE FUNCTIONING IN READING DISABILITY

M. Friedman, V. Welch, I. Fried, and G. Marsh

University of California at Los Angeles

This paper presents a theoretical and empirical analysis of reading disability using a framework motivated by current work in neuropsychology and cognitive psychology. Research on such clinical populations is important not only because of its obvious direct application, but also because the study of well-defined clinical populations has much to offer towards a basic understanding of brain-behavior relationships in human cognition.

The Nature of Reading Disability. A number of terms have been used to label the population of interest in this paper: developmental dyslexia, specific reading disability, learning disability, to name just a few. For simplicity, we will use the term reading disability (RD). These terms generally refer to children who (1) are of overall normal intelligence, (2) have no "hard" neurological signs or deficits, but (although receiving normal education) are (3) severely retarded as compared with their age group in the learning of basic academic skills--usually reading, spelling, and/or arithmetic. It is typically argued that RD "results from subtle dysfunction of the central nervous system, particularly with regard to the reception, transformation, and transmission of sensory data" (Clements, 1969, p. 32).

The most popular theories of RD are variations on a maturation lag of left hemisphere development and/or incomplete dominance hypothesis (Satz, 1976). These theories suggest that RD is due to slower and less complete differentiation of cerebral function. We have used this general theoretical framework to study some of the brain-behavior relationships in RD, particularly the nature of the specific and general cognitive deficits shown by RD children and the implications for remediation.

Hemispheric Functioning and Reading Disability. In an initial study (Guyer & Friedman, 1975), we explored the relationship between processing skills associated with left and right hemispheric functioning and RD. The study compared a group of 41 RD boys and normal control boys matched for age (7.7-12.7 years) and Verbal IQ (90-146 range). All of the boys were administered a set of tests chosen for their association with specialized hemispheric processing and standard achievement tests. The test battery included measures of visual and auditory verbal short- and long-term memory, visual and verbal closure, verbal concepts, and hand awareness and laterality. The main results were that the RD boys performed more poorly on

DOI: 10.4324/9781003164647- 32

left hemisphere tests of verbal skills and hand awareness, but were equivalent to the normal controls on right hemisphere non-verbal tasks. Velluntino (1977) and others have reported similar results.

Testing Implications for Remediation. Further studies explored the implications of these results for remediation. Visual imagery, as a coding strategy, has been shown to be related to right hemisphere functioning (Seamon & Gazzaniga, 1973). Since RD children appear to be better able to comprehend and remember visual material, we wondered whether imagery strategies might be used to bolster their academic skills. Guyer-Christie, Friedman and Miller (1977) examined different hemispherically based memory strategies in RD and normal children. The experimental task required the learning of word lists by either phonemic, taxonomic, or imagic association strategies. The phonemic (rhyming) strategy presumably depends largely on left hemisphere auditory-verbal skills. The taxonomic strategy, which required the grouping of nouns into semantic categories, could appear to depend on both imagic and verbal factors. The main results of this study were that both groups were able to use all of the strategies to increase the number of words recalled relative to unstructured control lists, and both groups recalled more words correctly using the imagic strategy. The phonemic and taxonomic strategies produced more intrusion errors than the imagic strategy, and this effect was greater for the RD children than the normal controls. The hypothesis of left hemisphere dysfunction in RD is consistent with these results. More important, however, the results for the imagic strategy suggest that the intact right hemisphere abilities might be used as a support system for the deficient left hemisphere skills.

A further study (Guyer-Christie, 1976) examined specific strategy training for both reading comprehension and spelling skills. In this study, two strategies were compared: the imagery strategy called for children to form visual images and describe them, to visualize spelling words, to draw pictures in response to sentences, and to visualize in response to a story. The verbal strategy required children to verbalize in response to sentences, and to silently verbalize in response to a story. A group of 48 RD children (8-14 years of age) were compared with a group of 41 normal children. As in the previous study, imagery strategies were found to be effective in improving memory and comprehension in both groups. However, the normal children were better able to use instructions to visualize words as aids to spelling them, while the RD children retained spelling words better when emphasis was placed on associating letters to sounds. Further analysis suggested that control children were actually using *two* strategies (imagery plus phonics) while the RD children did not spontaneously use phonics in addition to imagery instructions. In support of left hemisphere dysfunction, RD children scored considerably lower than the normal control children on a test requiring analytic comparison of speech sounds, a function associated with the left hemisphere. Scores on this auditory analysis test, which did not involve written material or verbalization, were as strongly correlated with each other.

We have concluded from these studies that imagery is of limited use in remediation and training of RD children, and that the interrelation of oral language and reading are such that remediation must attack that interface (Rozin & Gleitman, 1977).

Analysis of Reading Strategies. The results reviewed in the preceding section have led us to move from the study of global right and left hemisphere strategies to an analysis of strategies actually employed in reading and spelling (Marsh, Friedman, Desberg, & Saterdahl, 1981; Marsh, Friedman, Welch, & Desberg, 1980, 1981). Broadly speaking, our results here suggest that there is a developmental progression in the strategies actually used in reading and spelling. In the earliest stages of reading, a child learns to recognize a limited set of words by rote association and uses general language context to read unknown words rather than using the perceptual characteristics of the word itself. A second stage is the development of the understanding and usage of grapheme-phoneme rules. Reading in both of these early stages probably consists of accessing word meaning through speech recoding. In the third stage, phoneme-grapheme rules become automatized, and a more direct access to word meaning than through speech recoding develops. Novel, unknown words are read by analogy to known words already automatized into an internal lexicon. Skilled reading surely requires all of these strategies: language context, speech recoding, automatized reading, and reading by analogy.

Clinical Subtypes of Reading Disability. There is evidence for clinical subgroups of RD that can be related to this strategy classification. Although there are several competing formulations, Boder (1973) has proposed a classification that is particularly interesting to us. She has suggested the following categories of RD on the basis of patterns of reading and spelling errors: (1) *Dysphonetic*--poor grapheme-phoneme rule understanding; some sight vocabulary; spelling "by eye" but not "by ear"; (2) *Dyseidetic*--poor visual memory for words; reads analytically through phonemic analysis rather than whole words; very limited sight vocabulary; spells phonetically; (3) *Alexic*--most severely handicapped, has both dyseidetic and dysphonetic characteristics.

In terms of the strategy analysis outlined above, Boder's dysphonetic category can be interpreted as a dysfunction in the use of the grapheme-phoneme route to meaning, while the dyseidetic category may be a deficit in the automatized direct access to meaning. The alexic group appears to have deficits in both channels. Using a standardized reading and spelling test, Boder classified a carefully selected group of 107 RD children as follows: dysphonetic = 63%; dyseidetic = 9%; alexic = 22%.

Studies of adults with acquired dyslexia--reading disability due to brain damage--also support the notion of two relatively independent routes to meaning (Marshall & Newcombe, 1973). There are cases of acquired dyslexia which show sudden impairment of word attack skills and inability to read new words, but with a relatively intact sight vocabulary. Similarly, there are dyseidetic-like cases of acquired dyslexia in which word attack skills are intact, but there is a deterioration of fluent sight vocabulary. Left hemisphere lesions account for almost all reported cases of acquired dyslexia (see Ellis, 1984, for a recent discussion of acquired dyslexia and other issues discussed in this paper).

Neurophysiological Measures and Reading Disability. Research by Fried (1979) provides physiological evidence supporting Boder's classification based on reading and spelling patterns. He studied auditory event-related potentials (ERP) to word and to musical chord stimuli in a group of RD boys classified into Boder's clinical subtypes, and in a group of normal readers. The results are complex and difficult to summarize, but we will note two important findings: First, latency differences between ERPs to word vs. musical chord stimuli were attenuated in all of the RD subjects as compared to the normal controls. Figure 25.1 shows ERP examples in a normal and an alexic boy. Second, the normal readers tended to have greater word than musical chord waveform differences over their left as compared with right hemisphere. Similar differences were found for the dyseidetic and alexic groups. However, these differences for words and musical chords were absent for the dysphonetic group, again supporting a left hemisphere dysfunction interpretation of RD (at least for the dysphonetic group).

Attention in Reading Disability. Our discussion of RD has focused on the specific cognitive deficits associated with RD. However, it is frequently reported that RD children are also generally inattentive and distractible. The nature of the general attentional deficit seen in many RD children is illustrated in a study by Deikel and Friedman (1976). They compared the performance of groups of normal and RD children on a card-sorting task which required the classification of cards into categories on the basis of patterns of holes punched in the cards. After completing the task, they were tested for their memory of the details of the task as well as their memory for incidental, extraneous material that was on the cards they sorted, but clearly unrelated to the primary task. The results of this study are shown in Figure 25.2. The normal readers showed superior recall for the details of the primary task, but the RD children showed superior recall of the extraneous

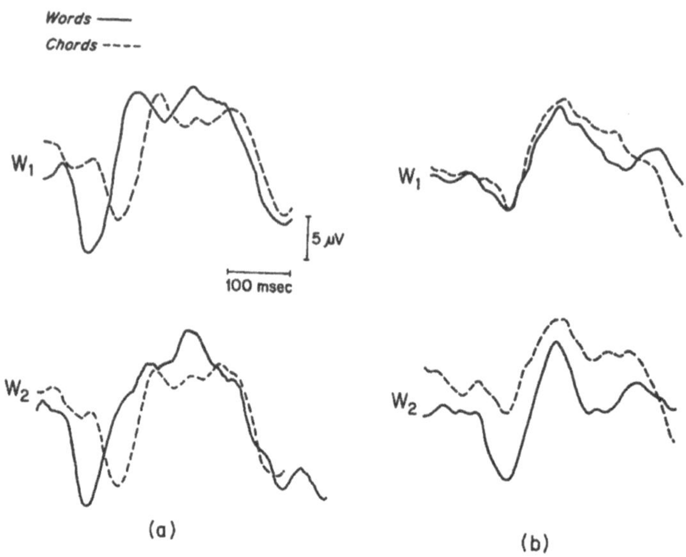

FIGURE 25.1. Auditory event-related potentials (ERPs) to word and musical-chord stimuli in a typical normal reader (a) and an "alexic" subject (b). W_1 overlies the Wernicke region of left hemisphere, W_2 overlies a homotopic area on the right. There is a clear shift in latency of the large positive-negative-positive potential between word and musical chord ERPs in the normal subject. This shift is of much smaller magnitude in the alexic subject (after Fried, 1979).

material. This inability of the RD children to focus on the relevant aspects of the task might be considered as immature behavior and indicative of a "maturational lag" in cognitive development.

But how might incomplete dominance or deficient left hemisphere functioning account for such attentional deficits? We have argued as follows (Friedman, Guyer-Christie, & Tymchuk, 1976): The brain may be regarded as a collection of semi-independent processing systems all operating on the same data, trying to make sense of experience and reach decisions (Gazzaniga, 1974). "Dominance" is the decision or control system that brings order and attentional control of these various information-processing activities. Vygotsky (1962) proposed a special role for language processing systems in normal development. He suggested that language and thought develop from preverbal thought through a series of stages, beginning with the use of language for communication and social purposes and then to a stage of egocentric speech where language acquires its "second signal properties." At this point, self-speech begins to obtain control over attentional processes, and as a newer phylogenetic ability, it inhibits control by older phylogenetic

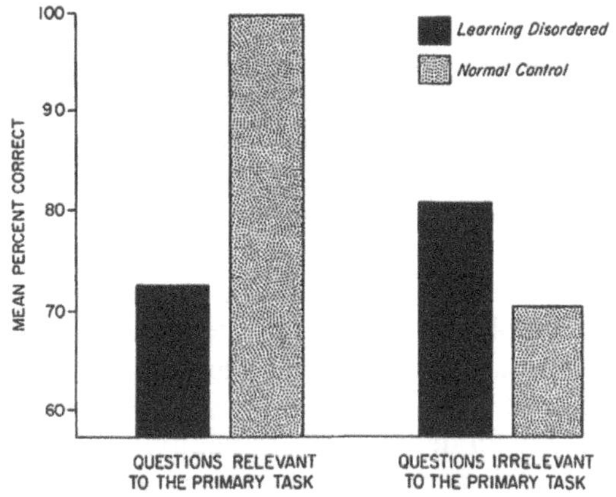

FIGURE 25.2. Post-test scores on questions relevant and irrelevant to the primary tasks for groups of Learning Disordered and Normal Control Children (after Deikel & Friedman, 1976).

systems. As applied to the RD child, deficient or immature left hemisphere function would lead to both deficient verbal functioning and attentional control.

Cognitive Style in Reading Disability. We might further conjecture that this asynchrony of development of the various brain processing systems might result in a particular *cognitive style* of information processing that would affect performance over a wide range of perceptual, learning, and problem-solving tasks. Thus, when task requirements and cognitive style are incompatible, a deficit in performance may be detectable. The construct of cognitive style has emerged in recent years as an important theoretical concept in child development and is critical for an understanding of RD children (Keogh, 1977).

The cognitive style dimension of particular interest here is Witkin's Field Articulation dimension (Witkin, Goodenough, & Altman, 1977). One of the principal tests used to define the concept is the Rod and Frame Test. This task requires the subject to align a rod vertically in the presence of a distorting visual frame with no other visual cues available. Successful performance requires the subject to use postural cues and to ignore the distracting visual frame. Those subjects who are able to ignore the frame and make few errors in aligning the rod are said to be "field-independent." According to Witkin, field independent people tend to be relatively focused and analytic in their approach to cognitive tasks. On the other hand, the "field dependent" person cannot ignore the distracting frame and tends to make errors in setting the rod. In general, the field dependent person tends to respond to the total field and to use a more global and perhaps less verbal information processing style. There is considerable evidence linking field articulation style with performance over a broad range of tasks. There is also evidence relating field articulation with brain functioning (Friedman et al., 1976; Witkin et al., 1977). The essential idea is that although the Rod and Frame Test appears to be a perceptual task, it is the attentional function of the left hemisphere which is important for successful performance, because the distracting visual information from the frame must be suppressed.

The theory of incomplete dominance of left hemisphere systems as a principal factor in RD would thus lead to an expectation that RD children would tend to be field dependent. We tested this hypothesis in the Guyer and Friedman (1975) study mentioned earlier. In addition to the battery of right and left hemisphere tests, we administered the Rod and Frame Test of Field Articulation. As predicted, we found that the RD children tended to be much more field dependent than the normal controls. Indeed, we found that with three measures of left hemisphere function which did not involve reading--the Rod and Frame, verbal long-term memory, and verbal closure--it was possible to correctly classify 82% of the boys into the normal or RD groups.

In related work on this issue, Welch (1981) examined some of the relationships of cognitive style with more general concepts of social and cognitive competence in RD and normal children. This study focused on the self-concept, behavioral, and affective correlates of learning disabled children through an analysis of their causal attributions. The major discriminators of group differences between learning disabled and control children in this study were the tendency to externalize failure, to report numerous indices of depression, and to exhibit problem classroom behaviors, particularly antisocial behaviors.

CONCLUSION

Our research was initially motivated by some rather simplistic notions about brain-behavior relationships. However, the melding together of concepts from neuropsychology and cognitive psychology and our empirical results have led to a detailed analysis of the components of cognitive performance and their neuropsychological foundations. Thus the results of this very conceptually oriented program have implications for both the identification, diagnosis, and treatment of reading disability and basic understanding of human cognition.

REFERENCES

Boder, E. (1973). Developmental dyslexia: A diagnostic approach based on three reading-spelling patterns. *Developmental Medicine and Child Neurology, 15*, 663-687.

Clements, S. (1969). A new look at learning disabilities. In L. Tarponol (Ed.), *Learning disabilities: Introduction to educational and medical management* (pp. 156-175). Springfield, IL: Charles C. Thomas.

Deikel, S., & Friedman, M. (1976). Selective attention in children with learning disabilities. *Perceptual and Motor Skills, 42*, 675-678.

Ellis, A. W. (1984). *Reading, writing, and dyslexia: A cognitive analysis.* London: Lawrence Erlbaum Associates.

Fried, I. (1979). Cerebral dominance and subtypes of developmental dyslexia. *Bulletin of the Orton Society, 29*, 101-112.

Friedman, M., Guyer-Christie, L., & Tymchuk, A. (1976). Cognitive style and specialized hemispheric processing learning disability. In R. Knights & D. Bakker (Eds.), *The neuropsychology of learning disorders: Theoretical approaches* (pp. 257-263). Baltimore: University Park Press.

Gazzaniga, M. (1974). Cerebral dominance viewed as a decision system. In S. Diamond & J. Beaumont (Eds.), *Hemispheric functions in the human brain* (pp. 367-382). London: Halstead.

Guyer, L., & Friedman, M. (1975). Hemispheric processing and cognitive styles in learning disabled and normal children. *Child Development, 46*, 658-668.

Guyer-Christie, L. (1976). *Tutoring strategies based on a hemispheric processing paradigm compared for learning disabled and normal children.* Unpublished dissertation, University of California, Los Angeles.

Guyer-Christie, L., Friedman, M., & Miller, J. (1977). *Memory strategies in learning-disabled and normal children.* Unpublished manuscript.

Keogh, B. (1977). Research on cognitive styles. In R. Kneedler & S. Tarver (Eds.), *Changing perspectives in special education* (pp. 183-195). Columbus, OH: Merrill.

Marsh, G., Friedman, M., Desberg, P., & Saterdahl, K. (1981). Comparison of reading and spelling strategies in normal and reading disabled children. In M. Friedman, J. P. Das, & N. O'Connor (Eds.), *Intelligence and learning* (pp. 363-367). New York: Plenum.

Marsh, G., Friedman, M., Welch, V., & Desberg, P. (1980). The development of strategies in spelling. In U. Frith (Ed.), *Cognitive processes in spelling* (pp. 339-353). London: Academic Press.

Marsh, G., Friedman, M., Welch, V., & Desberg, P. (1981). A cognitive-developmental approach to reading acquisition. In G. MacKinnon & T. Waller (Eds.), *Reading research: Advances in theory and practice* (pp. 199-221). New York: Academic Press.

Marshall, J., & Newcombe, F. (1973). Patterns of paralexia: A psycholinguistic approach. *Journal of Psycholinguistic Research, 2*, 175-199.

Rozin, P., & Gleitman, L. (1977). The structure and acquisition of reading. II. The reading process and acquisition of the alphabetic principle. In A Reber & D. Scarborough (Eds.), *Towards a psychology of reading* (pp. 176-201). Hillsdale, NJ: Erlbaum.
Satz, P. (1976). Cerebral dominance and reading disability: An old problem revisited. In R. Knights & D. Bakker (Eds.), *The neuropsychology of reading disorders: Theoretical approaches* (pp. 21-38). Baltimore: University Park Press.
Seamon, J., & Gazzaniga, M. (1973). Coding strategies and cerebral laterality effects. *Cognitive Psychology, 5*, 249-256.
Vellutino, F. (1977). Alternative conceptualization of dyslexia: Evidence in support of a verbal deficit hypothesis. *Harvard Educational Review, 47*, 334-353.
Vygotsky, L. S. (1962). *Thought and language*. Cambridge, MA: MIT Press.
Welch, V. O. (1981). *An attributional analysis of the relationship between learning disabilities and self-concept, affect, and behavior in children.* Dissertation, University of California, Los Angeles.
Witkin, H., Goodenough, D., & Altman, P. (1977). Psychological differentiation: Current status. *Research Bulletin 77-17*. Princeton, NJ: Educational Testing Service.

26 · LANGUAGE LATERALITY, SEX, AND STUTTERING: ERPs TO CONTEXTUAL MEANING

W. S. Brown, J. T. Marsh, R. E. Ponsford,
and L. E. Travis

Fuller Graduate School
and
University of California at Los Angeles

Available evidence, reviewed by McGlone (1980), gives some support to the hypothesis that females have a lower margin of hemispheric asymmetry for the processing of language than males, i.e., females have some degree of bilaterality in the hemispheric distribution of language function. Data from dichotic listening (Harshman, Remington, & Krashen, 1976; Lake & Bryden, 1976), tachistoscopic (Levy & Reid, 1976), and EEG studies (Ray, Morell, Frediani, & Tucker, 1976; Tucker, 1976) appear to support this assertion.

Similarly, it has long been postulated (Orton, 1927; Travis, 1931) that stuttering results from a failure to develop adequate hemispheric asymmetry for language functions. While the results have been somewhat contradictory, this theory has received at least some support from experiments using dichotic listening (Curry & Gregory, 1969; Quinn, 1972), articulatory tracking (Sussman & MacNeilage, 1975), and CNV (Zimmerman & Knott, 1973,1974).

Several reports from this laboratory (Brown, Marsh, & Smith, 1973, 1976, 1979; Marsh & Brown, 1977) demonstrate that differences in the waveforms of the EEG event-related potentials (ERPs) produced by the noun and verb meanings of a homophone (e.g., "fire") are greater for responses from lefthemisphere placements than homologous right hemisphere placements in normal right-handed males. ERP waveforms to different meanings of the same homophone word were more different for left hemisphere than right hemisphere loci in 17 of 21, 12 of 15, and 17 of 22 subjects in three different experiments. Although this proportion (80%) is less than the incidence of left hemisphere dominance for speech reported for right-handers with the Wada technique (Milner, 1974), it is at least as good as other non-invasive measures, e.g., dichotic listening and tachistoscopic techniques. Thus, while the original intent of our earlier work was to study ERP correlates of meaning in language, it seemed to us that the paradigm might have some utility as an additional non-invasive measure of hemispheric asymmetry of language processing. Additionally, our ERP technique offers the advantage of allowing inter-hemispheric comparison for more than a

 DOI: 10.4324/9781003164647- 33

single hemispheric pair of recording loci, as well as intra-hemispheric comparisons.

The study reported here utilized asymmetry of noun/verb ERP correlation as a measure of the hemispheric asymmetry in language-processing. Language dominance was studied in three groups: (a) normal right-handed males, (b) normal right-handed females, and (c) male right-handed stutterers.

METHOD

Twenty-nine adult (18-30 years old), right-handed individuals were used in this research. Of these subjects, 10 were males and 10 were females with no history of speech dysfluency or neurological disorder. The other 9 were male stutterers who were at that time involved in therapy for stuttering and classified by a speech pathologist as moderate to severe stutterers. To assure a reasonably homogeneous group relative to etiology of stuttering, only subjects whose stuttering began before the age of six years took part in this study. All subjects were assessed for handedness on the basis of a questionnaire (Raczkowski, Kalat, & Nebes, 1974).

The stimulus paradigm utilized the visually presented homophone "fire" in the phrases "fire is hot" and "fire the gun." Our previous research (Brown et al., 1973, 1976, 1979; Marsh & Brown, 1977) using these and other stimulus phrases has demonstrated that (1) ERP waveform differences are real and not related to hemispheric differences in artifact or signal-to-noise ratios of background EEG; (2) the differences are observed only when the specific meaning of the homophone word is unambiguous when presented; (3) ERP waveform differences are the result of meaning-related changes in specific ERP components; and (4) left-lateralized ERP differences are found for auditory and visual stimulus presentation.

Subjects in the present experiment were prepared by affixing four silver-disc electrodes to the scalp with Grass electrolyte paste. Electrodes were placed approximately over Broca's and Wernicke's areas on the left and over homologous areas on the right (F_7 and F_8 and 3 cm posterior and dorsal to T_3 and T_4). All channels were recorded with reference to linked earlobes. The subject was seated in a comfortable chair within a recording chamber facing a 4" x 8" translucent screen at eye level, approximately 4 feet from the chair. Visual stimuli were back-projected using a slide projector outside the soundproof recording chamber.

Stimuli were 35 mm slides containing white block letters on a black background, projected through a red filter to reduce flash EPs. The word "fire" appeared on one slide and the remaining phrase, "is hot" or "the gun," appeared on the following slide. Stimuli were presented for 0.23 sec with 0.77 sec between the stimulus word and the rest of the phrase, and 1.77 sec between phrases. A trigger pulse for use in signal-averaging was generated as the stimulus word was presented. Phrases were presented in blocks of 60 successive repetitions of the same phrase. The subject was informed before each block which phrase would be presented, and instructed to read each phrase silently as it appeared and to try to form an image appropriate to the meaning of the phrase.

The EEG was amplified and recorded on tape along with the stimulus pulse. Amplifier pass-band was set at 1-50 Hz (3 dB down). All channels of the EEG were also written on an oscillograph and carefully edited off-line, response by response, so that records contaminated by muscle or eye-movement artifact could be excluded from averaging. Artifact free responses to the word "fire" were digitized (128 samples/sec) and averaged (N=55-60) over a 500-msec epoch from the onset of the trigger pulse. ERPs to the word "fire" were averaged separately for the noun and verb meanings at each of the four recording loci. The two waveforms at a particular locus were compared using a Pearson product-moment correlation of the 64 digital values in each waveform (Brown et al., 1973). This yields an index of ERP waveform similarity, with low correlations indicating dissimilar ERPs for the noun and verb conditions. The correlation coefficients for the four recording leads were Z-transformed prior to further analysis to eliminate skewness. For each subject a separate laterality index for ERP correlations, R/(L+R), was then derived for anterior and posterior leads. Since a low correlation coefficient is an index of greater noun/verb ERP waveform difference, relative right dominance is indicated by values less than 0.5, and relative left dominance by values greater than 0.5.

RESULTS

The findings for the male reference group replicate the results of studies previously reported from this laboratory (Brown et al., 1973, 1976, 1979; Marsh & Brown, 1977). ERP waveforms from one subject to the two meanings of the stimulus word are shown in pairs for each scalp derivation at the top of Figure 26.1. Correlations between the two waveforms are indicated. As the figure shows, correlations are lower for left hemisphere derivations than for right, indicating greater dissimilarity of ERP waveforms over the left hemisphere. This effect is greater for anterior electrode derivations. That this subject reflects the group mean correlations is apparent from inspection of Table 26.1. For group means the difference is apparent only for anterior derivations. The distributions of laterality indices for each subject are shown in Figure 26.2. Nine of 10 subjects had lateralization scores greater than 0.5 for the anterior leads, but only 6 of 10 for the posterior leads. For the anterior derivation, t tests indicated that the lateralization indices differed significantly from 0.5 (t=2.57, df=9, p<.05).

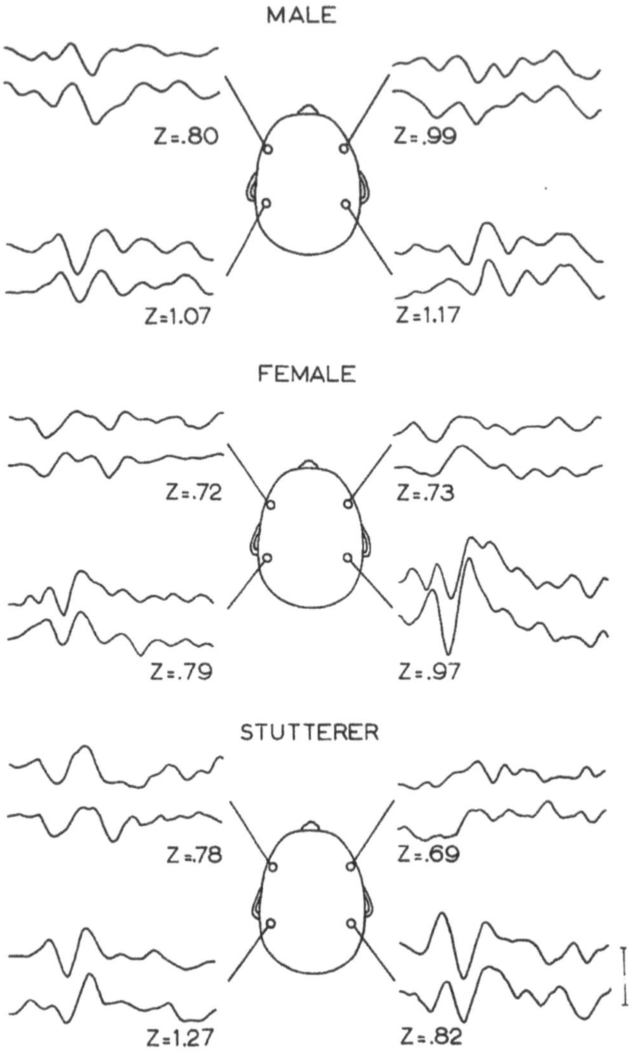

FIGURE 26.1. Average potentials evoked by the word "fire" recorded from four electrode loci in three individual subjects: a normal male, a normal female, and a male stutterer. The upper trace of each pair is related to the noun meaning of "fire" and the bottom trace is related to the verb meaning. Z-transformed correlations between the 64 digital values representing the noun and verb waveforms appear below each waveform pair. Vertical scale is 4 μV, positive up. Horizontal scale is 100 msec.

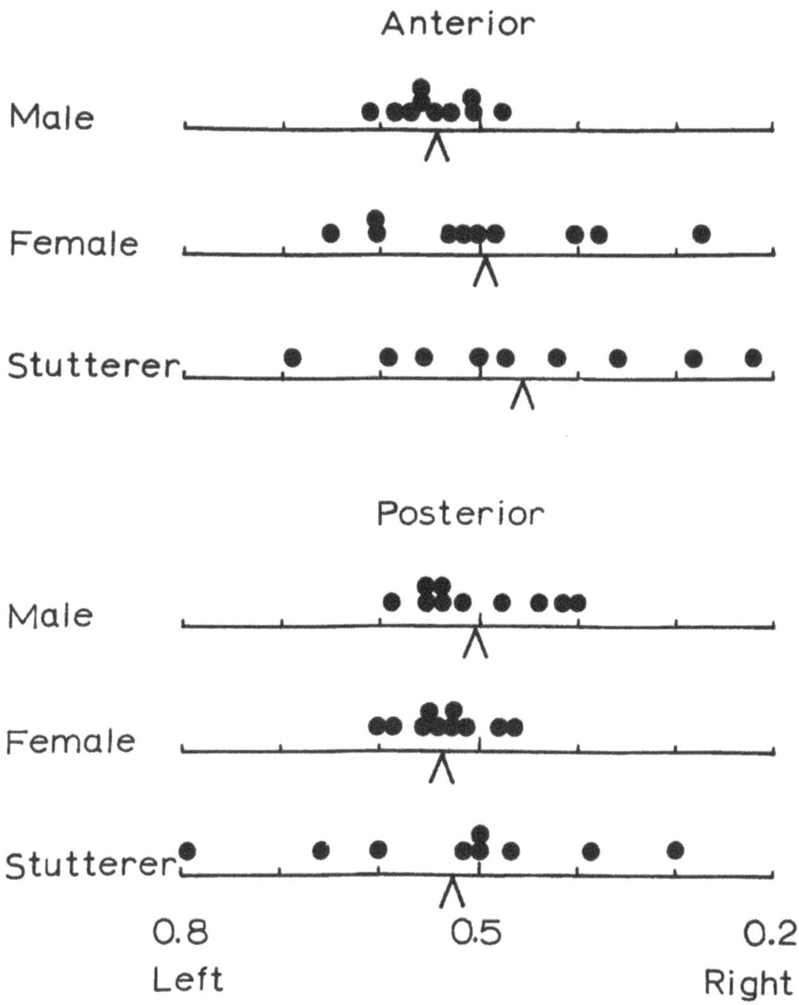

FIGURE 26.2. Laterality scores (R/(L+R)) for normal males and females and male stutterers computed from the noun-verb ERP waveform correlations. Values greater than 0.5 indicate left laterality of greatest ERP waveform difference. Distribution means are indicated by the arrow below each graph. Upper graphs are from the anterior homologous electrode pair, and the lower, from the posterior pair.

While females also show relative left lateralization, the pattern is different such that the greatest margin of lateralization is at the posterior leads (illustrated for one subject in the middle of Figure 26.1). This is confirmed by the mean correlations from Table 26.1. Eight of 10 female subjects had laterality scores greater than 0.5 at the posterior leads, and 6 of 10 for the anterior electrode pair. Laterality scores for the posterior leads were significantly different from 0.5 (t=2.50, df=9, p<.05).

The most striking difference between the males and females was the greater variability of laterality scores for the anterior lead in the female group (Figure 26.2). Range tests (F', or substitute F-ratio, Beyer, 1966) indicated significant differences between the anterior and posterior laterality scores for the females (F'=.250, p<.005). The range for the anterior female scores was also significantly greater than the range of male anterior scores (F'=.355, p<.005). In contrast, for posterior derivations the range of scores for females was significantly less than that for males (F'=.496, p<.05).

With respect to the stutterer group, the mean correlation value and mean laterality indices would suggest right laterality for the anterior derivation and left laterality for posterior. However, neither anterior nor posterior mean laterality scores differ significantly from 0.5.

The lack of significant laterality scores in the stutterer group reflects what is, in fact, the most striking aspect of these results, namely the extreme variability of laterality scores in this group. As can be seen from Figure 26.2, both anterior and posterior laterality scores in the stutterer group ranged from extreme left to extreme right laterality.

Differences between the stutterers and the male reference group in the range of laterality scores were significant for both the anterior (F'=.298, p<.005) and posterior (F'=.271, p<.005) leads. Differences in range between the female and stutterer groups were significant for the posterior (F'=.191, p<.005), but not for anterior leads.

In summary, the reference male group shows left lateralization only at the anterior leads, and little variance in either anterior or posterior laterality scores. Females show a significant left laterality and small variance for posterior derivations, but a wide range of laterality scores for anterior leads. Stutterers show an inconsistent pattern and a wide range of laterality scores at both anterior and posterior derivations.

The wide range of laterality scores within the stutterer group and to a lesser extent in the female group raises the question of the within-subject consistency of anterior and posterior laterality measures. As expected, the male reference group shows a high level of consistency (anterior/posterior rank order correlation, rho=.73, p<.05). Female subjects were less consistent (rho=.46) and stutterers are totally inconsistent (rho=.18).

TABLE 26.1. Means and Standard Deviations of Noun/Verb Correlations (Z).

		Anterior		Posterior	
		Left	Right	Left	Right
Males	X	.867	1.006*	1.045	1.042
	S.D.	.481	.512	.276	.230
Females	X	1.025	1.032	1.057	1.168
	S.D.	.394	.422	.220	.246
Stutterers	X	.887	.801	.942	1.029
	S.D.	.362	.442	.435	.468

* Difference between right and left hemisphere correlations significant at $p<.01$ (2-tail probability).

DISCUSSION

Any discussion of these results must necessarily be preceded by some answer to the question of whether the findings could simply be a function of consistently lower signal-to-noise ratios in the data obtained from stutterers. This could be the case if, for example, they were as a group more tense and anxious in the recording situation than control subjects. Lower signal-to-noise ratios could be expected to produce larger differences in average ERP waveforms and hence lower correlations. As is clear from mean correlation values in Table 26.1, the correlations for stutterers are in the same range as those for both the non-stutterer groups. A second and more convincing test of the question is the calculation of the discrimination index (Brown et al., 1973, 1976; Buchsbaum & Fedio, 1970). Briefly, this index involves a split-half reliability test to determine whether condition-related ERP differences exceed differences between replications of the same stimulus condition. This analysis showed that, in fact, for all four electrode derivations in the stutterer group the discrimination index is positive, i.e., condition-related (noun/verb) ERP differences exceed differences obtained from replication of the same condition. Further, these positive discrimination indices fall within the same range as those for both the female and male non-stutterer groups. Thus, the effects predicted on the basis of an assumption of lower signal-to-noise ratio for stutterers are not found in these data. Rather, the positive discrimination index values suggest that the laterality scores are based on correlations which reflect consistent ERP differences related to the processing of linguistic meaning.

Taken together these results suggest that the three groups are distinguished not so much by the magnitude of laterality scores but by their consistency within and between subjects. Thus, while the most extreme lateralization scores, both left and right, occur in the stutterer group, they are the most inconsistent, i.e., show the widest range between subjects and the lowest anterior/posterior correlation. The male reference group, in contrast, while showing only moderate degrees of lateralization, is characterized by little variation of laterality scores between subjects and highly consistent lateralization between anterior and posterior areas.

The difference in range of laterality scores between stutterers and the male reference group observed here is consistent with the electrophysiological data of Zimmerman and Knott (1974) which showed stutterers to differ from controls in having "variable inter-hemispheric relationships." Quinn (1972), using dichotic listening, also found stutterers to be more variable, but not less lateralized, in ear preference.

Females occupy an intermediate position between these groups, with tightly clustered and left lateralized scores only at posterior leads and some degree of dissociation of anterior and posterior laterality scores. This finding is congruent with the results reported by Tucker (1976) who found that during the performance of a vocabulary task females showed greater EEG desynchrony at posterior leads than did males. Tucker suggests that "while males showed greater specialization of the brain laterally, females seemed to show more differential usage in the rostral-caudal dimension" (p. 453). We have reported data from another study (Brown, Lehmann, & Marsh, 1980) which further support the notion of "more differential usage in the rostral-caudal direction in females." A study of the scalp-topography of meaning-related ERP differences in three groups of female subjects showed the largest noun/verb effects in the anterior-posterior plane. Similar studies of primarily male subjects (Brown et al., 1979) showed the greatest noun/verb effects in the right-left dimension.

Support for the concept of greater variability in the hemispheric asymmetry of females has come from the work of McGlone (1977, 1980). Based on a neuropsychological study of language functions in males and females with right or left hemispheric lesions, McGlone (1977) concludes that males, as a group, are less variable than females with respect to the cerebral organization of speech. Both the anterior laterality scores and the anterior/posterior correlations in the present data would seem to support this conclusion.

The relationship between stuttering and sex may be reflected in the anterior lateralization data reported here. Males are at far greater risk for both stuttering and dyslexia than females (Johnson & Myklebust, 1965; Milisen, 1971), yet females appear from our data to manifest frontal ERP patterns more like the stutterers than the non-stutterers. Thus, it could be hypothesized that the development of normal speech in right-handed females is not dependent upon left dominance for language in frontal brain areas; whereas a departure from a pattern of consistent left lateralization in the frontal areas of right-handed males is associated with abnormal speech development. Such a notion is congruent with McGlone's data (1980) indicating a lower incidence of aphasia from left brain damage in females than in males.

Three aspects of the data presented here should be emphasized. First, they suggest that the interrelationship between hemispheric dominance for language functions, sex, and stuttering is not a simple one which lends itself to one-dimensional conceptualization. At least, the anterior/posterior dimension of cortical processing should be taken into account in addition to the more commonly studied left-right dimension. For example, these findings do not support a notion that stuttering is associated with an insufficient margin of left asymmetry for language. Rather, it suggests that in males a lack of consistency in anterior and posterior asymmetries may be a more important factor. Second, we view the demonstrated relationships between the ERP index of laterality, sex, and stuttering to indicate simply an association among these variables which is not necessarily a causal relationship. For example, the fact that laterality scores of a few of our stutterers were similar to those of the non-stuttering males suggests that some third factor, perhaps early minimal brain damage, operated to produce stuttering which often, but not invariably, produces deviant laterality scores. Finally, these data emphasize the usefulness of EEG and evoked-potential methods in experimental neuropsychology. Dichotic listening, tachistoscopic viewing tasks, administration of sodium amytal, and tests of right- and left-hand dexterity or reaction time all yield results which treat the cerebral hemispheres as merely dichotomous neural masses. It is possible through recording EEG or evoked-potentials from four or more bilaterally symmetric electrode sites to measure brain processing in two spatial dimensions, rather than a single left-right dichotomy.

REFERENCES

Beyer, W. H. (Ed.). (1966). *CRC handbook of probability and statistics* (pp. 293-295). Cleveland: Chemical Rubber Co.

Brown, W. S., Lehmann, D., & Marsh, J. T. (1980). Linguistic meaning-related differences in evoked potential topography: English, Swiss-German and imagined. *Brain and Language, 11*, 340-353.

Brown, W. S., Marsh, J. T., & Smith, J. C. (1973). Contextual meaning effects on speech-evoked potentials. *Behavioral Biology, 9*, 755-761.
Brown, W. S., Marsh, J. T., & Smith, J. C. (1976). Evoked potential waveform differences produced by the perception of different meanings of an ambiguous phrase. *Electroencephalography and Clinical Neurophysiology, 41*, 113-123.
Brown, W. S., Marsh, J. T., & Smith, J. C. (1979). Principal component analysis of ERP differences related to the meaning of an ambiguous word. *Electroencephalography and Clinical Neurophysiology, 46*, 709-714.
Buchsbaum, M., & Fedio, P. (1970). Hemispheric differences in evoked potentials to verbal and nonverbal stimuli in the left and right visual fields. *Physiology and Behavior, 5*, 207-210.
Curry, F. K. W., & Gregory, H. R. (1969). The performance of stutterers on dichotic listening tasks thought to reflect cerebral dominance. *Journal of Speech and Hearing Research, 12*, 73-82.
Harshman, R., Remington, R., & Krashen, S. (1976). Sex, language, and the brain: Adult sex differences in lateralization. In D. O. Walter, L. Rogers, & J. M. Sinzi-Freed (Eds.), *Conference on human brain function*. Los Angeles: Brain Information Service/BRI Publications Office.
Johnson, D. J., & Myklebust, H. R. (1965). Dyslexia in childhood. In J. Hellmuth (Ed.), *Learning Disorders* (Vol. 1, pp. 259-292). Seattle: Bernie Staub and Jerome Hellmuth.
Lake, D. A., & Bryden, M. P. (1976). Handedness and sex differences in hemispheric asymmetry. *Brain and Language, 3*, 266-282.
Levy, J., & Reid, M. (1976). Variations in writing posture and cerebral organization. *Science, 194*, 337-339.
Marsh, J. T., & Brown, W. S. (1977). Evoked potential correlates of meaning in the perception of language. In J. Desmedt (Ed.), *Language and hemispheric specialization in man: Cerebral event related potential* (pp. 60-72). Basel: Karger.
McGlone, J. (1977). Sex differences in the cerebral organization of verbal functions in patients with unilateral brain lesions. *Brain, 100*, 775-793.
McGlone, J. (1980). Sex differences in human brain asymmetry: A critical survey. *The Behavioral and Brain Sciences, 3*, 215-263.
Milisen, R. (1971). The incidence of speech disorders. In L. E. Travis (Ed.), *Handbook of speech pathology and audiology* (pp. 619-633). New Jersey: Prentice-Hall.
Milner, B. (1974). Hemispheric specialization: Scope and limits. In F. O. Schmitt & F. G. Worden (Eds.), *The neurosciences third study program* (pp. 75-89). Cambridge: MIT Press.
Orton, S. T. (1927). Studies in stuttering. *Archives of Neurology and Psychiatry, 18*, 671-672.

Quinn, P. (1972). Stuttering, cerebral dominance and the Dichotic Word Test. *Medical Journal of Australia, 2*, 639-643.
Raczkowski, D., Kalat, J. W., & Nebes, R. (1974). Reliability and validity of some handedness questionnaire items. *Neuropsychologia, 12*, 43-47.
Ray, W. J., Morell, M., Frediani, A. W., & Tucker, D. (1976). Sex differences and lateral specialization of hemispheric functioning. *Neuropsychologia, 14*, 391-394.
Sussman, H., & MacNeilage, P. (1975). Hemispheric specialization for speech production and perception in stutterers. *Neuropsychologia, 13*, 19-26.
Travis, L. E. (1931). *Speech pathology*. New York: Appleton Century.
Tucker, D. M. (1976). Sex differences in hemispheric specialization for synthetic visuospatial functions. *Neuropsychologia, 14*, 447-454.
Zimmerman, G. N., & Knott, J. R. (1973). Slow potentials preceding speech in stutterers and normal speakers. *Electroencephalography and Clinical Neurophysiology, 36*, 216.
Zimmerman, G. N., & Knott, J. R. (1974). Slow potentials of the brain related to speech processing in normal speakers and stutterers. *Electroencephalography and Clinical Neurophysiology, 27*, 599-607.

27 · EVOKED POTENTIALS RELATED TO DECISION CONFIDENCE

J. W. Rohrbaugh, L. M. Chalupa, and D. B. Lindsley

National Institute for Alcohol Abuse and Alcoholism
University of California at Davis
and
University of California at Los Angeles

A familiar feature of psychophysical judgments is their remarkable trial-by-trial variability, in both actual and apperceived performance. We have sought underlying neural correlates of this variability by recording visual event-related potentials (ERPs) and selectively averaging them on the basis of performance and subjective variables. We used a visual discrimination task which entailed the comparison of two line stimuli. The task was structured in such a way as to allow observation of a variety of EEG phenomena preceding and following the discriminanda. An indicator of predisposing EEG conditions was provided by the contingent negative variation (CNV). The CNV was first described by Walter, Cooper, Aldridge, McCallum, and Winter (1964) and often is held to index preparatory activities. Post-stimulus analyses are focused on the primary ERP components which can, in some cases, bear on the integrity of sensory input and on the endogenous P300 wave, which reflects stimulus evaluation processes (Donchin, Ritter, & McCallum, 1978).

METHOD

ERP and judgment data were obtained from eight male volunteers with normal or corrected to normal vision. The data from two of these subjects were discarded because of insufficient data for some of the error and low-confidence response categories.

The subjects were seated comfortably in an electrically shielded and sound-attenuated experimental room, and were instructed to direct their gaze towards a computer-controlled CRT. Aside from a small luminous fixation dot in the center of the CRT and a small amount of illumination over the remainder of the CRT screen, the room was completely darkened. Extraneous sounds were masked by continuous white noise delivered over a speaker communication system.

Each trial contained three stimuli, separated by interstimulus intervals (ISIs) of 2 sec. The first stimulus was a brief tone pip that served as a warning stimulus. The 2K Hz tone was rectangularly gated for a duration of

DOI: 10.4324/9781003164647- 34

30 msec and delivered over binaural insert earphones at an approximate intensity of 90 dB SPL. After the first ISI of 2 sec, a short horizontal line was flashed for 25 msec, centered on the fixation point. At a viewing distance of 75 cm this line subtended 0.13^{o} angle. Following a second ISI of 2 sec, a horizontal line was again presented in the same location for 25 msec. This second line served as the comparison stimulus and on 50% of the trials was the same length as the previously exposed standard line. On the remaining 50% of the trials the comparison line was longer than the standard line. Task difficulty was manipulated by varying the length of the comparison line in separate sessions: 20% longer in the Hard condition, 25% longer in the Medium condition, and 30% longer in the Easy condition. Subjects judged whether the two lines were the same or different in length, and indicated their confidence in their judgments on a 3-point confidence scale ranging from 1 (certain) to 3 (guessing).

Subjects participated in two sessions, the first of which was a practice session designed to stabilize their use of confidence ratings. Subjects were exposed to all experimental conditions and were encouraged to adopt a standard confidence framework, to which they could adhere throughout the Hard, Medium and Easy difficulty conditions. During the second session, subjects received, in balanced order, six blocks of 40 trials each, two blocks under each difficulty level. (Subsequent analyses combined all data under each difficulty level.) Within each block of trials, the same and different length comparison lines were presented 20 times each in random order. Trials were separated from one another by an average inter-trial interval of 8.5 sec (range 7 to 10 sec). Subjects were cautioned to maintain their fixation on the fixation dot and to withhold any blinks or eye movements (as well as the verbal response) until well into the inter-trial interval. Data acquisition was preceded by a 10-min period of dark adaptation.

EEG was recorded from central (C_z), parietal (P_3), and occipital (O_1) sites, referred to linked earlobes, with Grass 6A1B DC amplifiers, modified to yield a 12-sec time-constant. After cleansing the scalp site with alcohol, Beckman miniature silver/silver chloride electrodes were affixed with Bentonite paste. The electrooculogram (EOG) was recorded similarly from a pair of electrodes about the orbital ridge. EEG and EOG data were stored on FM tape for off-line computer analysis. All trials that showed significant EOG activity were excluded from analyses.

RESULTS

Shown in Figure 27.1 are grand-averaged waveforms from the P_3 electrode, under the various experimental conditions. The records were digitized at a rate of 27.3 msec per point. The trends described here were apparent in individual subjects at all electrode sites. The waveforms showed

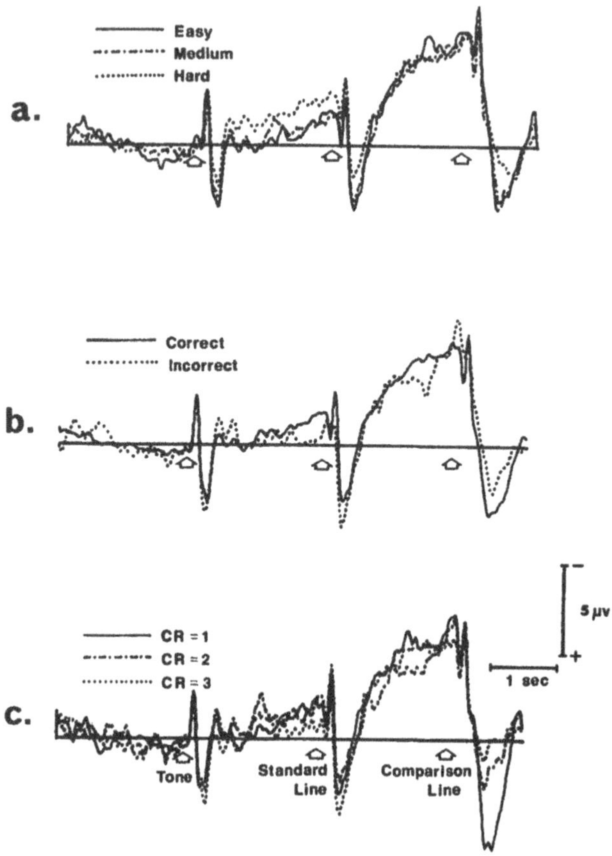

FIGURE 27.1. Shown in (a) are visual evoked potentials associated with all trials in the Easy, Medium and Hard difficulty conditions. In (b) the averages are computed separately for trials having correct and incorrect judgments, and are averaged over the three difficulty conditions. In (c) the potentials for the Medium difficulty condition have been computed on the basis of the confidence rating associated with the judgment, ranging from most confident (CR=1) to least confident (CR=3). Confidence-related variations in P300 following the comparison line, as shown in (c), were statistically significant. The waveforms in all cases are averaged over 6 subjects, and are from the P_3 electrode.

clear evoked potentials to the warning and line stimuli and also showed CNVs during the intervals separating stimuli. The maximum CNV was attained immediately before the comparison line. The comparison line is accompanied also by a large positive P300 component, which was appreciably larger than that elicited by the standard line.

In Figure 27.1a waveforms are presented for the Easy, Medium and Hard difficulty conditions. Judgment performance varied considerably over the three difficulty levels, averaging 95% correct in the Easy condition, 88% in the Medium condition, and 65% in the Hard condition. These differences in performance were significant ($F(2,10)=56.7, p<.001$), as determined in a repeated-measures Analysis of Variance (ANOVA) including as factors both the difficulty level and judgment confidence (confidence data are considered below).

As is apparent in Figure 27.1a, however, the CNV does not show any changes over the difficulty levels. Analyses of the CNV were based on the averaged amplitude values during the 500-msec period immediately preceding the onset of the comparison line. A three-way ANOVA considering the factors of Difficulty, Confidence and Electrode site found that no significant CNV effects were associated with the factors of Difficulty ($F(2,10)=0.2$) or the interaction of Difficulty with Electrode site ($F(4,20)=0.3$).

A separate three-way ANOVA was performed on the P300 component following the comparison line, measured as the averaged amplitude during the 350-650 msec post-stimulus relative to the average measure during the 500 msec immediately preceding the comparison line. P300 tended to become progressively smaller in amplitude as difficulty increased; however, this apparent trend did not attain statistical significance ($F(2,10)=0.5$).

In Figure 27.1b the waveforms (averaged over all difficulty levels) have been computed separately for correct and incorrect judgments. There are no effects in the CNV associated with correctness ($F(1,5)=0.1$), as determined in a separate ANOVA based on the factors of Correctness, Difficulty, and Electrode site. The P300 measure averaged 21% smaller for incorrect judgments, although this apparent effect was not statistically significant ($F(1,5)=4.8, p=.08$).

Figure 27.1c displays averages (from the Medium difficulty condition) that are computed on the basis of confidence rating associated with the judgment. Performance data indicated that subjects used the ratings in a consistent and meaningful manner. The mean confidence rating progressed from 1.6 in the Easy condition to 2.0 in the Medium difficulty condition, and to 2.2 in the Hard condition. Accuracy varied commensurately as a function of confidence ratings ($F(2,10)=14.9, p<.005$), dropping from 89% for the highest confidence rating of 1, to 82% for a rating of 2, and 71% for a rating of 3. While there was a trend toward a general easing in confidence standard during the Medium and Hard conditions (as manifest in a slightly increased error rate for the respective confidence categories), the trend was not statistically significant (interaction of confidence and difficulty: ($F(4,20)=1.9$).

Again, no significant effects were present for the CNV, although there was a trend over all electrode sites and difficulty levels for the CNV to be largest preceding low confidence ratings ($F(2,10)=3.3, p=.08$). However, large and systematic effects were associated with confidence in the P300 measure. This effect was especially prominent in the records from the O_1 electrode, where the P300 amplitude decreased to 67% as confidence decreased from 1 to 2, and the amplitude further decreased to 46% at a confidence rating of 3. The ANOVA showed these effects of confidence on P300 to be highly significant ($F(2,10)=19.6$, $p<.001$), and to interact significantly with electrode site ($F(4,20)=4.2$, $p<.05$). Confidence did not interact with difficulty level ($F(4,20)=1.5$). The magnitude of the changes in P300 related to confidence suggests that the confidence factor encompasses and may account for the smaller P300 variations associated with difficulty and correctness, as shown in Figure 27.1a and b.

Finally, it should be noted that a detailed analysis of the primary components following the comparison line (as identified in averages based on shorter epochs encompassing only the response to the comparison line) permitted the measurement for each subject of a positive wave peaking at about 75 msec, and a subsequent negative wave peaking at about 130 msec. ANOVAs based on respective base-to-peak and peak-to-peak measures for the components revealed no systematic or significant effects associated with any experimental variable.

DISCUSSION

The sensitivity shown by P300 to judgment confidence is in striking contrast to the CNV, which shows no significant changes related to difficulty, correctness, or confidence. In general, the CNV literature shows an equally discouraging pattern of results, with some studies finding essentially negative results (e.g., Jarvilehto & Fruhstorfer, 1970; Warren, 1974), others finding a positive correlation between CNV amplitude and performance (e.g., Hillyard, Squires, Bauer, & Lindsay, 1971; Wilkinson & Haines, 1970), and still others finding a negative correlation (Delse, Marsh, & Thompson, 1972). This inconsistency has led to proposals that additional factors are involved in CNV production (McCarthy & Donchin, 1978) or to formulations (e.g., Rohrbaugh, Syndulko, & Lindsley, 1976) emphasizing motor-related elements (although the motorial consequences of visual discrimination as in the present case would be very difficult indeed to specify).

P300, on the other hand, is shown clearly to vary with judgment performance, most notably with spontaneous fluctuations in confidence. These findings are in harmony with other data relating P300 to judgment behavior, much of which have been obtained from near-threshold detection paradigms in which criterion has been shown to be an important variable (e.g., Donchin, 1968; Paul & Sutton, 1972; Squires, Hillyard, & Lindsay, 1973; Squires, Squires, & Hillyard, 1975a, 1975b). Our data combine with accounts for these and other investigators (Kutas, McCarthy, & Donchin, 1977;

Rohrbaugh, Donchin, & Eriksen, 1974; Ruchkin & Sutton, 1976) in stressing the importance that must be attached to the idiosyncratic roles whereby subjects receive and appreciate information. P300 seems clearly to reflect the quality of the associated evaluation or decision processes--particularly the confidence with which such decisions are made.

REFERENCES

Delse, F. C., Marsh, G. R., & Thompson, L. W. (1972). CNV correlates of task difficulty and accuracy of pitch discrimination. *Psychophysiology, 9,* 53-62.

Donchin, E. (1968). Average evoked potentials and uncertainty resolution. *Psychonomic Science, 12,* 103.

Donchin, E., Ritter, W., & McCallum, W. C. (1978). Cognitive psychophysiology: The endogenous components of the ERP. In E. Callaway & S. H. Koslow (Eds.), *Event related brain potentials in man* (pp. 349-411). New York: Academic Press.

Hillyard, S. A., Squires, K. C., Bauer, J. W., & Lindsay, P. H. (1971). Evoked potential correlates of auditory signal detection. *Science, 172,* 1357-1360.

Jarvilehto, T., & Fruhstorfer, H. (1970). Differentiation between slow cortical potentials associated with motor and mental acts in man. *Experimental Brain Research, 11,* 309-317.

Kutas, M., McCarthy, G., & Donchin, E. (1977). Augmenting mental chronometry: The P300 as an index of stimulus evaluation time. *Science, 197,* 792-795.

McCarthy, G., & Donchin, E. (1978). Brain potentials associated with structural and functional visual matching. *Neuropsychologia, 16,* 571-585.

Paul, D. D., & Sutton, S. (1972). Evoked potential correlates of response criteria in auditory signal detection. *Science, 177,* 362-364.

Rohrbaugh, J. W., Donchin, E., & Eriksen, C. W. (1974). Decision making and the P300 component of the cortical evoked response. *Perception and Psychophysics, 15,* 368-374.

Rohrbaugh, J. W., Syndulko, K., & Lindsley, D. B. (1976). Brain wave components of the contingent variation in humans. *Science, 191,* 1055-1057.

Ruchkin, D. S., & Sutton, S. (1976). *Equivocation and P300 amplitude.* Paper presented before the 4th International Congress on Event Related Slow Potentials of the Brain, Hendersonville, NC.

Squires, K. C., Hillyard, S. A., & Lindsay, P. H. (1973). Vertex potentials evoked during auditory signal detection: Relation to decision criteria. *Perception and Psychophysics, 10,* 445-452.

Squires, K. C., Squires, N. K., & Hillyard, S. A. (1975a). Vertex evoked potentials in a rating-scale detection task: Relation to signal probability. *Behavioral Biology, 13*, 31-34.

Squires, K. C., Squires, N. K., & Hillyard, S. A. (1975b). Decision-related cortical potentials during an auditory signal detection task with cued observation intervals. *Journal of Experimental Psychology: Human Perception and Performance, 1*, 268-279.

Walter, W. G., Cooper, R., Aldridge, V. J., McCallum, W. C., & Winter, A. L. (1964). Contingent negative variation: An electric sign of sensori-motor association and expectancy in the human brain. *Nature, 203*, 380-384.

Warren, C. (1974). *The contingent negative variation and late evoked potential as a function of task difficulty and short-term memory load*. Unpublished doctoral dissertation, University of Illinois.

Wilkinson, R. T., & Haines, E. (1970). Evoked response correlates of expectancy during vigilance. *Acta Psychologica, 33*, 402-413.

28· AN INTERPRETATION OF HUMAN CLASSICAL CONDITIONING OF ELECTRODERMAL ACTIVITY

I. Maltzman and M. Pendery

University of California at Los Angeles

Garcia and others (e.g., Garcia, McGowan, & Green, 1972) have demonstrated that the laws of classical conditioning differ depending upon the nature of the CS and UCS. There is a biological "belongingness" (Thorndike, 1931) such that illness aversions are more readily conditioned to interoceptive than exteroceptive stimuli. Temporal contiguity between CS and UCS is of much greater importance when exteroceptive stimuli are paired than when interoceptive stimuli are paired. Generality of the laws of classical conditioning are constrained in that they do not apply in the same manner to all stimuli and to all CS-UCS combinations.

A thesis of this paper is that there is another constraint on the laws of classical conditioning. They do not apply in the same manner to all response systems. They do not apply in the same manner to all of the measures commonly employed as the dependent variable in classical conditioning studies. Phasic electrodermal activity, the GSR, has been the most commonly employed measure of classical conditioning in humans in the past decade, clearly replacing eyeblink conditioning in popularity. But phasic electrodermal activity is also the most commonly employed measure of the orienting reflex (OR), suggesting that studies of classical conditioning using phasic electrodermal activity as the dependent measure are actually studying the OR. Laboratory experiments on classical conditioning are studying aspects of attention as reflected by variations in the GSR index of the OR (Maltzman, 1979a, 1979b).

Initial conditions for the occurrence of an OR are generally considered to belong to two general classes of events: (1) novelty or stimulus change (Berlyne, 1960; Sokolov, 1960, 1963), and (2) significance of the stimulus (Luria & Vinogradova, 1959; Maltzman, 1979a, 1979b; Razran, 1961; Sokolov, 1963). Significance of a stimulus may be (a) biologically determined (Razran, 1961), (b) imparted by the experimenter's instructions (Luria & Vinogradova, 1959; Maltzman & Raskin, 1965, 1979; Sokolov, 1963), (c) induced as the result of past learning, or (d) discovered during the course of a subject's problem-solving activity. Such problem-solving activity characteristically occurs in laboratory studies of classical conditioning with college students serving as subjects (Pendery & Maltzman, 1977).

 DOI: 10.4324/9781003164647- 35

Sokolov (1963, p. 56) has asserted that the GSR probably best reflects the laws of the OR. If this is the case, the GSR should also express principles of the OR when it is taken as a CR in a classical conditioning experiment. Its presence as the dependent variable in a conditioning paradigm should not change its nature. The GSR-CR, the electrodermal change meeting the experimental criteria of a CR, should also be an expression of an OR. But this can only be the case if the OR is induced by the significance or signal value of stimuli as well as the more commonly considered stimulus change. It has been known for many years that the GSR and other expressions of an OR can be induced, heightened, or its habituation retarded, by providing stimuli with signal value (Latash, 1968; Luria, 1973; Luria & Homskaya, 1970; Luria & Vinogradova, 1959; Maltzman & Raskin, 1965; Sokolov, 1960, 1963). It is apparent that the same process occurs in classical conditioning (Pendery & Maltzman, 1977). According to our interpretation there is only one important procedural difference between the usual habituation study of the signal value of stimuli and classical conditioning experiments as ordinarily conducted. In the conditioning experiment the participant discovers the signal value of stimuli whereas in the habituation experiment the signal value is imparted by the experimenter's instructions. "Verbal conditioning" of the OR is the process mediating occurrence of the heightened electrodermal activity in both cases.

The purpose of this paper is to consider the implications of habituation studies of the GSR to signal and nonsignal stimuli for the interpretation of the results from classical conditioning experiments.

Significance or signal value as imparted by instructions can readily be demonstrated as a powerful variable influencing the magnitude of the GSR. A brief description of such an experimental demonstration follows. Further details may be found elsewhere (Maltzman, 1971; Maltzman & Raskin, 1965, 1979).

EXPERIMENT I

Subjects. A total of 140 college students served in four different groups.

Procedure. Subjects in an Associate group were instructed to implicitly free associate to the word "LIGHT" each time that they heard it; a Pedal Press group was instructed to press a foot pedal every time they heard "LIGHT"; a Count group was asked to implicitly count the number of times that they heard "LIGHT"; and a Control group was asked to simply sit quietly and listen to the words.

Twenty different neutral, unrelated, prerecorded words were presented via stereophonic headphones followed without interruption by 14 presentations of "LIGHT" interspersed among neutral filler words. Palmar GSRs were recorded throughout the experiment. They were scored in the interval 0.5 - 5.5 sec following each presentation of the first 20 words, the critical word "LIGHT," and the immediately preceding (C_1) and following (C_2) control words.

RESULTS AND DISCUSSION

Figure 28.1 presents the results for each of the three experimental groups and the control group. Mean magnitude of the GSR to LIGHT (L) is depicted along with the mean magnitude of the GSR to the preceding C_1, and following C_2 control words on each of the 14 presentations of LIGHT.

It is apparent that assignment of signal value to a stimulus by the experimenter may result in a marked selective increase in the GSR to that stimulus. We hypothesize that the selective increase is a reflection of the evocation of what we call a voluntary OR (Maltzman, 1979b; Pendery & Maltzman, 1977). It is a voluntary OR because it presumably is a

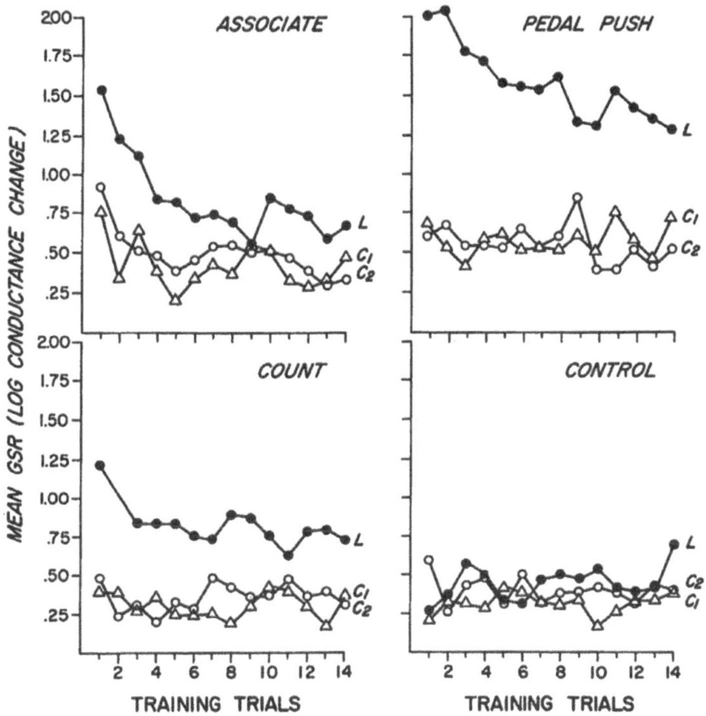

FIGURE 28.1. Mean magnitude of the GSR to the critical word LIGHT (L) and the immediately preceding (C_1) and following (C_2) control words for the control and experiment groups (Maltzman, 1971).

consequence of verbal conditioning. It cannot be a consequence of simple stimulus change or novelty since LIGHT was no different than any other word in this regard. Stimulus change or novelty may induce an involuntary OR which may be considered an unconditioned response to the stimulus change in question. Involuntary ORs also may be influenced by instructions and the state of the organism at the moment (Maltzman, 1979a; Pendery & Maltzman, 1977). But elicitation of an involuntary OR is less a consequence of verbal regulation than is the voluntary OR. Peripherally, in terms of the GSR, voluntary and involuntary ORs do not seem to be differentiable. There is evidence, however, that the two differ in morphology (Luria, 1973) as well as functional characteristics (Maltzman, 1979b; Maltzman, Gould, Barnett, Raskin, & Wolff, 1979).

As indicated in Figure 28.1 the signal value of a stimulus may have a marked effect on the magnitude of the GSR. Our hypothesis is that the usual laboratory experiment on classical conditioning of the GSR also involves evocation of an OR by the signal value imparted to the CS. There is one important difference between the usual study of habituation and the classical conditioning experiment. In the conditioning experiment the participant discovers the signal value of the stimulus whereas in the habituation experiment the signal value is imparted by the experimenter's instructions.

Results reported by Pendery and Maltzman (1977) from an experiment on verbal conditioning of the GSR support the preceding conceptions of stimulus significance, the voluntary OR, and its role in classical conditioning. A brief description of the experiment follows (further details may be found in Pendery & Maltzman, 1977).

EXPERIMENT II

Subjects. A total of 105 college students served in two principal groups.

Procedure. A habituation series of 20 words prerecorded on magnetic tape was first presented via stereophonic headphones. They were followed without interruption by six conditioning trials consisting of the CS word "PLANT" followed in 10 sec by an innocuous 0.5 sec 1000 Hz 70 dB tone. Conditioning trials were interspersed among different innocuous unrelated words. Previous experiments had demonstrated that the innocuous tone does not evoke a differential GSR and that conditioning does not occur with such a tone serving as a UCS (Maltzman et al., 1979).

Classical conditioning can be obtained with an innocuous stimulus of the above sort by transforming the situation into a forewarned reaction time paradigm. A standard classical conditioning group, here called a Discovery group, was instructed to respond to the tone as quickly as possible by lifting their foot from a pedal. Under such circumstances, a large GSR will occur to the tone-pedal-lift complex, and reliable differential conditioning will follow (Maltzman, Gould, Barnett, Raskin, & Wolff, 1977, 1979). An instructed group was verbally conditioned and was informed of the CS word (PLANT) that would always precede the tone to which they were to respond. Two-thirds of the Instructed students received the tone after every word in the experiment in order to ensure the credibility of the instructions, that in fact the tone was innocuous. Thus, both principal groups received a critical word PLANT followed by an innocuous tone where this word-tone pair was interspersed among other unrelated words. Both groups were instructed to overtly respond to the tone as quickly as possible upon hearing it. However, the Instructed group was told that the critical tone was always preceded by the word PLANT whereas the Discovery group was not informed of this CS-UCS contingency. As in the usual classical conditioning experiment, these latter subjects must learn the contingency in order to manifest a differential conditioned response, a differentially larger GSR to the critical word PLANT as compared to preceding and following control words.

Palmar GSRs were recorded continuously throughout the experiment. At the conclusion of the experiment students in the Discovery group were presented with a recognition test designed to determine whether or not they could state which word was followed by the tone. If they could report the CS-UCS contingency, they were further queried concerning their recall of the trial on which they first discovered the contingency.

RESULTS AND DISCUSSION

Analysis of the magnitude of the GSR evoked in the interval 0.5 - 5.5 sec following the CS word and the control filler words immediately preceding and following each trial indicated that reliable verbal and classical conditioning occurred (Pendery & Maltzman, 1977). Figure 28.2 depicts the differential responsivity of the Instructed group and those students in the Discovery group who verbalized the CS-UCS contingency.

It is apparent that the maximum GSR for the Instructed students occurred on the first trial, prior to the presentation of the UCS or imperative stimulus. Differential verbal conditioning of the GSR was produced in the absence of a noxious UCS. It is apparent also that the maximum CR attained by the Discovery students approximates the magnitude of the CR displayed by the Instructed students on that trial. Examination of individual conditioning curves for the Discovery students indicated that the curves tend to be of all-or-none nature (Pendery & Maltzman, 1977). Their maximum GSR tends to occur on the trial these students reported that they first discovered the CS-UCS contingency. The group growth curve depicted in Figure 28.2 is a product of averaging over students who discovered the contingency on different trials. When the magnitude of their maximum CR is examined, it

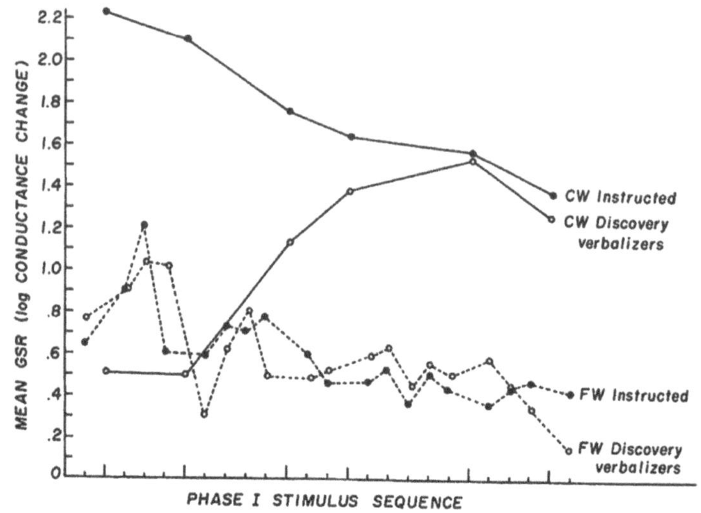

FIGURE 28.2. Mean magnitude of the GSR to the critical (CW) and filler (FW) words in Phase 1 of acquisition for the Instructed group and the verbalizer Discovery group (Pendery & Maltzman, 1977).

approximates the magnitude of the first response obtained by the Instructed group. Most students in the latter group manifested their largest CR on the first trial.

DISCUSSION

Results obtained for the Discovery and Instructed students suggest that conditioning of a measure of the OR such as the GSR is basically of the same nature in verbal and in classical conditioning as it is usually conducted in the laboratory. It represents the occurrence of a voluntary OR to a significant stimulus. In one case significance was imparted by the experimenter's instructions. In the other case it was discovered. Conventional classical conditioning of a measure of the OR such as the GSR is a problem solving discovery process. If the student is successful in problem solving, a large OR occurs following the CS reflected as a large GSR which is taken as the CR. Discriminative semantic condition is a relatively difficult problem situation and different subjects require varying numbers of trials to discover the CS-UCS contingency. Averaging over subjects produces the group growth curve. Individual subjects do not manifest the gradual acquisition process. Simple conditioning with only a CS+ or discriminative conditioning with a CS+ and a single CS- rarely manifest a growth curve because all students solve the problem quickly. Continued conditioning trials yield a habituation curve much like the response curve to the critical word produced by the Instructed group. The Instructed group, in turn, is much like the experimental groups in the habituation experiment. They also orient maximally to the first presentation of the significant stimulus following habituation to a list of nonsignal words.

Assertions that simple conditioning is a nonassociative process reflecting dishabituation of a previously habituated OR (Stewart, Stern, Winokur, & Fredman, 1961) or the occurrence of an OR to a new stimulus complex, the CS-UCS compound (Stern & Walrath, 1977) are in error. They fail to consider that there may be voluntary as well as involuntary ORs and the "conditioning" has occurred on essentially one trial for the individual subject. Subsequent trials yield habituation, inhibition with reinforcement (Hull, 1943), because habituation is a characteristic feature of ORs in predictable stimulus situations.

Voluntary and Involuntary ORs. Our interpretation of classical conditioning of the GSR rests upon the assumption that there are different kinds of ORs, voluntary as well as involuntary (Maltzman, 1979b). Indices of the OR occur not only as a consequence of stimulus change, but also as the consequence of significant stimuli, stimuli that have signal value (Luria, 1973; Sokolov, 1963).

Luria and his associates have presented extensive evidence suggesting that the morphology, physiology, and ontogeny of voluntary and involuntary ORs differ (e.g., Luria, 1973). Evidence in support of the distinction stems largely from Luria's experimental clinical research. He and his associates have found consistent differences among patients in their GSR, digital vasomotor responses and the spectral analysis of their EEG activity under signal and nonsignal stimulus conditions. Patients with lesions in the prefrontal cortex and patients with lesions in the occipital and parietal regions show abnormal ORs to simple nonsignal tones or light stimuli as reflected by the above measures. Patients tend to show very rapid habituation, e.g., of the GSR, no GSRs to speak of to nonsignal stimuli, or in some cases, they are abnormally resistant to habituation. Imparting stimuli with signal value by instructing the patients to perform a motor response to each stimulus or to judge which of the stimuli is of greater duration produces marked differences in ORs depending upon the locus of the patient's lesion. Providing the stimuli with signal value normalizes the OR obtained in patients with lesions in the occipital and parietal regions. Signal value usually increases the magnitude of autonomic measures of the OR. Significant stimuli do not normalize indices of ORs obtained in patients with lesions in the prefrontal cortex.

Näätänen (1979) has analyzed the different conditions under which the concept of an OR has been applied and has concluded that it is reasonable to differentiate between voluntary and involuntary ORs. His examination of the distribution of evoked potentials to significant and nonsignificant stimuli suggests that voluntary ORs tend to manifest disproportionately larger EP amplitudes to significant stimuli in the frontal areas whereas involuntary ORs to stimulus change appear to display their largest EP amplitudes in sensory specific areas.

Results of the above kind suggest that the prefrontal cortex is particularly important in the regulation of voluntary ORs, indicating that ORs to significant and nonsignificant stimuli are organized differently in the cerebral cortex. It follows that discovery of the signal value of the CS in classical conditioning induces a voluntary OR that is dependent upon the prefrontal cortex for its regulation. Such an OR may also be induced by the experimenter's instructions as in verbal conditioning.

Different Forms of Classical Conditioning. Not all classical conditioning in adults need involve a preliminary problem solving stage. Campbell, Sanderson, and Laverty (1964) have demonstrated that classical conditioning of the GSR can occur following a single traumatic experience. Razran's (1955) research suggests that verbal awareness of the CS-UCS contingency may interfere with classical conditioning of the salivary response. On the other hand, Brandeis and Lubow (1975) have obtained highly significant differential conditioning of the GSR in the absence of verbal awareness. They examined verbal awareness by means of a careful assessment of the subjects' ability to recognize the relevant contingencies in the experiment. These illustrative experimental results indicate that the typical cognitive approach which attempts to explain classical conditioning in terms of awareness (Brewer, 1974) and insists that successful classical conditioning always involves verbal awareness is inadequate to the task of coping with the array of available data.

Theoretical Bases for Voluntary and Involuntary ORs. We assume that there are different bases for the occurrence of involuntary and voluntary ORs and for conditioned voluntary and involuntary ORs. Involuntary ORs occur on the basis of a mismatch, much in the manner emphasized by Sokolov (1963). A neuronal model of a stimulus is established on the basis of past stimulation. A mismatch between the neuronal model of past stimulation and present stimulation results in the occurrence of an OR. Since the neuronal model based upon past stimulation persists, it has been learned. Occurrence of an OR due to the mismatch is dependent upon the previous acquisition of the neuronal model. In a sense therefore, the typical involuntary OR is a conditioned involuntary OR. Such conditioning is most clearly seen in the response to the omitted UCS during extinction or test trials (Grings,

Lockhart, & Dameron, 1962). We assume that once the significance of the CS is discovered, it is a signal for the UCS, a neuronal model may be established for the CS-UCS contingency. A mismatch occurs when the UCS is omitted, giving rise to a large response at the point in time where the UCS would have occurred. This is taken as a conditioned response (Stern, 1972). The basis for such a conditioned response is quite different from the conditioned response obtained in the CS-UCS interval in long trace or delayed conditioning. Here, as previously indicated, the response occurs because of the discovered signal value of the CS. But discovery of the significant CS cannot be the consequence of the establishment of a neuronal model. The neuronal model can only occur after the significance of the CS is established. We hypothesize that the physiological basis for significance is the same whether in the usual habituation experiment, in verbal conditioning experiments where it is established by instructions, or in the usual classical conditioning situation where it is discovered. It is a consequence of a dominant focus of excitability established primarily in the prefrontal cortex. A conception of the dominant focus, basic to Pavlov's theory of classical conditioning (Asratyan, 1953), has been studied extensively in the Soviet Union (Livanov, 1977; Rusinov, 1973), but its potential importance has not received the attention it warrants (John, 1967; Maltzman, 1979a, 1979b, Maltzman, Langdon, & Feeney, 1970; Razran, 1971).

We hypothesize that instructions to establish a stimulus as significant do so by inducing a dominant focus of excitability. When an appropriate stimulus occurs the dominant focus is activated and induces an OR. Selective attention and transmission of information occur as a consequence (Maltzman et al., 1970). As Bechterev (1932) pointed out, the dominant focus is the basis for mental sets and determining tendencies, traditional psychological concepts used to account for the selective transmission of information and behavior. In classical conditioning discovery of the significance of the CS as a signal for the UCS results in the establishment of a dominant focus for the CS and "communication" between the dominant foci for the CS and UCS (Livanov, 1977). An OR to the CS now occurs.

According to the present interpretation, classical conditioning of the GSR in the usual laboratory experiment involves the development of a differential voluntary OR to a significant stimulus. This is the initial phase of a multi-phase process when a skeletal motor response is being conditioned. After the OR has occurred differentially to the CS, because of the predictable stimulus situation, a CS-UCS contingency, the OR will begin to habituate. Only after the initial phase of differential orienting has occurred can an appropriate conditioned motor response develop (Smythies, 1970). Selective orienting (discrimination) must occur before response differentiation (avoidance responding), for example, is consistently manifested. The usual laboratory classical conditioning experiment with humans employing a measure of the OR as the index of a CR is studying only the former process, discrimination. Its principles, those of orienting and attention, are different from the usual principles involved in conditioning a skeletal motor response.

REFERENCES

Asratyan, E. A. (1953). *I. P. Pavlov: His life and works*. Moscow: Foreign Languages Publishing House.

Bechterev, V. M. (1932). *General principles of human reflexology*. New York: International Publishers. (Republished: New York: Arno Press, 1973).

Berlyne, D. E. (1960). *Conflict, arousal and curiosity*. New York: McGraw-Hill.

Brandeis, R., & Lubow, R. E. (1975). Conditioning without awareness-again. *Bulletin of the Psychonomic Society, 5*, 36-38.

Brewer, W. F. (1974). There is no convincing evidence for operant or classical conditioning in adult humans. In W. B. Weimer & D. S. Palermo (Eds.), *Cognition and the symbolic processes* (pp. 1-42). New York: Wiley.

Campbell, D., Sanderson, R. E., & Laverty, S. G. (1964). Characteristics of a conditioned response in human subjects during extinction trials following a single traumatic conditioning trial. *Journal of Abnormal and Social Psychology, 68*, 627-639.

Garcia, J., McGowan, B. K., & Green, K. F. (1972). Biological constraints on conditioning. In A. Black & W. F. Prokasy (Eds.), *Classical conditioning* (pp. 3-27). New York: Appleton-Century.

Grings, W. W., Lockhart, R. A., & Dameron, L. E. (1962). Conditioning autonomic responses of mentally subnormal individuals. *Psychological Monographs, 76* (39, Whole No. 558).

Hull, C. L. (1943). *Principles of behavior*. New York: Appleton-Century.

John, E. R. (1967). *Mechanisms of memory*. New York: Academic Press.

Latash, L. P. (1968). *Hypothalamus, adaptive activity, and the electroencephalogram*. Moscow: Nauka.

Livanov, M. N. (1977). *Spatial organization of cerebral processes*. New York: Wiley.

Luria, A. R. (1973). *The working brain*. New York: Basic Books.

Luria, A. R., & Homskaya, E. D. (1970). Frontal lobes and the regulation of arousal processes. In D. Mostofsky (Ed.), *Attention: Contemporary theory and analysis* (pp. 303-330). New York: Appleton-Century-Crofts.

Luria, A. R., & Vinogradova, S. (1959). An objective investigation of the dynamics of semantic systems. *British Journal of Psychology, 50*, 89-105.

Maltzman, I. (1971). The orienting reflex and thinking as determiners of conditioning and generalization to words. In H. H., Kendler & J. T. Spence (Eds.), *Essays in neobehaviorism: A memorial volume to Kenneth W. Spence* (pp. 89-111). New York: Appleton-Century-Crofts.

Maltzman, I. (1979a). Orienting reflexes and significance: A reply to O'Gorman. *Psychophysiology, 16*, 274-282.

Maltzman, I. (1979b). Orienting reflexes and classical conditioning in humans. In H. D. Kimmel, E. H. van Olst, & J. F. Orlebeke (Eds.), *The orienting reflex in humans* (pp. 323-351). Hillsdale, NJ: Lawrence Erlbaum Associates.

Maltzman, I., Gould, J., Barnett, O. J., Raskin, D. C., & Wolff, C. (1977). Classical conditioning components of the orienting reflex to words using innocuous and noxious unconditioned stimuli under different conditioned stimulus-unconditioned stimulus intervals. *Journal of Experimental Psychology: General, 106*, 171-212.

Maltzman, I., Gould, J., Barnett, O. J., Raskin, D. C., & Wolff, C. (1979). Habituation of the GSR and digital vasomotor components of the orienting reflex as a consequence of task instructions and sex differences. *Physiological Psychology, 7*, 213-220.

Maltzman, I., Langdon, B., & Feeney, D. (1970). Semantic generalization without prior conditioning. *Journal of Experimental Psychology, 83*, 73-75.

Maltzman, I., & Raskin, D. C. (1965). Effects of individual differences in the orienting reflex on conditioning and complex processes. *Journal of Experimental Research in Personality, 1*, 1-16.

Maltzman, I., & Raskin, D. C. (1979). Selective orienting and habituation of the GSR as a consequence of overt and covert activity. *Physiological Psychology, 7*, 204-208.

Näätänen, R. (1979). Orienting and evoked potentials. In H. D. Kimmel, E. H. van Olst, & J. F. Orlebeke (Eds.), *The orienting reflex in humans* (pp. 61-75). Hillsdale, NJ: Lawrence Erlbaum Associates.

Pendery, M., & Maltzman, I. (1977). Instructions and the orienting reflex in "semantic conditioning" of the galvanic skin response in an innocuous situation. *Journal of Experimental Psychology: General, 106*, 120-140.

Razran, G. (1955). A direct laboratory comparison of Pavlovian conditioning and traditional associative learning. *Journal of Abnormal and Social Psychology, 51*, 649-652.

Razran, G. (1961). The observable unconscious and the inferable conscious in current Soviet psychophysiology: Interoceptive conditioning, semantic conditioning, and the orienting reflex. *Psychological Review, 68*, 81-147.

Razran, G. (1971). *Mind in evolution*. Boston: Houghton Mifflin.

Rusinov, V. S. (1973). *The dominant focus*. New York: Consultants Bureau.

Smythies, J. R. (1970). *Brain mechanisms and behaviour* (2nd ed.). New York: Academic Press.

Sokolov, E. N. (1960). Neuronal models and the orienting reflex. In A. B. Brazier (Ed.), *The central nervous system and behavior.* New York: Josiah Macy, Jr., Foundation.

Sokolov, E. N. (1963). *Perception and the conditioned reflex.* New York: Macmillan.

Stern, J. A. (1972). Physiological response measures during classical conditioning. In N. S. Greenfield & R. A. Sternbach (Eds.), *Handbook of psychophysiology* (pp. 197-227). New York: Holt, Rinehart & Winston.

Stern, J. A., & Walrath, L. C. (1977). Orienting responses and conditioning of electrodermal responses. *Psychophysiology, 14*, 334-342.

Stewart, M. A., Stern, J. A., Winokur, G., & Fredman, S. (1961). An analysis of GSR conditioning. *Psychological Review, 68*, 60-67.

Thorndike, E. L. (1931). *Human learning.* New York: Century.

29 · A POSSIBLE CENTRAL NERVOUS SYSTEM CORRELATE (CNV AMPLITUDE) OF RELATIVELY HIGH SOCIAL STATUS

P. R. Barchas, W. S. Jose II, B. Payne, B. S. Kopell, and W. T. Roth

Stanford University
and
University of South Carolina

Lindsley's conception of the relationships between brain and behavior has long envisioned sociology as a contributing discipline (Lindsley, 1957, p. 49). This study presents data suggesting that alterations in the way a person is linked into the social status structure of interpersonal interaction has a central nervous system correlate. The particular status structure alteration in this research changes the subject's belief in his competence on a specific task, both in relation to a "partner" and in relation to "national standards." This produces a change in the subject's expectations for performance on the task, which in turn has been shown to produce behavioral effects on such social processes as deference, power, influence, and reward distribution in small group interaction (Barchas, 1976; Berger & Conner, 1969; Berger, Fisek, Norman, & Zelditch, 1977; Berger, Rosenholtz, & Zelditch, 1980).

The hypothesis of this study is that the manipulation of subjects into a high competence state relative to another actor produces a heightened contingent negative variation (CNV). This hypothesis was developed by noting a convergence in two distinct research areas. It has been shown in social psychological studies that a competence manipulation of the type used here alters the social behaviors mentioned above. Of particular interest to our inquiry is Conner's (Conner, 1977) finding that when subjects are manipulated into a high competence state, they react faster (i.e., their response latency is lower) than subjects who are manipulated into a low competence state. Observations from research on the CNV indicate that alternations in the magnitude of the CNV are also associated with response latencies; shorter latencies are associated with greater magnitude of the CNV. Both high status and CNV are hypothesized to be related to attention. We therefore postulate that the high competence manipulation will produce an increase in CNV magnitude.

 DOI: 10.4324/9781003164647- 36

METHODS

Measures of the CNV were taken both before and after the high competence manipulation, and mean scores were compared using each subject as his own control. In addition, subjects for whom the high competence manipulation was strong and effective were compared with those for whom it was not.

Each male undergraduate college student subject is told that this is a two-part study. He is instructed that in the first part, he/she and his/her "partner" (a confederate) will work separately while EEG recordings are made; and that in the second part of the study, he/she and his/her "partner" are to work together as a team on problems similar to those in the first part of the study. It was explained that the first part of the study consisted of a series of problems and that, after that series, scores would be reported to both the subject and his/her "partner" and related to "national standards" for this test--in reality this was a fictitious ability which is sufficiently believable to produce the status manipulation. The subject made a decision for each problem and recorded his/her answer by pressing the appropriate button on the arm of his/her chair. After the first set of problems, the subject was told that he/she scored 37 correct out of 40, placing him/her in the "above average" category on the "national standards" for this test; and his/her "partner" scored 9 correct out of 40, placing him/her in the "poor" category. The subject was then given another set of problems to work on. Those subjects who accepted and believed the scores were considered to have a higher evaluation of self than another person relative to the task at hand.

While the subject was working on the two slide series, electrical activity of the brain was recorded on FM tape. The electroencephalogram (EEG) was recorded from the vertex (C_z). The electrooculogram (EOG) was also recorded. A pin electrode was used at the vertex and disc electrodes elsewhere (Roth et al., 1976).

Each problem was projected tachistoscopically from a slide. For each slide presentation, the subject first heard a tone which indicated that the slide was about to appear. One second after the tone, the slide was shown for one second. If the subject had not made a decision one and one-half seconds after the slide went off, he/she heard a white noise to remind him/her to respond immediately. This situation is represented in Figure 29.1. The CNV occurs in the interval between the first tone and the onset of the slide as shown in the figure.

Analog signals for both the EEG and EOG in the time-locked interval between tone and slide presentation were recorded for each trial on magnetic tape. Cumulative waveforms were generated for each slide series using a Fabritek signal averaging computer. The averaged signals were then graphed on an X-Y plotter. It is known that eye movements contaminate the EEG

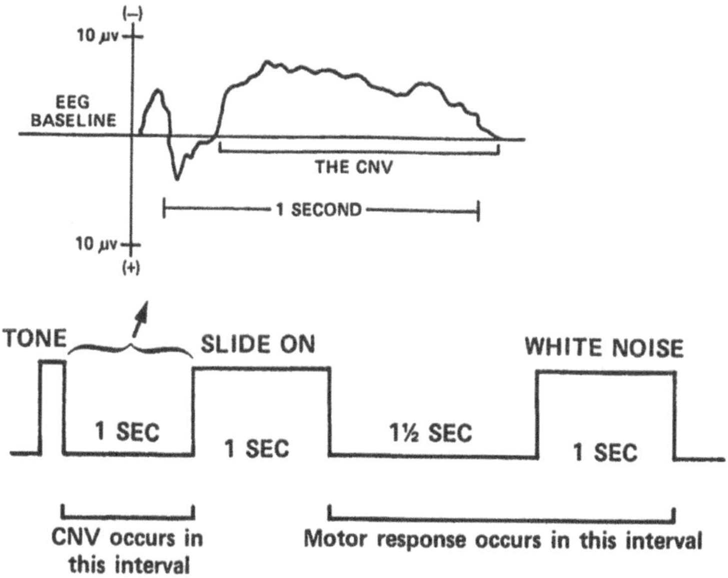

FIGURE 29.1. Schematic representation of sequence of events for each problem trial, with diagram of typical CNV wave.

record; therefore, a procedure for removing eye movement artifact from the CNV record was followed (Corby & Kopell, 1972). The measure for the CNV was the maximum amplitude of the CNV in microvolts during the 500- to 1000-msec period after the warning tone.

In a debriefing interview immediately after the study, subjects were classified according to the impact of the manipulation. A subject was considered to have a higher evaluation of self than another person with regard to competency on the task only if all of the following requirements were met: (1) subject believed there was another person going through the study with him/her and that he/she would be working with that person in the later phase of the study, (2) subject believed there were right and wrong answers to the test, (3) subject believed the scores reported for both himself/herself and his/her partner were true, and (4) the subject remembered his/her score on the test relative to his/her partner. Only if all of the above criteria were met was the subject categorized as having been strongly and effectively manipulated. Subjects who did not meet all these criteria were classified as having been only weakly manipulated. This category of subjects is representative of subjects whose behavior is ordinarily eliminated from analysis in typical social psychological experiments, due to violation of scope conditions. However, for this study they make an important comparison group because they have been through exactly the same physical procedures as the target group. Further, it is of interest in terms of generalizability of behavioral findings to know whether there are physiological differences in normal subjects who typically are not included in analyses for reasons of not meeting manipulation conditions of a study.

RESULTS

Table 29.1 presents the CNV amplitude measures for all subjects. For all subjects who were strongly manipulated (who accepted and believed all aspects of the manipulation), the CNV amplitude in all cases increased in the post-manipulation session. A one-tailed *t* test for paired samples indicates that the increase in CNV amplitude from pre- to post-manipulation is significant at the .0005 level. No significant differences were found between pre- and post-manipulation measures for subjects who were not strongly manipulated (subjects who did not accept and believe all aspects of the manipulation). Subjects who were strongly manipulated had an average CNV amplitude before manipulation of -4.50 μV, and -7.76 μV after manipulation. Subjects who were not strongly manipulated had a pre-manipulation CNV amplitude average of -9.08 μV, and -8.40 μV after manipulation. Five of five subjects who were strongly manipulated conformed to the hypothesis of increased CNV amplitude after manipulation; only one of the five who were not strongly manipulated showed an increase in CNV amplitude after manipulation into relatively high status.

DISCUSSION

The data yield two interpretations. It may be the case, as we have argued, that the competence manipulation, when strong and effective, increases the magnitude of the CNV. This interpretation is strengthened by the increased CNV amplitude for the subjects who were strongly manipulated and no significant increase for those who were not strongly manipulated. However, it also may be the case that we have two physiologically distinct subgroups of subjects in this sample. This interpretation is consistent with the observation that the mean CNV amplitude before manipulation for the strongly manipulated subjects is -4.50 μV, while for the weakly manipulated subjects it is -9.08 μV. This difference, while not statistically significant, suggests that subjects who do not respond to this and similar manipulations, and are therefore usually eliminated from analysis in behavioral social psychological studies, may be a physiologically distinct subgroup. Thus it is not clear at this point whether the weakly manipulated subjects failed to show increased CNV amplitude because of the weakness of the manipulation, or

TABLE 29.1. CNV amplitude for subjects who were strongly manipulated (those who believed and accepted all parts of the manipulation) and subjects who were weakly manipulated (those who did not believe and accept all parts of the manipulation).

Subject Identification (Does not reflect ordering of runs)	CNV Amplitude Before Manipulation	CNV Amplitude After manipulation	Difference (Before - After)
Subjects who were strongly manipulated*			
	-7.00	-10.50	+3.50
B	+1.20	-1.47	+2.67
C	-6.12	-8.83	+2.71
D	-5.62	-8.87	+3.75
E	-4.94	-9.11	+4.17
Mean	-4.50	-7.76	
Subjects who were weakly manipulated**			
F	-9.18	-6.38	-2.80
G	-3.00	-15.25	+12.25
H	-16.41	-9.05	-7.36
I	-6.66	-5.64	-1.02
J	-10.17	-5.71	-4.26
Mean	-9.08	-8.40	

*Paired t statistic for pre- vs. post-manipulation difference of means: t = 11.76, df = 4, p<.0005.
**Paired statistic for pre- vs. post-manipulation difference of means: t = 0.20, df = 4, non-significant.

whether they are a physiologically distinct subgroup who are, for physiological or psychological reasons, resistant to such manipulations. At this point it seems reasonable to suggest that there may be an interaction effect between amplitude of the CNV and susceptibility to the effects of social status changes such as the one manipulated here.

The discussion above should not obscure the focal point. Whether the data are taken to reveal two physiologically distinct subgroups or not, there is evidence of a positive relationship between CNV amplitude and manipulation into a high competence state relative to another individual. We conclude that the high competence manipulation did in fact have the effect of increasing the amplitude of the CNV for those subjects classified as strongly manipulated. This conclusion is further supported by consideration of the effects of habituation and fatigue which would act so as to diminish the size of the CNV. These factors would operate so as to produce an effect opposite to that which was found.

In further clarification of the posited relationship, studies are needed which incorporate a larger number of subjects in each cell and which have both a no-manipulation condition and a condition in which the subject is manipulated into a low state relative to another actor, although that design deserves careful human subjects consideration. In both experimental conditions, criteria for successful manipulation should be applied, and both groups, believers and nonbelievers, retained for analysis. A few studies using EEG measured shifts in lateralized brain activity in similar paradigms have produced positive correlations which are interpretable in terms of attentional distribution, and conscious and preconscious perception (Barchas, Crissman, Ford, & Wilkes, 1984; Barchas & Fisek, 1984; Barchas & Perlaki, 1986; Colomy & Barchas, 1984; Harris, 1981).

While other studies have shown that differences in social status are related to biochemical events (Back & Bogdonoff, 1964; Barchas & Barchas, 1977; Rose, Gordon, & Bernstein, 1972), this is the first demonstration we are aware of which has explicitly related a basic social processes to a central nervous system event for which there is consensus with regard to its meaning. By using a physiological measure of brain activity, we have been able to tap into postulated internal processes, shifts in attention, thought to be associated with interpersonal behavior (for a review see Carver & Scheier, 1981). In particular, these data show that a manipulation central to investigations of an elementary social process known to affect a range of interactive behaviors also affects a dimension of brain activity in humans. The argument that many of the behaviors and interactions known to be associated with position in a social structure are outcomes of distribution of attention has received at least partial support.

ACKNOWLEDGMENT

This work was supported by the Harry Frank Guggenheim Foundation and Office of Naval Research Grant No. N00014-79C0796.

REFERENCES

Back, K. W., & Bogdonoff, M. D. (1964). Plasma lipid responses to leadership conformity and deviation. In P. H. Leiderman & D. Shapiro (Eds.), *Psychobiological approaches to social behavior* (pp. 24-42). Stanford University Press.

Barchas, P. R. (1976). Physiological sociology: Interface of sociological and biological processes. *Annual Review of Sociology, 2*, 233-299.

Barchas, J., & Barchas, P. (1977). Social behavior and adrenal medullary function in relation to psychiatric disorders. In D. Hamburg, E. Usdin, & J. Barchas (Eds.), *Neuroregulators and psychiatric disorders* (pp. 99-102). New York: Oxford University Press.

Barchas, P. R., Crissman, S. J., Ford, J. B., & Wilkes, C. D. (1984). Social agreements, disagreements and relative hemispheric activity. In P. R. Barchas & S. P. Mendoza (Eds.), *Social cohesion: Essays toward a sociophysiological perspective* (pp. 151-162). Westport, CN: Greenwood Press.

Barchas, P. R., & Fisek, M. H. (1984). Hierarchical differentiation in newly formed groups of rhesus and humans. In P. R. Barchas (Ed.), *Social hierarchies: Essays toward a sociophysiological perspective* (pp. 23-44). Westport, CN: Greenwood Press.

Barchas. P. R., & Perlaki, K. M. (1986). Processing of preconsciously acquired information measured by hemispheric asymmetry and selection accuracy. *Journal of Behavioral Neuroscience, 100*, 343-349.

Berger, J., & Conner, T. L. (1969). Performance expectation and behavior in small groups. *Acta Sociologica, 12*, 186-198.

Berger, J., Fisek, M. H., Norman, R. Z., & Zelditch, M., Jr. (1977). *Status characteristics and social interaction: An expectation states approach*. New York: Elsevier.

Berger, J., Rosenholtz, S. J., & Zelditch, M., Jr. (1980). Status organizing processes. *Annual Review of Sociology, 6*, 479-508.

Carver, C. S., & Scheier, M. F. (1981). *Attention and self-regulation: A control-theory approach*. New York: Springer-Verlag.

Colomy, P. & Barchas, P. R. (1984). Social roles and hemispheric laterality. In P. R. Barchas & S. P. Mendoza (Eds.), *Social cohesion: Essays toward a sociophysiological perspective* (pp. 163-185). Westport, CN: Greenwood Press.

Conner, T. L. (1977). Performance expectation and the initiation of problem solving attempts. *Journal of Mathematical Sociology, 5*, 187-198.

Corby, J., & Kopell, B. S. (1972). Differential contribution of blink and vertical eye movements as artifacts in EEG recordings. *Psychophysiology, 9*, 640-644.

Harris, W. A. H. (1981). *A physiological investigation of status*. Unpublished doctoral dissertation, Stanford University.
Lindsley, D. B. (1957). Psychophysiology and motivation. In M. R. Jones (Ed.), *Nebraska Symposium on Motivation* (Vol. 5, pp. 44-105). Lincoln, NE: University of Nebraska Press.
Rose, R., Gordon, T., & Bernstein, I. (1972). Plasma testosterone levels in the male rhesus: Influences of sexual and social stimuli. *Science, 178*, 643-645.
Roth, W. T., Krainz, P. L., Ford, J. M., Tinklenberg, J. R., Rothbart, R. M., Kopell, B. S. (1976). Parameters of temporal recovery of the human auditory evoked potential. *Electroencephalography and Clinical Neurophysiology, 40*, 623-632.

APPENDIX A.

Donald Benjamin Lindsley

Born:	Brownhelm (Lorain County) Ohio, December 23, 1907
Married:	Ellen Ford, Iowa City, Iowa, August 16, 1933
Children:	David Ford Lindsley, Margaret (Lindsley) Block, Robert Kent Lindsley, Sara Ellen (Lindsley) Lyons

Academic and Professional

1929	A.B.	Wittenberg College, Springfield, Ohio	(Major: Psychology)
1930	M.A.	University of Iowa, Iowa City, Iowa	(Major: Psychology)
1932	Ph.D.	University of Iowa, Iowa City, Iowa	(Major: Psychology; Minor: Physiology)

1932-33	Instructor in Psychology, University of Illinois, Urbana-Champaign, Illinois
1933-35	National Research Council Fellow, Department of Physiology, Harvard Medical School, and Department Neuropsychiatry, Massachusetts General Hospital, Boston, Massachusetts
1935-38	Research Associate in Psychobiology, Western Reserve University Medical School, and the Brush Foundation, Cleveland, Ohio
1938-46	Assistant Professor, Psychology, Brown University, Providence, Rhode Island, and Director, Psychology Department and Neurophysiology Laboratories, Bradley Hospital, East Providence
1943-45	Director, World War II Research Project on Radar Operation, Yale University Contract with the Office of Scientific Research and Development; Headquarters, Army Signal Corps School, Camp Murphy, Hobe Sound, Florida and U.S. Army Air Forces Base, Boca Raton Field, Florida
1946-1951	Professor of Psychology, Northwestern University, Evanston, Illinois
1951-77	Professor, University of California, Los Angeles: Psychology and Pediatrics (1951-53); Psychology and Psychiatry (1953-56); Psychology and Physiology (1956-77); Chairman, Department of Psychology (1959-62); Member, Brain Research Institute (1961-77)
1977-	**Professor Emeritus, University of California, Los Angeles**
1949	Visiting Professor, Columbia University (Summer)
1950	Visiting Professor, University of California, Los Angeles (Summer)
1961	Visiting Professor, University of Hawaii (Summer)

Honorary Degrees

Sc.D.	Brown University, 1958
D.Sc.	Wittenberg University, 1959
Sc.D.	Trinity College, Hartford, Conn., 1965
D.Sc.	Loyola University, Chicago, 1969
Ph.D.	*Honoris Causa*, Johannes Gutenberg University, Mainz, West Germany, 1977

DONALD BENJAMIN LINDSLEY

Honors: Awards, Elections, Lectureships, etc.

Elected to Society of Experimental Psychologists, 1942
Presidential Certificate of Merit for World War II Effort, 1948 (signed by Harry S. Truman)
Elected to the National Academy of Sciences, 1952
William James Lecturer, Harvard University, Fall Semester, 1958
Guggenheim Fellow, 1959
Distinguished Scientific Contribution Award, American Psychological Association, 1959
University Research Lecturer, University of California, Los Angeles, 1960
First Walter B. Pillsbury Lecturer, Cornell University, 1963
Elected to the American Academy of Arts and Sciences, Boston, 1965
Scientific Achievement in Psychology Award, California Psychological Association, 1977
Annual Donald B. Lindsley Prize in Behavioral Neuroscience, for best Ph.D. thesis in the field for the preceding year, established by the Society for Neuroscience and funded by the Grass Foundation
Awarded Honorary Membership, with Great Distinction, by Western EEG Society, 1978
Awarded Honorary Membership in American Electroencephalographic Society, 1980
Elected Honorary Fellow in American Electroencephalographic Society, 1982
Distinguished Scientific Achievement Award, Society for Psychophysiological Research, 1984
Award: Fellow of the UCLA School of Medicine, For Great Contributions to Medicine, 1986
Elected to Foreign Membership, Finnish Academy of Science and Letters, 1987
Distinguished Graduate Award, Department of Psychology, The University of Iowa, 1987

Membership in Professional Societies

American Psychological Association (1932; Fellow, 1937): Numerous committees and Offices)
American Physiological Society (1937)
Society for Experimental Biology and Medicine (1937)
American Psychosomatic Society (1944): Council, 1956-59
American Electroencephalographic Society (1947): President, 1964-65
American Association for the Advancement of Science (1943): Vice President, Chairman, Sec. I (1954), J (1977)
Central Electroencephalographic Society (1946): President 1949
Western Electroencephalographic Society (1951): President 1957-58
Midwestern Psychological Association (1932): President 1951
Western Psychological Association (1951): President 1959-60
Society for Research in Child Development (1952)
American Academy for Cerebral Palsy (Honorary, 1953)
Psychonomic Society (1959): Board of Governors, 1960-65
American Academy of Neurology (1956)
IBRO-UNESCO - International Brain Research Organization (1960): Council, 1960-63
National Association for Retarded Children (1958): Scientific Advisory Board, 1958-71
Pavlovian Society of America (1968): Membership Chairman, 1976-77

APPENDIX A.

Major Offices, Committees, etc.,

American Psychological Association

President, Division 6, Physiological and Comparative Psychology (1947)
Representative to National Research Council (1948-51)
Member, Program Committee (1948-51); Chairman (1951)
Member Program Committee, Division of Clinical and Abnormal Psychology (1947-50)
Chairman, Standards and Training Committee (1949-51)
Member, Education and Training Board (1951-53)
Member, Policy and Planning Board (1955-57)
Member, Board of Scientific Affairs (1957-59)
Representative to the American Association for the Advancement of Science (1967-69)

National Academy of Sciences

Chairman, Psychology and Physiology Panel, Undersea Warfare Committee (1947-49)
Member, Undersea Warfare Committee, National Research Council (1951-64)
Chairman, Psychology Section, NAS (1958-62)
Member, Committee on Governmental Relations (1961-62)
Member, Committee on Science and Public Policy (COSPUP) (1962-64)
Member, Committee on Brain Sciences (1967-71)
Member, Space Science Board (1967-71)
Chairman, Committee on Long-Duration Mission in Space (Human Factors), Space Science Board (1967-72)
Member, Life Science Committee, Space Science Board (1967-72)
Member, Space Medicine Committee, Space Science Board (1969-72)
Member, Ad Hoc Committee on Publications Review (1969)

American Electroencephalographic Society

Charter Member (1947)
Council Member (1954-57)
President (1964-65)

International Brain Research Organization - IBRO/UNESCO

Charter and Organizing Member (1958)
Council Member (1958-63)
Treasurer (1967-72)

Military and Governmental Agencies

Member, Aviation Psychology Committee, National Research Council - Civil Aeronautics Administration (1942-51)
Member, Scientific Advisory Board, U.S. Air Forces (1947-49)
Chairman, Human Resources Professional Advisory Committee, U.S. Air Forces (1948-49)
Undersea Warfare Committee, NRC - Liaison with Office of Naval Research (1951-64)
Member, Central Veterans Administration Advisory Committee, Psychiatry and

Neurology, Veterans Administration, Washington (1956-58)

Govemmental Granting Agenicies (Service on Committees or Panels)

National Science Foundation, Advisory Panel, General Experimental Psychology (1952-54)
United States Public Health Service, Mental Health Study Section, NIMH (1951-54)
United States Public Health Service, Neurology Study Section, NINDB (1958-63)
United States Public Health Service, Behavioral Sciences Training Panel, NIGMS (1965-69); Chairman, (1966-69)
AIBS-NASA Behavioral Biology Panel (1966-71)

State of California: Special Appointments

Member, Science and Technology Council, Advisory to the State of California Legislative Assembly, a pioneering venture funded by the National Science Foundation (1969-70)
Member, Selection Jury of the California Museum of Science and Industry to select the California Scientist of the Year (1970-75); Chairman of Jury (1975)

Foundations

Trustee, the Grass Foundation (1958-)
Member, Board of Directors and Scientific Advisory Board, Brain Research Foundation, Chicago (1954-57)
Consultant Reviewer, The Guggenheim Foundation (1963-70)
Educational Advisory Board, The Guggenheim Foundation (1970-78)

Editorial Board and Consultant Editorships

Consulting Editor, *Journal of Experimental Psychology* (1947-68)
Consulting Editor, *Journal of Comparative and Physiological Psychology* (1952-62)
Consulting Editor, *Journal of Personality* (1958-62)
Editorial Board, *International Journal of Physiology and Behavior* (1965-)
Board of Co-Editors, *Experimental Brain Research* (1965-77)
Editorial Board, *Journal of Developmental Psychobiology* (1967-)
Board of Editors, *Science* (1974-76)
Advisory Board, *Archiv. für Psychologie* (1978-)

APPENDIX B.

Ph.D. Degrees Sponsored by Donald B. Lindsley

Northwestern University (1946-51)

1947 **James E. Birren,** Psychology
1948 **Howard L. Siple,** Psychology
1950 **Robert J. Ellingson,** Psychology
1950 **Maime J. Jones,** Speech
1950 **Walter D. Obrist,** Psychology
1951 **Dorothy H. Eichorn,** Psychology
1951 **Julius Wishner,** Psychology
1951 **William B. Knowles,** Psychology

University of California, Los Angeles (1951-84)

1952 **Robert B. Voas,** Psychology
1953 **Murray Korngold,** Psychology
1954 **Robert W. Lansing,** Psychology
1955 **Jack L. Michael,** Psychology
1956 **Robert F. Snowden,** Psychology
1958 **Robert M. Gerard,** Psychology
1960 **Roy S. Griffiths,** Psychology
1962 **Mitchell L. Kietzman,** Psychology
1962 **Robert C. Boyle,** Psychology
1963 **Maurice B. Sterman,** Psychology
1964 **Emanuel Donchin,** Psychology
1964 **Orval T. Ellsworth,** Physiology
1964 **Gary C. Galbraith,** Psychology
1965 **Guenter H. Rose,** Psychology
1965 **Jo Ann Helmy,** Psychology
1966 **Joel W. Adkins,** Psychology
1966 **Lester G. Fehmi,** Psychology
1966 **Paul Spong,** Psychology
1967 **Ann S. Alpern,** Psychology
1967 **Sidney R. Roth,** Psychology
1967 **James E. Skinner,** Psychology
1968 **Constance W. Kovar,** Psychology
1968 **Ken Nakayama,** Psychology
1968 **John A. Satterberg,** Psychology
1968 **Jerry D. Wicke,** Psychology
1968 **Chuong C. Huang,** Physiology
1969 **Walter L. Salinger,** Psychology
1969 **Thomas T. Norton,** Psychology
1969 **Carol K. Peck,** Psychology
1970 **Nancy A. Hoisman,** Psychology
1970 **Thomas L. Davies,** Psychology
1971 **Jay E. Gould,** Psychology
1971 **Stephen Weissbluth,** Psychology
1972 **Marshall D. Buck,** Psychology
1972 **Karl Syndulko,** Psychology
1973 **David M. Seales,** Psychology
1974 **James R. Coleman,** Psychology
1974 **Kent M. Perryman,** Psychology
1976 **James C. Smith,** Psychology
1977 **Brad C. Motter,** Psychology
1980 **Thomas F. Sanquist,** Psychology
1984 **Richard S. Coyle,** Psychology

Postdoctoral fellows and visiting scientists sponsored by Donald B. Lindsley

1955-57 **James Olds,** Ph.D.
Harvard University
1956-57 **Jack D. Barchas,** M.D.
Yale University
1956-58 **Paul Bach-Y-Rita,** M.D.
University of Mexico
1956-58 **Robert G. Eason,** Ph.D.
University of Missouri
1956-58 **Walter Isaac,** Ph.D.
Ohio State University
1956-58 **Bruce Konigsmark,** M.D.
Stanford University
1956-58 **Arthur S. Schwartz,** Ph.D.
State University of New York, Buffalo
1956-58 **Edward Schwartz,** Ph.D.
Northwestern University
1957-58 **Alexander K. Bartoshuk,** Ph.D.
McGill University
1957-58 **Gershon Berkson,** Ph.D.
George Peabody College for Teachers

DEGREES, FELLOWS AND VISITING SCIENTISTS

1957-59 **Joaquin M. Fuster**, M.D.
University of Barcelona, Spain
1957-58 **Everett J. Wyers**, Ph.D.
University of California, Berkeley
1957-59 **Ann Richardson**, Ph.D.
Bryn Mawr College
1958-60 **Robert E. Farley**, Ph.D.
University of Washington
1958-60 **Frederick P. Gault**, Ph.D.
Indiana University
1958-60 **Carlos Guzman-Flores**, M.D.
Universidad Nacional Autonoma de Mexico
1958-60 **Nicolas Buendia**, M.D.
University of Bogota, Columbia
1960-61 **Clint B. Anderson**, Ph.D.
Indiana University
1960-61 **Andre J. Angyan**, M.D., Ph.D.
University of Pecs, Hungary
1960-61 **Robert S. Feldman**, Ph.D.
University of Michigan
1960-62 **Richard E. Keesey**, Ph.D.
Brown University
1960-62 **Ulker T. Keesey**, Ph.D.
Brown University
1960-61 **David A. Rodgers**, Ph.D.
University of Chicago
1960-62 **James Whitehouse**, Ph.D.
University of Colorado
1961-63 **Ernest S. Barratt**, Ph.D.
University of Texas
1961-63 **Norman M. Weinberger**, Ph.D.
Western Reserve University
1962-63 **Otto-Joachim Grüsser**, M.D.
University of Freiburg
1962-63 **Ursula H. F. Grüsser-Cornels**, M.D.
University of Marburg
1962-63 **Joseph B. Sidowski**, Ph.D.
University of Wisconsin
1962-63 **Richard W. Sroges**, Ph.D.
Northwestern University
1963-64 **Emanuel Donchin**, Ph.D.
University of California, Los Angeles
1963-64 **Charles M. Fair**, Guggenheim Fellow
1963-65 **Marcos Velasco-Campos**, M.D.
University of Mexico
1963-65 **Manfred Haider**, M.D., Ph.D.
University of Vienna
1964-65 **William M. Smith**, Ph.D.
Princeton University
1965 **Guenter H. Rose**, Ph.D.
University of California, Los Angeles
1965-67 **Solon B. Holstein**, Ph.D.
University of Missouri
1965-66 **Risto Näätänen**, Ph.D.
University of Helsinki
1965-67 **Graciela Olmos**, M.D.
University of Mexico, Guadalajara
1967-68 **Lester G. Fehmi**, Ph.D.
University of California, Los Angeles
1967-68 **Ilan Spector**, Ph.D.
University of Paris
1967-68 **Jean-Claude Roy**, D. Sc.
University of Paris
1967-68 **Paul V. Simonov**, M.D.
University of Moscow, USSR
1967-69 **James E. Skinner**, Ph.D.
University of California, Los Angeles
1967-69 **Madeleine L. Schlag-Rey**, Ph.D.
University of Bruxelles
1968-70 **Chuong G. Huang**, M.D., Ph.D.
University of Taiwan
University of California, Los Angeles
1968-71 **Harvey Anchel**, Ph.D.
McMaster University
1968-71 **Jerry D. Wicke, Ph.D.:,**
University of California, Los Angeles
1968-72 **Glenn F. Wilson**, Ph.D.
University of Arizona
1969 **Carol K. Peck**, Ph.D.
University of California, Los Angeles
1969-72 **Walter L. Salinger**, Ph.D.
University of California, Los Angeles
1970-75 **Leo M. Chalupa**, Ph.D.
Queens College, CUNY
1971-73 **Angelica W. Macadar**, M.D.
University of Uruguay, Montevideo
1972-77 **Charles L. Wilson**, Ph.D.
State University of New York, Stony Brook
1972-73 **Michel Serdaru**, M.D.
University of Paris
1972-74 **Diana S. Woodruff**, Ph.D.
University of Southern California
1972-78 **John W. Rohrbaugh**, Ph.D.
University of Illinois
1972-73 **Shoji Kitajima**, Ph.D.
Hokkaido University, Japan
1973 **Yuri Uryvaev**, M.D.
First Moscow Medical School, USSR

APPENDIX B.

1973-74 **Duane E. Shuttlesworth,** Ph.D.
University of North Carolina
1974-75 **David N. Young,** Ph.D.
University of California, Los Angeles
1974-76 **Jesus P. Machado-Salas,** M.D., Ph.D.
University of Mexico
University of California, Los Angeles
1974-76 **Stephen C. Coffin,** Ph.D.
Stanford University
1974-77 **Kent M. Perryman,** Ph.D.
University of California, Los Angeles
1976 **Noriaki Hirasuna,** Ph.D.
State University of New York, Stony Brook
1976-78 **Henry V. Soper,** Ph.D.
University of Connecticut
1977-78 **Henry Michalewski,** Ph.D.
Simon Fraser University, Vancouver
1977-78 **Lawrence Majovski,** Ph.D.
Fuller Theological Seminary, Pasadena, CA
1980 **Thomas F. Sanquist,** Ph.D.
University of California, Los Angeles
1981-83 **Jiang Qiyuan,** M.D.
Beijing Medical School, Beijing, China
1982-84 **Alfonso Lee Gonzalez**
Military Medical School, Mexico City, D. F

APPENDIX C.

Publications of Donald B. Lindsley

1. Travis, L. E., & Lindsley, D. B. (1931). The relation of frequency and extent of action currents to intensity of muscular contraction. *Journal of Experimental Psychology, 14*, 359-381.
2. Herren, R. Y., & Lindsley, D. B. (1931). Central and peripheral latencies in some tendon reflexes of the rat. *American Journal of Physiology, 99*, 167-171.
3. Travis, L. E., & Lindsley, D. B. (1933). An action current study of handedness in relation to stuttering. *Journal of Experimental Psychology, 16*, 258-270.
4. Lindsley, D. B. (1933). Some neuro-physiological sources of action current frequencies. *Psychological Monographs, 44*, 33-60.
5. Lindsley, D. B. (1934). Inhibition as an accompaniment of the knee jerk. *American Journal of Physiology, 109*, 181-191.
6. Anderson, F. M., & Lindsley, D. B. (1935). Action potentials from intercostal muscles before and after unilateral peumonectomy. *Journal of Laboratory and Clinical Medicine, 20*, 623-628.
7. Lindsley, D. B. (1935). Characteristics of single motor unit responses in human muscles during various degrees of contraction. *American Journal of Physiology, 113*, 88.
8. Herren, R. Y., & Lindsley, D. B. (1935). A note concerning cerebral dominance in the rat. *Journal of Genetic Psychology, 47*, 469-472.
9. Lindsley, D. B. (1935). Electrical activity of human motor units during voluntary contraction. *American Journal of Physiology, 114*, 90-99.
10. Lindsley, D. B. (1935). Myographic and electromyographic studies of myasthenia gravis. *Brain, 58*, 470-482.
11. Herren, R. Y., Travis, L. E., & Lindsley, D. B. (1936). The effect of lesions in the central nervous system of the rat upon reflex time. *Journal of Comparative Neurology, 63*, 241-249.
12. Rosenblueth, A., Lindsley, D. B., & Morison, R. S. (1936). A study of some decurarizing substances. *American Journal of Physiology, 115*, 53-68.
13. Lindsley, D. B., & Curnen, E. C. (1936). An electromyographic study of myotonia. *Archives of Neurology and Psychiatry* (Chicago), *35*, 253-269.
14. Lindsley, D. B. (1936). Electromyographic studies of neuromuscular disorders. *Archives of Neurology and Psychiatry* (Chicago), *36*, 128-157.
15. Lindsley, D. B. (1936). Brain potentials in children and adults. *Science, 84*, 354.
16. Lindsley, D. B., & Rubenstein, B. B. (1937). Relationship between brain potentials and some other physiological variables. *Proceedings of the Society for Experimental Biology and Medicine, 35*, 558-563.
17. Rubenstein, B. B., & Lindsley, D. B. (1937). Relation between human vaginal smears and body temperature. *Proceedings of the Society for Experimental Biology and Medicine, 35*, 618-619.
18. Lindsley, D. B., & Sassaman, W. H. (1938). Autonomic activity and brain potentials associated with "voluntary" control of the pilomotors (MM. Arrectores Pilorum). *Journal of Neurophysiology, 1*, 342-349.
19. Lindsley, D. B. (1938). Foci of activity of the alpha rhythm in the human electroencephalogram. *Journal of Experimental Psychology, 23*, 159-171.
20. Lindsley, D. B. (1938). Electrical potentials of the brain in children and adults. *Journal of General Psychology, 19*, 285-306.

21. Lindsley, D. B. (1938). "Brain waves." In W. L. Valentine (Ed.), *Experimental foundations of general psychology* (pp. 312-315). New York: Farrar & Rinehart.
22. Lindsley, D. B., & Bradley, C. (1939). Electroencephalography as an aid to understanding certain behavior disorders of childhood. *Zeitschrift für Kinderpsychiatrie, 6*, 33-37.
23. Lindsley, D. B. (1939). A longitudinal study of the occipital alpha rhythm in normal children: Frequency and amplitude standards. *Journal of Genetic Psychology, 55*, 197-213.
24. Lindsley, D. B., & Hunter, W. S. (1939). A note on polarity potentials from the human eye. *Proceedings of the National Academy of Sciences* (Washington), *25*, 180-183.
25. Lindsley, D. B. (1940). Bilateral differences in brain potentials from the two cerebral hemispheres in relation to laterality and stuttering. *Journal of Experimental Psychology, 26*, 211-225.
26. Lindsley, D. B., & Cutts, K. K. (1940). Electroencephalograms of "constitutionally inferior" and behavior problem children: Comparison with those of normal children and adults. *Archives of Neurology and Psychiatry* (Chicago), *44*, 1199-1212.
27. Lindsley, D. B. (1940). The ontogenetic development of brain potentials in human subjects. *Yearbook of National Society Studies in Education, 39*, 127-130.
28. Lindsley, D. B., & Cutts, K. K. (1941). Clinical and electroencephalographic changes in a child recovering from encephalitis. *Archives of Neurology and Psychiatry* (Chicago), *45*, 156-161.
29. Lindsley, D. B. (1942). Review: Bartley, S. H. Vision: A study of its basis. *Psychological Bulletin, 39*, 126-127.
30. Lindsley, D. B., Finger, F. W., & Henry, C. E. (1942). Some physiological aspects of audiogenic seizures in rats. *Journal of Neurophysiology, 5*, 185-198.
31. Lindsley, D. B., & Henry, C. E. (1942). The effect of drugs on behavior and the electroencephalograms of children with behavior disorders. *Psychosomatic Medicine, 4*, 140-149.
32. Lindsley, D. B. (1942). Heart and brain potentials of human fetuses *in utero*. *American Journal of Psychology, 55*, 412-416.
33. Lindsley, D. B. (1942). Physiological psychology. In R. Seashore (Ed.), *Fields of psychology* (chaps. 7-11, pp. 85-168). New York: Holt.

During World War II, as Director of a National Defense Research Committee Project on Radar Operation and Training under the Office of Scientific Research and Development (Project SC-70, NS-146, Applied Psychology Panel, NDRC), Dr. Lindsley edited or collaborated with other project members in the writing of 64 memoranda, research reports or manuals from 1943-45. Projects and studies were carried out in Army, Navy, Marine and Air Force installations mainly along the East Coast and in California. All of the publications were classified during the war.

---

34. Lindsley, D. B. (1944). Electroencephalography. In J. McV. Hunt (Ed.), *Personality and the behavior disorders* (Vol. 2, pp. 1037-1103). New York: Ronald Press.
35. Lindsley, D. B., & Wendt, G. R. (1944). *Studies in motion sickness (Series A, II): An investigation into the relationship of the electroencephalogram to motion sickness susceptibility.* Civil Aeronautics Administration, Division of Research, Washington, DC Report No. 40, 25-30.

36. Lindsley, D. B. (1945). Review: Goldstein, K. After-effects of brain injuries in war. *American Journal of Psychology, 58*, 281-283.
37. Lindsley, D. B. (1946). Review: Brock, S. The basis of clinical neurology. *Psychological Bulletin, 43*, 598-599.
38. Lindsley, D. B. (1948). Studying neuropsychology and bodily functions. In T. G. Andrews (Ed.), *Methods of psychology* (chap. 15, pp. 417-458). New York: Wiley.
39. Sprague, J. M., Schreiner, L. H., Lindsley, D. B., & Magoun, H. W. (1948). Reticulospinal influences on stretch reflexes. *Journal of Neurophysiology, 11*, 501-508.
40. Lindsley, D. B. (1949). Review: Halstead, W. C. Brain and intelligence. *Psychological Bulletin, 46*, 173-176.
41. Lindsley, D. B. (1949). *Psychophysiology: Military research and application of studies of vision, hearing, motor functions and environmental stress.* Section VI, 15 p. Armed Forces Familiarization Course in Military Psychology, American Psychological Association, Washington, DC.
42. Lindsley, D. B., Schreiner, L. H., & Magoun, H. W. (1949). An electromyographic study of spasticity. *Journal of Neurophysiology, 12*, 197-205.
43. Schreiner, L. H., Lindsley, D. B., & Magoun, H. W. (1949). Role of brain stem facilitatory systems in maintenance of spasticity. *Journal of Neurophysiology, 12*, 207-216.
44. Peterson, E. W., Magoun, H. W., McCulloch, W. S., & Lindsley, D. B. (1949). Production of postural tremor. *Journal of Neurophysiology, 12*, 371-384.
45. Lindsley, D. B. (1949). Preface to *Human factors in undersea warfare*. Washington, DC: Committee on Undersea Warfare, National Research Council. Served as Chmn. Panel on Psychology and Physiology which planned and organized this volume, 541 pp.
46. Lindsley, D. B., Bowden, J. W., & Magoun, H. W. (1949). Effect upon the EEG of acute injury to the brain stem activating system. *Electroencephalography and Clinical Neurophysiology, 1*, 475-486.
47. Lindsley, D. B., Schreiner, L. H., Knowles, W. B., & Magoun, H. W. (1950). Behavioral and EEG changes following chronic brain stem lesions in the cat. *Electroencephalography and Clinical Neurophysiology, 2*, 483-498.
48. Lindsley, D. B. (1950). Sleep. In *Collier's encyclopedia* (Vol. 17, p. 639). New York: F. P. Collier & Son.
49. Lindsley, D. B. (1950). Emotions and the electroencephalogram. In M. L. Reymert (Ed.), *Feelings and emotions: The Mooseheart symposium* (chap. 19, pp. 238-246). New York: McGraw-Hill.
50. Lindsley, D. B. (1951). Psychosomatics: Behavior disorders in children. In *Fourth Annual Guidance Workshop Publication No. 524*, Division of Higher Education (Advisement Service, Los Angeles City Schools), Los Angeles, June.
51. Lindsley, D. B. (1951). Emotion. In S. S. Stevens (Ed.), *Handbook of experimental psychology* (chap. 14, pp. 473-516). New York: Wiley.
52. Lindsley, D. B. (1952). Obituary: Robert Holmes Seashore, 1902-1951. *American Journal of Psychology, 65*, 114-116.
53. Lindsley, D. B. (1952). Brain stem influences on spinal motor activity. *Proceedings of the Association for Research in Nervous and Mental Diseases, 30*, 174-195. (1952). Also in P. Bard (Ed.), *Patterns of organization in the central nervous system* (chap. 9). Baltimore: Williams & Wilkens.
54. Lindsley, D. B. (1952). Psychological phenomena and the electroencephalogram. *Electroencephalography and Clinical Neurophysiology, 4*, 443-456.

55. Lindsley, D. B. (1953). Effect of photic stimulation in visual pathways from retina to cortex. *Science, 117*, 469.
56. Verzeano, M., Lindsley, D. B., & Magoun, H. W. (1953). Nature of recruiting response. *Journal of Neurophysiology, 16*, 183-195.
57. Lindsley, D. B. (1953). Response to photic stimulation in visual pathways from retina to cortex. *American Psychologist, 8*, 510.
58. Lindsley, D. B. (1954). Review: Walter, W. C., The living brain. *Psychological Bulletin, 51*, 317-318.
59. Lindsley, D. B. (1954). Electrical response to photic stimulation in visual pathways of the cat. *Electroencephalography and Clinical Neurophysiology, 6*, 690-691.
60. Lindsley, D. B. (1954). Psychology. In E. A. Spiegel (Ed.), *Progress in neurology and psychiatry, 9*, 391-416.
61. Lindsley, D. B. (1955). Symposium de Talamo (Discussion). *VI Congreso Latino Americano de Neurocirugia, 6*, 947-949. (Montevideo, Uruguay)
62. Lindsley, D. B., & Griffiths, R. S. (1955). Differential response of eye and optic pathways to intensity and wave-length. *Science, 122*, 976.
63. Lindsley, D. B. (1955). Higher functions of the nervous system. *Annual Review of Physiology, 17*, 311-338.
64. Lindsley, D. B. (1955). The psychology of lie detection. In G. J. Dudycha (Ed.), *Psychology for law enforcement officers* (pp. 89-125). Springfield, IL: Thomas.
65. Lindsley, D. B. (1956). Physiological psychology. *Annual Review of Psychology, 7*, 323-348.
66. Lansing, R. W., Schwartz, E., & Lindsley, D. B. (1956). Reaction time and EEG activation. *American Psychologist, 11*, 433.
67. Lindsley, D. B., & Lansing, R. W. (1956). Flicker and two-flash fusional thresholds. *American Psychologist, 11*, 433.
68. Blinn, K. A., & Lindsley, D. B. (1956). Color sensitivity in photoepilepsy. *Electroencephalography and Clinical Neurophysiology, 8*, 166-167.
69. Lindsley, D. B., & Jones, M. (1956). Clinical and EEG correlations in cerebral palsy children and adults. *Electroencephalography and Clinical Neurophysiology, 8*, 168.
70. Lindsley, D. B. (1956). Basic perceptual processes and the EEG. *Psychiatry Research Reports, 6*, 161-170.
71. Lindsley, D. B. (1957). Psychological aspects of consciousness. In *Clinical Neurosurgery: Proceedings of the Congress of Neurological Surgeons, Los Angeles, 1955* (pp. 175-186). Baltimore: Williams & Wilkens.
72. Lindsley, D. B. (1957). Psychophysiology and motivation. In M. R. Jones (Ed.), *Nebraska Symposium on Motivation* (pp. 44-105). Lincoln, NE: University of Nebraska Press.
73. Lindsley, D. B. (1958). Psychophysiology and perception. In R. Patton (Ed.), *The description and analysis of behavior* (pp. 48-91). (University of Pittsburgh Tenth Annual Conference on Current Trends in Psychology.) Pittsburgh: University of Pittsburgh Press.
74. Lindsley, D. B. (1958). The reticular system and perceptual discrimination. In H. H. Jasper et al. (Eds.), *International Symposium on Reticular Formation of the Brain* (pp. 513-534). Boston: Little, Brown.
75. Lindsley, D. B. (1958). The brain and nervous system as an integrating mechanism. In F. L. Ruch (Ed.), *Psychology and life* (pp. 558-562). Chicago: Scott, Foresman.
76. Lindsley, D. B., & Emmons, W. H. (1958). Perceptual blanking, evoked potentials and perception time. *Electroencephalography and Clinical Neurophysiology, 10*, 359.

77. Lindsley, D. B., & Emmons, W. H. (1958). Perception time and evoked potentials. *Science, 127*, 1061.
78. Lansing, R. W., Schwartz, E., & Lindsley, D. B. (1959). Reaction time and EEG activation under alerted and nonalerted conditions. *Journal of Experimental Psychology, 58*, 1-7.
79. Lindsley, D. B. (1960). Attention, consciousness, sleep and wakefulness. In J. Field (Ed.), *Handbook of physiology: Neurophysiology III* (chap. LXIV, pp. 1553-1593). Washington, DC: American Physiological Society.
80. Lindsley, D. B. (1961). Encephalographic and psychological aspects of the kernicterus child. In C. A. Swinyard (Ed.), *Kernicterus and its importance in cerebral palsy* (pp. 247-253). Springfield, IL: Thomas.
81. Lindsley, D. B. (1961). The reticular activating system and perceptual integration. In D. E. Sheer (Ed.), *Electrical stimulation of the brain* (pp. 331-349). (Houston Symposium on Brain Stimulation.) Austin, TX: University of Texas Press.
82. Guzman-Flores, C., Buendia, N., & Lindsley, D. B. (1961). Cortical and reticular influences upon evoked responses in dorsal column nuclei. *Federation Proceedings, 20*, 330.
83. Lindsley, D. B. (1961). Common factors in sensory deprivation, sensory distortion, and sensory overload. In P. Solomon et al. (Eds.), *Sensory deprivation* (chap. 12, pp. 174-194). Cambridge, MA: Harvard University Press.
84. Schwartz, A. S., & Lindsley, D. B. (1961). Critical flicker frequency and photic following in the cat. *Federation Proceedings, 4, No. 3*, 105.
85. Lindsley, D. B. (1961). Electrophysiology of the visual system and its relation to perceptual phenomena. In M. A. B. Brazier (Ed.), *Brain and behavior* (Vol. I, pp. 359-392). Washington, DC: American Institute of Biological Sciences.
86. Caveness, W. F., van Wagenen, G., & Lindsley, D. B. (1961). Comparison of monkey and human EEG development from birth to puberty. *Electroencephalography and Clinical Neurophysiology, 13*, 105.
87. Guzman-Flores, C., Buendia, N., Anderson, C., & Lindsley, D. B. (1962). Cortical and reticular influences upon evoked responses in dorsal column nuclei. *Experimental Neurology, 5*, 37-46.
88. Lindsley, D. F., Wendt, R. H., Fugett, R., Lindsley, D. B., & Adey, W. R. (1962). Diurnal activity cycles in monkeys under prolonged visual-pattern deprivation. *Journal of Comparative and Physiological Psychology, 55*, 633-640.
89. Keesey, R. E., & Lindsley, D. B. (1962). Duration of stimulation and the reinforcing properties of hypothalamic stimulation. *American Psychologist, 17*, 375.
90. Donchin, E., Wicke, J. D., & Lindsley, D. B. (1963). Cortical evoked potentials and perception of paired flashes. *Science, 141*, 1285-1286.
91. Guzman-Flores, C., Gault, F. P., Anderson, C., & Lindsley, D. B. (1963). Pyramidal influences upon potentials evoked in sensory nuclei. *Boletin Inst. Estudios Medicos y Biologicas, Mexico, 21*, 65-75.
92. Haider, M., Spong, P., & Lindsley, D. B. (1964). Attention, vigilance and cortical evoked-potentials in humans. *Science, 145*, 180-182.
93. Lindsley, D. B. (1964). Brain development and behavior: Historical introduction. In W. A. & H. E. Himwich (Eds.), *The developing brain: Progress in brain research* (Vol. 9, pp. 1-5). Amsterdam: Elsevier.
94. Lindsley, D. B. (1964). The ontogeny of pleasure: Neural and behavioral development. In R. G. Heath (Ed.), *The role of pleasure in behavior* (pp. 3-22). New York: Hoeber.

95. Lindsley, D. B., Wendt, R. H., Lindsley, D. F., Fox, S. S., Howell, J., & Adey, W. R. (1964). Diurnal activity, behavior and EEG responses in visually deprived monkeys. *Annals of the New York Academy of Sciences, 117*, 564-588. Also in H. E. Whipple (Ed.), *Photo-neuro-endocrine effects in circadian systems, with particular reference to the eye*. New York: New York Academy of Sciences.

96. Wicke, J. D., Donchin, E., & Lindsley, D. B. (1964). Visual evoked potentials as a function of flash luminance and duration. *Science, 146*, 83-85.

97. Lindsley, D. B., & Harrell, T. W. (1964). History of the Western Psychological Association. *American Psychologist, 19*, 290-291.

98. Schwartz, A. S., & Lindsley, D. B. (1964). Critical flicker frequency and photic following in the cat. *Boletin Inst. Estudios Medicos y Biologicas, Mexico, 22*, 249-262.

99. Haider, M., & Lindsley, D. B. (1964). Microvibrations in man and dolphin. *Science, 146*, 1181-1183.

100. Lindsley, D. B. (1964). Electrocortical correlates of temporally ordered visual phenomena. *Acta Psychologica, 23*, 201-202.

101. Haider, M., Spong, P., & Lindsley, D. B. (1964). Cortical evoked potentials during visual vigilance task performance. *Electroencephalography and Clinical Neurophysiology, 17*, 714.

102. Spong, P., Haider, M., & Lindsley, D. B. (1964). Average evoked potentials during fluctuations of attention in humans. *Electroencephalography and Clinical Neurophysiology, 17*, 714-715.

103. Weinberger, N. M., & Lindsley, D. B. (1964). Behavioral and electroencephalographic arousal to contrasting novel stimulation. *Science, 144*, 1355-1357.

104. Donchin, E., & Lindsley, D. B. (1964). Retroactive brightness enhancement with paired flashes. *American Psychologist, 19*, 515.

105. Lindsley, D. B. (1964). (English Editor) *Kh. S. Koshtoyants, Essays on the history of physiology in Russia* (361 pp.). Washington, DC: American Institute of Biological Sciences. Translated from Russian by David P. Boder, Kristan Hanes, & Natalie O'Brien. (Available from American Psychological Association, Washington, DC)

106. Lindsley, D. B. (1965). (English Editor) *I. M. Sechenov, Autobiographical notes* (174 pp.). Washington, DC: American Institute of Biological Sciences. Translated from Russian by Kristan Hanes. (Available from American Psychological Association, Washington, DC)

107. Lindsley, D. B. (1965). (English Editor) *L. G. Voronin, A. N. Leontiev, A. R. Luria, E. N. Sokolov, & O. S. Vinogradova (Eds.), Orienting reflex and exploratory behavior* (462 pp.). Washington, DC: American Institute of Biological Sciences. Translated from Russian by Vsevolod Shmelev and Kristan Hanes. (Available from American Psychological Association, Washington, D. C.)

108. Weinberger, N. M., Velasco, M., & Lindsley, D. B. (1965). Effects of lesions upon thalamically induced electrocortical desynchronization and recruiting. *Electroencephalography and Clinical Neurophysiology, 18*, 369-377.

109. Weinberger, N. M., Velasco, M., & Lindsley, D. B. (1965). Differential effects of reinforced and non-reinforced stimuli upon electrocortical desynchronization and recruiting responses. *Psychonomic Science, 2*, 129-130.

110. Donchin, E., & Lindsley, D. B. (1965). Retroactive brightness enhancement with brief paired flashes of light. *Vision Research, 5*, 59-70.

111. Spong, P., Haider, M., & Lindsley, D. B. (1965). Selective attentiveness and cortical evoked potentials to visual and auditory stimuli. *Science, 148*, 395-397.

112. Lindsley, D. B., & Rose, G. H. (1965). Development of visually evoked responses in kittens. *Federation Proceedings, 24*, 274.
113. Velasco, M., Weinberger, N. M., & Lindsley, D. B. (1965). A unitary arousal system revealed by blocking of recruiting responses. *Electroencephalography and Clinical Neurophysiology, 18*, 517.
114. Donchin, E., & Lindsley, D. B. (1965). Cortical evoked potentials and reaction times. *Electroencephalography and Clinical Neurophysiology, 18*, 523.
115. Rose, G., & Lindsley, D. B. (1965). Longitudinal development of evoked potentials in kittens. *Electroencephalography and Clinical Neurophysiology, 18*, 525.
116. Spong, P., Haider, M., & Lindsley, D. B. (1965). Selective attention and cortical evoked potentials. *Electroencephalography and Clinical Neurophysiology, 18*, 523.
117. Rose, G. H., & Lindsley, D. B. (1965). Visually evoked electrocortical responses in kittens: Development of specific and nonspecific systems. *Science, 148*, 1244-1246.
118. Velasco, M., & Lindsley, D. B. (1965). Role of orbital cortex in regulation of thalamocortical electrical activity. *Science, 149*, 1375-1377.
119. Lindsley, D. B., Fehmi, L. G., & Adkins, J. W. (1965). Electrophysiological correlates of visual perception in man and monkey. *Sixth International Congress of Electroencephalography and Clinical Neurophysiology, 6*, 237-238.
120. Donchin, E., & Lindsley, D. B. (1965). Visually evoked response correlates of perceptual masking and enhancement. *Electroencephalography and Clinical Neurophysiology, 19*, 325-335.
121. Donchin, E., & Lindsley, D. B. (1966). Average evoked potentials and reaction times to visual stimuli. *Electroencephalography and Clinical Neurophysiology, 20*, 217-223.
122. Rose, G. H., & Lindsley, D. B. (1966). The differentiation of visually evoked cortical response components in kittens. *Electroencephalography and Clinical Neurophysiology, 21*, 404.
123. Lindsley, D. B. (1966). Psychology and the future of man. In *Reflections on the Future - The life sciences* (pp. 31-42). Proceedings of the Convocation at Trinity College. Hartford, CT: Trinity College Press.
124. Thompson, R. F., Lindsley, D. B., & Eason, R. G. (1966). Physiological psychology. In J. B. Sidowski (Ed.), *Experimental methods and instrumentation in psychology* (pp. 117-182). New York: McGraw-Hill.
125. Lindsley, D. B. (1967). Electroencephalography. In *International encyclopedia of the social sciences*. New York: Crowell Collier and Macmillan.
126. Weinberger, N. M., Velasco, M., & Lindsley, D. B. (1967). The relationship between cortical synchrony and behavioral inhibition. *Electroencephalography and Clinical Neurophysiology, 23*, 297-305.
127. Skinner, J. E., & Lindsley, D. B. (1967). Electrophysiological and behavioral effects of blockade of the nonspecific thalamo-cortical system. *Brain Research, 6*, 95-118.
128. Lindsley, D. B., & Lumsdaine, A. A. (1967). (Eds.). *Brain function and learning* (364 pp.). Brain Function, Vol. 4. Berkeley and Los Angeles: University of California Press.
129. Clemente, C. D., & Lindsley, D. B. (1967). (Eds.). *Aggression and defense: Neural mechanisms and social patterns* (361 pp.). Brain Function, Vol. 5. Berkeley and Los Angeles: University of California Press.
130. Velasco, M., Skinner, J. E., & Lindsley, D. B. (1967). Blocking of electrocortical activity mediated by a thalamo-cortical system by lesions in the forebrain and rostral diencephalon. *Electroencephalography and Clinical Neurophysiology, 22*, 292.

131. Skinner, J. E., & Lindsley, D. B. (1967). Reversible blocking of a thalamo-cortical system by cryogenic probes in acute and chronic cat preparations. *Electroencephalography and Clinical Neurophysiology, 22*, 292.

132. Fehmi, L. G., Adkins, J. W., & Lindsley, D. B. (1967). Electrophysiological correlates of visual discrimination in monkeys. *Electroencephalography and Clinical Neurophysiology, 22*, 292-293.

133. Lindsley, D. B., Fehmi, L. G., & Adkins, J. W. (1967). Visually evoked potentials during perceptual masking in man and monkey. *Electroencephalography and Clinical Neurophysiology, 23*, 79.

134. Lindsley, D. B. (1967). (English Editor) V. N. Chernigovskiy, *Interoceptors* (904 pp.). Washington, DC: American Psychological Association. Translated from Russian by George Onischenko. (Available from American Psychological Association, Washington, DC)

135. Skinner, J. E., & Lindsley, D. B. (1967). The effect of cryogenic blocking of the diffuse thalamocortical recruiting system upon visual evoked potentials and behavior. *Electroencephalography and Clinical Neurophysiology, 23*, 79.

136. Lindsley, D. B. (1968). Foreward to the English Edition of S. M. Blinkov and I. I. Glezer, *The human brain in figures and tables*. (Translated from Russian by Basil Haigh.) New York: Plenum Press-Basic Books.

137. Weinberger, N. M., Nakayama, K., & Lindsley, D. B. (1968). Electrocortical recruiting responses during classical conditioning. *Electroencephalography and Clinical Neurophysiology, 24*, 16-24.

138. Weinberger, N. M., Yeudall, L., & Lindsley, D. B. (1968). EEG correlates of reinforced behavioral inhibition. *Psychonomic Science, 10*, 11-12.

139. Rose, G. H., & Lindsley, D. B. (1968). Development of visually evoked potentials in kittens: Specific and nonspecific responses. *Journal of Neurophysiology, 31*, 607-623.

140. Skinner, J. E., & Lindsley, D. B. (1968). Reversible cryogenic blockade of neural function in the brain of unrestrained animals. *Science, 161*, 595-597.

141. Velasco, M., Skinner, J. E., Asaro, K. D., & Lindsley, D. B. (1968). Thalamocortical systems regulating spindle bursts and recruiting responses. I. Effect of cortical ablations. *Electroencephalography and Clinical Neurophysiology, 25*, 463-470.

142. Lindsley, D. B., & Riesen, A. (1968). (Eds.). Biological substrates of development and behavior. In *Perspectives on human deprivation: Biological, psychological and sociological* (chap. IV, pp. 229-269). Bethesda, MD: National Institute of Child Health and Human Development, U.S. Department of Health, Education and Welfare.

143. Lindsley, D. B., & Livingston, R. B. (1968). Investigations on learning in the sloth. *Conditional Reflex, 3*, 141-142.

144. Lindsley, D. B. (1968). Growth, maturation and development. In *Perspectives on human deprivation: Biological, psychological and sociological* (pp. 231-240). Bethesda, MD: National Institute of Child Health and Human Development, U.S. Department of Health, Education and Welfare.

145. Skinner, J. E., & Lindsley, D. B. (1968). The effects of blocking the nonspecific thalamocortical system upon the sensory evoked response in the visual cortex in the cat. *Electroencephalography and Clinical Neurophysiology, 24*, 393.

146. Spong, P., & Lindsley, D. B. (1968). Cortical evoked responses and attentiveness in man: Differential effects of selective attentiveness and general alertness level. *Electroencephalography and Clinical Neurophysiology, 24*, 396-397.

147. Lindsley, D. B. (1968, 1967). Developmental brainstorming. *Contemporary Psychology, 13*, 627-628. (Review of Esther Milner: *Human neural and behavioral development: A relational inquiry, with implications for personality* (p. 393). Springfield, IL: Thomas.
148. Velasco, M., Weinberger, N. M., & Lindsley, D. B. (1968). Effect of thalamo-cortical activation on recruiting responses: I. Reticular stimulation. *Acta Neurologica Latinoamerica, 14*, 99-115.
149. Velasco, M., & Lindsley, D. B. (1968). Effect of thalamo-cortical activation on recruiting responses: II. Peripheral and central neural stimulation. *Acta Neurologica Latinoamerica, 14*, 188-199.
150. Velasco, M., & Lindsley, D. B. (1968). Effect of thalamo-cortical activation on recruiting responses: III. Reticular lesions. *Acta Neurologica Latinoamerica, 14*, 271-282.
151. Adkins, J. W., Fehmi, L. G., & Lindsley, D. B. (1969). Perceptual discrimination in monkeys: Retroactive visual masking. *Physiology and Behavior, 4*, 255-260.
152. Fehmi, L. G., Adkins, J. W., & Lindsley, D. B. (1969). Electrophysiological correlates of visual perceptual masking in monkeys. *Experimental Brain Research, 7*, 299-316.
153. Wicke, J. D., & Lindsley, D. B. (1969). Apparatus for randomly determining visual stimulus positions using a 'forced-choice' psychophysical procedure. *American Journal of Psychology, 82*, 410-413.
154. Skinner, J. E., & Lindsley, D. B. (1969). Effect of cryogenic blockade of a nonspecific thalamo-cortical synchronizing system upon sensory evoked potentials. *Electroencephalography and Clinical Neurophysiology, 26*, 333-334.
155. Wilson, G. F., & Lindsley, D. B. (1969). Differential effects of repetitive visual stimulation on alpha rhythm and average visual evoked potentials. *Electroencephalography and Clinical Neurophysiology, 27*, 704-705.
156. Lindsley, D. B. (1969). Average evoked potentials - achievements, failures and prospects. In E. Donchin & D. B. Lindsley (Eds.), *Average evoked potentials: Methods, results and evaluations* (pp. 1-43). Washington, DC: NASA.
157. Donchin, E., & Lindsley, D. B. (1969). (Eds). *Average evoked potentials: Methods, results and evaluations* (400 pp.). Washington, DC: NASA.
158. Lindsley, D. B. (1969). Neurophysiological mechanisms of attention and perception: Their role in information processing in the visual system. In D. P. Kimble (Ed.), *Readiness to remember* (pp. 437-507). (Proceedings of Third Conference on Learning, Remembering and Forgetting.) New York: Gordon and Breach.
159. Schlag-Rey, M., & Lindsley, D. B. (1970). Effect of prefrontal lesions on trained anticipatory visual attending in cats. *Physiology and Behavior, 5*, 1033-1041.
160. Lindsley, D. B. (1970). The role of nonspecific reticulo-thalamo- cortical systems in emotion. In P. Black (Ed.), *Physiological correlates of emotion* (pp. 147-188). New York: Academic Press.
161. Young, F. A., & Lindsley, D. B. (1970). (Eds.). *Early experience and visual information processing in perceptual and reading disorders* (533 pp.). Washington, DC: National Academy of Sciences.
162. Lindsley, D. B., & Young, F. A. (1970). Introduction. In F. A. Young & D. B. Lindsley (Eds.), *Early experience and visual information processing in perceptual and reading disorders* (pp. 1-13). Washington, DC: National Academy of Sciences.
163. Wilson, G. F., & Lindsley, D. B. (1971). Detachable head holder for human subjects. *Psychophysiology, 7*, 281-282.
164. Lindsley, D. B. (1971). (English Editor) S. M. Blinkov & N. A. Smirnov, *Brain displacements and deformations* (218 pp.). New York: Plenum.

165. Kietzman, M. L., Boyle, R. C., & Lindsley, D. B. (1971). Perceptual masking: Peripheral vs. central factors. *Perception and Psychophysics, 9*, 350-352.
166. Skinner, J. E., & Lindsley, D. B. (1971). Enhancement of visual and auditory evoked potentials during blockade of the nonspecific thalamo-cortical system. *Electroencephalography and Clinical Neurophysiology, 31*, 1-6.
167. Salinger, W. L., & Lindsley, D. B. (1971). The suppression-recovery effect and visual adaptation in the cat. *Vision Research, 11*, 1435-1444.
168. Salinger, W. L., & Lindsley, D. B. (1971). Response latency as a function of flash rate. *Society for Neuroscience Abstracts*, 49.
169. Norton, T. T., & Lindsley, D. B. (1971). Visual behavior after bilateral superior colliculus lesions in kittens and cats. *Federation Proceedings, 30*, 615.
170. Anchel, H., & Lindsley, D. B. (1972). Differentiation of two reticulo-hypothalamic systems regulating hippocampal activity. *Electroencephalography and Clinical Neurophysiology, 32*, 209-226.
171. Peck, C., & Lindsley, D. B. (1972). Average evoked potential correlates of two-flash perceptual discrimination in cats. *Vision Research, 12*, 641-652.
172. Lindsley, D. B. (1972). (Translation Editor) A. N. Sokolov, *Inner speech and thought* (283 pp.). New York: Plenum.
173. Salinger, W. L., & Lindsley, D. B. (1972). The suppression-recovery effect in relation to stimulus repetition and rapid light adaptation. *Vision Research, 12*, 1897-1905.
174. Lindsley, D. B. (1972). (Editor and Organizer) *Human factors in long-duration space flight* (272 pp.). Washington, DC: National Academy of Sciences.
175. Lindsley, D. B. (1972). Introduction. In D. B. Lindsley (Ed.), *Human factors in long-duration space flight* (pp. 1-14). Washington, DC: National Academy of Sciences.
176. Lindsley, D. B. (1972). Summary and concluding remarks. In C. D. Clemente, D. P. Purpura, & F. E. Mayer (Eds.), *Sleep and the maturing nervous system* (pp. 443-460).
177. Chalupa, L. M., Anchel, H., & Lindsley, D. B. (1972). Modification of photic pulvinar responses by local cooling of cortical and subcortical visual structures. *Federation Proceedings, 31*, 84.
178. Peck, C. K., & Lindsley, D. B. (1972). Unit and evoked potentials in cat optic tract to paired light flashes and their relationship to perceptual discrimination. *Society for Neuroscience Abstracts*, 98.
179. Chalupa, L. M., Anchel, H., & Lindsley, D. B. (1972). Visual input to the pulvinar via lateral geniculate, superior colliculus and visual cortex in the cat. *Experimental Neurology, 36*, 449-462.
180. Lindsley, D. B. (1972). Two thousand years of pondering brain and behavior. In H. Messel & S. T. Butler (Eds.), *Brain mechanisms and the control of behaviour* (pp. 207-234). Sydney, Australia: Shakespeare Head Press. Also in G. C. Galbraith, M. L. Kietzman, & E. Donchin (Eds.), (1987) *Neurophysiology and psychophysiology: Experimental and clinical applications* (pp. 6-27). Hillsdale, NJ: Lawrence Erlbaum Associates.
181. Lindsley, D. B. (1972). Ontogenetic development of brain and behaviour. In H. Messel & S. T. Butler (Eds.), *Brain mechanisms and the control of behaviour* (pp. 274-325). Sydney, Australia: Shakespeare Head Press.
182. Huang, C., & Lindsley, D. B. (1973). Polysensory responses and sensory interaction in pulvinar and related posterolateral nuclei in cat. *Electroencephalography and Clinical Neurophysiology, 34*, 265-280.
183. Peck, C. K., & Lindsley, D. B. (1973). Single unit and evoked potential responses in cat optic tract to paired light flashes. *Experimental Brain Research, 16*, 371-382.

184. Chalupa, L. M., Anchel, H., & Lindsley, D. B. (1973). Effects of cryogenic blocking of pulvinar upon visually evoked responses in the cortex of the cat. *Experimental Neurology, 39*, 112-122.

185. Lindsley, D. B., Seales, D. M., & Wilson, G. F. (1973). Changes in the late components of visual evoked potentials with visual information processing. *Society for Neuroscience Abstracts*, 422.

186. Chalupa, L. M., Macadar, A. W., & Lindsley, D. B. (1973). Differential modification of responses of tonic and phasic lateral geniculate units to visual flash stimuli after click-flash pairing. *Society for Neuroscience Abstracts*, 299.

187. Salinger, W. L., & Lindsley, D. B. (1973). Patterns of unit activity in optic tract of cat during suppression-recovery effect: Relationship to high intensity effect. *Vision Research, 13*, 2121-2127.

188. Macadar, A. W., & Lindsley, D. B. (1973). A search for the brainstem origin of two hypothalamic-hippocampal systems mediating hippocampal theta activity and desynchronization. *Society for Neuroscience Abstracts*, 282.

189. Skinner, J. E., & Lindsley, D. B. (1973). A nonspecific mediothalamic-frontocortical system: Its influence on electrocortical activity and behavior. In K. H. Pribram & A. R. Luria (Eds.), *Psychophysiology of the frontal lobes* (pp. 185-234). New York: Academic Press.

190. Chalupa, L. M., Macadar, A. W., & Lindsley, D. B. (1973). Spontaneous and evoked activity in the pulvinar and lateralis posterior of the cat. *Federation Proceedings, 32*, 339.

191. Lindsley, D. B. (1973). EEG and autonomic relationships. In G. R. Gullickson (Ed.), *The psychophysiology of Darrow* (pp. 35-40). New York: Academic Press.

192. Gould, J. E., Chalupa, L. M., & Lindsley, D. B. (1974). Modification of pulvinar and geniculo-cortical evoked potentials during visual discrimination learning in monkeys. *Electroencephalography and Clinical Neurophysiology, 36*, 639-649.

193. Macadar, A. W., Chalupa, L. M., & Lindsley, D. B. (1974). Differentiation of brain stem loci which affect hippocampal and neocortical electrical activity. *Experimental Neurology, 43*, 499-514.

194. Lindsley, D. B., & Wicke, J. D. (1974). The electroencephalogram: Autonomous electrical activity in man and animals. In R. F. Thompson & M. M. Patterson (Eds.), *Bioelectric recording techniques*. Part B. *Electroencephalography and human brain potentials* (pp. 3-83). New York: Academic Press.

195. Chalupa, L. M., Macadar, A. W., & Lindsley, D. B. (1974). Response of pulvinar and lateralis posterior units to visual and auditory input in the cat: Interaction with cortex, lateral geniculate and superior colliculus. *Electroencephalography and Clinical Neurophysiology, 36*, 210.

196. Lindsley, D. B., Wilson, G. F., & Seales, D. M. (1974). Evoked potentials and information processing. *Electroencephalography and Clinical Neurophysiology, 36*, 213.

197. Coleman, J. R., & Lindsley, D. B. (1974). Hippocampal and neocortical EEG changes during free and operant behavior. *Society for Neurosciences Abstracts*, 175.

198. Wilson, C. L., & Lindsley, D. B. (1974). Effects of hypothalamic stimulation upon septal pacemaker mechanisms influencing hippocampal theta rhythm. *Society for Neurosciences Abstracts*, 479.

199. Perryman, K. M., & Lindsley, D. B. (1974). Visually evoked responses in pulvinar, lateral geniculate and visual cortex to patterned and unpatterned stimuli in squirrel monkey. *Society for Neurosciences Abstracts*, 371.

200. Chalupa, L. M., Rohrbaugh, J., Gould, J. E., & Lindsley, D. B. (1974). Evoked potential and reaction time correlates in monkeys during a simultaneous visual discrimination task. *Society for Neurosciences Abstracts*, 164.
201. Velasco, M., Skinner, J. E., Asaro, K. D., & Lindsley, D. B. (1975). Thalamo-cortical systems regulating spindle bursts and recruiting responses. II. Effect of thalamic lesions. *Acta Neurologica Latinoamerica, 21*, 31-39.
202. Seales, D., & Lindsley, D. B. (1975). Modification of human visually evoked potentials by complexity and cerebral lateralization of information processing. *Electroencephalography and Clinical Neurophysiology, 38*, 101.
203. Atwell, C. W., & Lindsley, D. B. (1975). Development of visually evoked responses and visually guided behavior in kittens: Effects of superior colliculus and lateral geniculate lesions. *Developmental Psychobiology, 8*, 465-478.
204. Chalupa, L. M., Macadar, A. W., & Lindsley, D. B. (1975). Response plasticity of lateral geniculate neurons during and following pairing of auditory and visual stimuli. *Science, 190*, 290-292.
205. Chalupa, L. M., Coyle, R., & Lindsley, D. B. (1975). Contrasting effects of inferior and medial-lateral pulvinar lesions on visual pattern discrimination learning in monkeys (Macaca nemestrina). *Society for Neurosciences Abstracts, 1*, 50.
206. Wilson, C. L., Motter, B., & Lindsley, D. B. (1975). Effects of cryogenic blockade of dorsal fornix and fimbria upon hippocampal theta rhythm and septal cellular activity. *Society for Neuroscience Abstracts, 1*, 544.
207. Lindsley, D. B., & Wilson, C. L. (1975). Brainstem-hypothalamic system influencing hippocampal activity and behavior. In R. L. Isaacson & K. H. Pribram (Eds.), *The hippocampus* (Vol. 2, pp. 247-277). New York: Plenum.
208. Coleman, J. R., & Lindsley, D. B. (1975). Hippocampal electrical correlates of free behavior and behavior induced by stimulation of two hypothalamic-hippocampal systems in the cat. *Experimental Neurology, 49*, Part 2, 506-528.
209. Rohrbaugh, J. W., Syndulko, K., & Lindsley, D. B. (1976). Brain wave components of the contingent negative variation in humans. *Science, 191*, 1055-1057.
210. Lindsley, D. B. (1976). Mental retardation: Historical and psychophysiological perspective. In R. Karrer (Ed.), *Developmental psychophysiology of mental retardation* (pp. 3-38). Springfield, IL: Thomas.
211. Chalupa, L. M., Rohrbaugh, J. W., Gould, J. E., & Lindsley, D. B. (1976). Cortical and subcortical visual evoked potential correlates of reaction time in monkeys. *Journal of Comparative and Physiological Psychology, 90*, 119-126.
212. Chalupa, L. M., Coyle, R. S., & Lindsley, D. B. (1976). Effect of pulvinar lesions on visual pattern discrimination in monkeys. *Journal of Neurophysiology, 39*, 354-369.
213. Wilson, C. L., Motter, B. C., & Lindsley, D. B. (1976). Influences of hypothalamic stimulation upon septal and hippocampal electrical activity in the cat. *Brain Research, 107*, 55-68.
214. Velasco, M., Skinner, J. E., & Lindsley, D. B. (1976). Thalamo-cortical systems regulating spindle bursts and recruiting responses. III. Effects of lesions in the forebrain and rostral diencephalon. *Acta Neurologica Latinoamerica, 22*, 1-9.
215. Wilson, C. L., Uryvaev, Y. V., & Lindsley, D. B. (1976). Diencephalic stimulation affecting hippocampal and neocortical electrical activity in cats. *Federation Proceedings, 35, No. 3*, 483.
216. Perryman, K. M., Lindsley, D. F., & Lindsley, D. B. (1976). Pulvinar unit responses associated with eye movements in squirrel monkeys. *Society for Neuroscience Abstracts, 2*, Part II, 280.

217. Wilson, C. L., Hirasuna, N., & Lindsley, D. B. (1976). Brain stem sources of hippocampal theta rhythm investigated by cryogenic blockade of hypothalamic pathways in the cat. *Society for Neuroscience Abstracts, 2*, Part I, 399.

218. Perryman, K. M., & Lindsley, D. B. (1977). Visual responses in geniculo-striate and pulvino-striate systems to patterned and unpatterned stimuli in squirrel monkeys. *Electroencephalography and Clinical Neurophysiology, 42*, 157-177.

219. Coleman, J. R., & Lindsley, D. B. (1977). Behavioral and hippocampal electrical changes during operant learning in cats and effects of stimulating two hypothalamic-hippocampal systems. *Electroencephalography and Clinical Neurophysiology, 42*, 309-331.

220. Motter, B. C., & Lindsley, D. B. (1977). Response properties in the nucleus lateralis posterior of the cat associated with visual orientation. *Society for Neuroscience Abstracts, 3*, 570.

221. Syndulko, K., & Lindsley, D. B. (1977). Motor and sensory determinants of cortical slow potential shifts in man. In J. E. Desmedt (Ed.), *Attention, voluntary contraction and event-related potentials. Progress in clinical Neurophysiology* (Vol. 1, pp. 97-131). Basel: Karger.

222. Rohrbaugh, J. W., Syndulko, K., & Lindsley, D. B. (1978). Cortical slow negative waves following non-paired stimuli: Effects of task factors. *Electroencephalography and Clinical Neurophysiology, 45*, 551-567.

223. Rohrbaugh, J. W., Syndulko, K., & Lindsley, D. B. (1979). Cortical slow negative waves following non-paired stimuli: Effects of modality, intensity and rate of stimulation. *Electroencephalography and Clinical Neurophysiology, 46*, 416-427.

224. Lindsley, D. B. (1979). Does the EEG reflect space- and time-coded cerebral functions? *Contemporary Psychology, 24*, 824. (Review of M. N. Livanov: *Spatial organization of cerebral processes* (1977). New York: Halstead Press).

225. Soper, H. V., Zweizig, S., Gilman, T., & Lindsley, D. B. (1979). Performance of Nemestrina monkeys with bilateral inferotemporal cortex lesions in a simultaneous visual pattern discrimination task demanding of attention. *Society for Neuroscience Abstracts, 5*, 121.

226. Rohrbaugh, J. W., Syndulko, K., Sanquist, T. F., & Lindsley, D. B. (1980). Synthesis of the contingent negative variation brain potential from noncontingent stimulus and motor elements. *Science, 208*, 1165-1168.

227. Sanquist, T. F., Rohrbaugh, J. W., Syndulko, K., & Lindsley, D. B. (1980). An event-related potential analysis of coding processes in human memory. In H. H. Kornhuber & L. Deecke (Eds.), *Motivation, motor and sensory processes in the brain. Progress in Brain Research* (Vol. 54, pp. 655-660). Amsterdam: Elsevier/North Holland Biomedical Press.

228. Sanquist, T. F., Rohrbaugh, J. W., Syndulko, K., & Lindsley, D. B. (1980). Electrocortical signs of levels of processing: Perceptual analysis and recognition memory. *Psychophysiology, 17*, 568-576.

229. Lindsley, D. F., Perryman, K. M., & Lindsley, D. B. (1980). Brain mechanisms of attention and perception. In R. F. Thompson, L. H. Hicks, & V. B. Shvyrkov (Eds.), *Neural mechanisms of goal-directed behavior and learning* (pp. 387-396). New York: Academic Press.

230. Perryman, K. M., Lindsley, D. F., & Lindsley, D. B. (1980). Pulvinar neuron responses to spontaneous and trained eye movements and to light flashes in squirrel monkeys. *Electroencephalography and Clinical Neurophysiology, 49*, 152-161.

APPENDIX C.

231. Sanquist, T. F., Beatty, J. T., & Lindsley, D. B. (1981). Slow potential shifts of human brain during forewarned reaction. *Electroencephalography and Clinical Neurophysiology, 51*, 639-649.
232. Serdaru, M., Rohrbaugh, J., Syndulko, K., & Lindsley, D. B. (1981). Visual evoked potentials during classical eyeblink conditioning. *Society for Neuroscience Abstracts, 7*, 750.
233. Lindsley, D. B. (1982). Neural mechanisms of arousal, attention and information processing. In J. Orbach (Ed.), *Neuropsychology after Lashley* (pp.315-407). Hillsdale, NJ: Lawrence Erlbaum Associates.
234. Lindsley, D. B. (1985). Brain potentials and complexity of information processing. In W. D. Froelich, G. Smith, J. G. Draguns, & U. Hentschel (Eds.), *Psychological processes in cognition and personality* (pp. 231-245). New York: Hemisphere Publishing.
235. Lindsley, D. B. (1984). *Behavioral neuroscience: vertebrate brain and behavior.* (Booklet of 13 essays and bibliographies (24 pp.) associated with 1984 calendar of neuroscience illustrations). Quincy, MA: Grass Instrument Company.
236. Rohrbaugh, J. W., Chalupa, L. M., & Lindsley, D. B. (1987). Evoked potentials related to decision confidence in a visual discrimination task. In G. C. Galbraith, M. L. Kietzman, & E. Donchin (Eds.), *Neurophysiology and psychophysiology: Experimental and clinical applications* (pp. 326-332). Hillsdale, NJ: Lawrence Erlbaum Associates.
237. Wilson, C. L., Hirasuna, N., & Lindsley, D. B. (1987). Brainstem-limbic systems and behavior. In G. C. Galbraith, M. L. Kietzman, & E. Donchin (Eds.), *Neurophysiology and psychophysiology: Experimental and clinical applications* (pp. 57-67). Hillsdale, NJ: Lawrence Erlbaum Associates.
238. Lindsley, D. B. (1987). Activation, arousal, alertness and attention. In G. Adelman (Ed.), *Encyclopedia of neuroscience (Vol 1)*, (pp. 3-6). Boston: Birkhäuser/Springer-Verlag.

AUTHOR INDEX

AUTHOR INDEX

AUTHOR INDEX

AUTHOR INDEX

AUTHOR INDEX

AUTHOR INDEX

AUTHOR INDEX

SUBJECT INDEX

SUBJECT INDEX

SUBJECT INDEX

SUBJECT INDEX

SUBJECT INDEX

For Product Safety Concerns and Information please contact our
EU representative GPSR@taylorandfrancis.com Taylor & Francis
Verlag GmbH, Kaufingerstraße 24, 80331 München, Germany